Cooperative breeding refers to a social system in which individuals other than the parents provide care for the offspring. In addition to alloparental care, two further characteristics are common among species exhibiting cooperative breeding: delayed dispersal and delayed reproduction. Among vertebrates, cooperative breeding is expressed most prominently in birds and mammals. The book explores the phenomenon in a wide variety of mammals, including rodents, primates, viverrids, and carnivores. Comparative studies of cooperative breeding provide important tests for the origin and maintenance of sociality in complex groups. Understanding the behavioral and physiological mechanisms underlying cooperative breeding yields insights into the fundamental building blocks of social behavior in animal societies.

Although several recent volumes have summarized the state of our knowledge of the ecology and evolution of cooperative breeding in birds, *Cooperative Breeding in Mammals* is the first book devoted to these issues in mammals, and it will appeal to zoologists, ecologists, evolutionary biologists, and those interested in animal behavior.

COOPERATIVE BREEDING IN MAMMALS

COOPERATION E BREEDING IN MAMMALS

COOPERATIVE BREEDING IN MAMMALS

Edited by

NANCY G. SOLOMON
Miami University, Oxford, Ohio

JEFFREY A. FRENCH
University of Nebraska at Omaha

CAMBRIDGE
UNIVERSITY PRESS

CAMBRIDGE UNIVERSITY PRESS
Cambridge, New York, Melbourne, Madrid, Cape Town, Singapore, São Paulo

Cambridge University Press
The Edinburgh Building, Cambridge CB2 8RU, UK

Published in the United States of America by Cambridge University Press, New York

www.cambridge.org
Information on this title: www.cambridge.org/9780521454919

First published 1997
This digitally printed version 2007

A catalogue record for this publication is available from the British Library

Library of Congress Cataloguing in Publication data
Cooperative breeding in mammals / edited by Nancy G. Solomon, Jeffrey
A. French
   p. cm.
Includes index.
ISBN 0–521–45491–3 (hc)
1. Mammals – Behavior. 2. Cooperative breeding in animals.
I. Solomon, Nancy G. II. French, Jeffrey A.
QL739.3.C665 1996
599.056–dc20   96–17982
                                                                        CIP

ISBN 978-0-521-45491-9 hardback
ISBN 978-0-521-03828-7 paperback

To Bruce and Mary

Contents

Contributors

David H. Abbott
Wisconsin Regional Primate
  Research Center and
Department of Obstetrics and
  Gynecology
University of Wisconsin
Madison, WI 53715

Cheryl S. Asa
St. Louis Zoo
Forest Park
St. Louis, MO 63110

C. Sue Carter
Department of Zoology
University of Maryland
College Park, MD 20742

Scott R. Creel
Field Research Center for Ecology
  and Ethology
Rockefeller University
Tyrrel Road, RR2 Box 38B
Millbrook, New York 12545

Christopher G. Faulkes
Institute of Zoology
Zoological Society of London
London NW1 4RY,
England, U.K.

Jeffrey A. French
Department of Psychology and
  Nebraska Behavioral Biology
  Group
University of Nebraska at Omaha
Omaha NE 68182

Lowell L. Getz
Department of Ecology, Ethology
  and Evolution
University of Illinois at Urbana-
  Champaign
Urbana, IL 61801

Heribert Hofer
Max-Planck-Institut für
  Verhaltensphysiologie
D-82319 Seewiesen
Germany

Eileen A. Lacey
Museum of Vertebrate Zoology
Department of Integrative
  Biology
University of California
Berkeley, CA 94270

Susan E. Lewis
Department of Biology
Carroll College

100 North East Avenue
Waukesha, WI 53186

Jeffrey R. Lucas
Department of Biological Sciences
Purdue University
West Lafayette, IN 47907

Patricia D. Moehlman
Wildlife Conservation Society
Bronx, NY 10460

Ronald L. Mumme
Department of Biology
Allegheny College
Meadville, PA 16335

Anne E. Pusey
Department of Ecology, Evolution
   and Behavior
1987 Upper Buford Circle
University of Minnesota
St. Paul, MN 55108

R. Lucille Roberts
Department of Zoological Research

National Zoological Park
Smithsonian Institution
Washington, DC 20008

Paul W. Sherman
Section of Neurobiology and
   Behavior
Cornell University
Ithaca, NY 14853

Nancy G. Solomon
Department of Zoology
Miami University
Oxford, OH 45056

Suzette D. Tardif
Department of Biological Sciences
Kent State University
Kent, OH 44242

Peter M. Waser
Department of Biological Sciences
Purdue University
West Lafayette, IN 47907

Preface

As with most projects of this nature, this book began innocently. At the 1989 Animal Behavior Society meetings at Northern Kentucky University, only a handful of papers were presented on cooperative breeding and alloparental care in mammals. Two of these contributions were by the editors of this volume. At the time, much of the excitement generated by new research on cooperative breeding, either empirical or theoretical, was coming primarily from ornithologists. Woolfenden and Fitzpatrick's monograph on the Florida scrub jay and Brown's review of helping behavior in birds had been published in the previous five years. Two additional volumes, Stacey and Koenig's edited volume on long-term field studies of cooperative breeding in birds and Koenig and Mumme's monograph on the acorn woodpecker, were to be published in the immediate future. Yet it was surprising, at least to us, that, with the exception of a couple of reviews in scientific periodicals, no systematic survey or compilation of issues on cooperative breeding in mammals existed at the time.

A simple conversation was held between the editors at the 1989 meetings as we walked across a campus quad at Northern Kentucky University. The conversation, the gist of which was "someone ought to organize something," led to a proposal to the Animal Behavior Society to conduct a symposium on the topic. We are grateful to the ABS and to Lynne Houck and the other members of the program committee for approving this request and sponsoring our symposium at the 1992 ABS meetings in Kingston, Ontario. At this meeting, we gathered together eight distinguished presenters in a symposium that stretched over two days. By all estimates, the symposium was a great success and the presenters played to a full house at the Kingston meetings.

We received, from many in the field whom we respect, strong positive feedback on both the organization and the content of the symposium, and the question of publication was raised. Primarily at the urging of Paul Sherman, we explored the possibility of producing an edited volume that would summarize the proceedings of the symposium. Although initially hesitant, we felt, in light of the dearth of a single compendium on cooperative breeding in mammals and the favorable reaction of our colleagues, that it would be a worthwhile venture. Two additional goals directed our philosophy toward publication. First, we felt that the book would serve the discipline best if each of the contributions covered broad territory, made new contributions to the understanding of the origins and maintenance of cooperative breeding, and synthesized a large body of literature. In other words, our insistence to authors was that their chapters be more than just summaries of the past five years of their research program. Second, we had been limited by time constraints at ABS to limit the number of participants, and as a consequence several important long-term research programs were not included in that program. Therefore, we took the opportunity to expand the coverage to more taxa and to gain a better representation of scientists addressing both proximate and ultimate issues in cooperative breeding. The expanded volume now contains 12 content-oriented chapters on alloparental care, delayed dispersal, and reproductive suppression in mammals from scientists who are among the most active in this field.

As with any project of this magnitude, many people deserve credit for their unsung contributions along the way. The hosts of the ABS meetings in Kingston, Laurene Ratcliffe and Kathy Wynne-Edwards, provided a supportive environment in which to hold the symposium. We also are grateful to Dr. Robin Smith of Cambridge University Press for patience and guidance along the way during the preparation of the book. John Vandenbergh provided the senior editor with a supportive environment at North Carolina State University in which to pursue the early stages of this project. Rick Lee, at Miami University, graciously provided advice on editing a symposium volume.

We also would like to thank all of the individuals who reviewed manuscripts in draft stages for us. These include the contributors to the book, each of whom critiqued one or more of the chapters. In addition, the following persons provided helpful critiques and commentaries on chapters in the book: Andrew Baker, Stanton Braude, Nancy Marusha Creel, Steve Emlen, J. P. Hearn, John Gittleman, M. S. Gosling, Jennifer Jarvis, Walt Koenig, Dan Leger, A. S. I. Loudon, David Macdonald, Betty McGuire, Jo Manning, Craig Packer, Hudson Reeve, Anthony Rylands, Ann Rypstra, Wendy Saltzman, Colleen Schaffner, John Schneider, Steve Schoech, Nancy Schultz-Darken, Tessa Smith, Sandra Vehrencamp, and Rosie Woodroffen.

For the editors, Beth McDonald at the University of Nebraska at Omaha and Darla Lipscomb at Miami University of Ohio provided excellence in editorial assistance in the final, obsessive-compulsive phase of the preparation of the volume. Nancy Solomon was supported by funds from the National Institutes of Health (MH 52471–01), and Jeff French was supported by funds from the National Science Foundation (IBN 92–09528 and OSR 92–55225) during the preparation of the volume. Thanks to Denise for keeping the marmoset colony going and to Colleen for keeping everything else afloat during this time. Finally, deep gratitude is expressed by the editors to support from friends and especially loved ones, whose unflagging enthusiasm kept spirits alive. Thanks to all.

1

The Study of Mammalian Cooperative Breeding

NANCY G. SOLOMON and JEFFREY A. FRENCH

1.1 Introduction

Behavioral biologists and comparative psychologists have long been interested in species that display a social system labeled "cooperative breeding," in which members of the social group assist in rearing young that are not their own offspring. The definitional hallmarks of cooperative breeding concentrate on the distinctive attributes of these care-giving individuals, and they include (1) delayed dispersal from the natal group, (2) reproductive suppression, and (3) care for others' offspring. The individuals engaging in care of young may be nonbreeding adults or subadults (usually called helpers, auxiliaries, or alloparents), or they may be reproductive adults sharing in the care of young with other breeders in the social group. In the former, all three of the previously mentioned characteristics are seen, whereas in the latter, typically only the third characteristic is required for groups to be considered cooperative breeders. The types of care given to young by alloparents is remarkably diverse and includes patterns classified as both depreciable parental care (defined by Clutton-Brock 1991 as those patterns that change as a function of the number of offspring that are cared for; e.g., provisioning with food) and nondepreciable ones (those patterns that do not vary with the number of offspring that are cared for; e.g., burrow defense).

Cooperative breeding has received considerable attention from biologists, especially subsequent to the publication of Hamilton's landmark paper on the evolution of sociality (Hamilton 1964). Before Hamilton's formalization of inclusive fitness theory, alloparental behavior appeared to be an enigma, since the self-sacrifice of delayed reproduction and the provision of assistance to nondescendant offspring were at odds with maximizing individual fitness. Brown (1987) considered helping behavior to be the "principal test case" (italics in original) for evaluating Hamilton's notions of the importance of inclusive fitness. Early studies of cooperative breeding focused almost exclu-

1

sively upon the question of the adaptive significance of alloparental behavior. Recently, however, a number of alternative adaptive hypotheses (e.g., Emlen et al., 1991) have been proposed and have received some empirical support in the avian literature. Furthermore, a provocative nonadaptive hypothesis for alloparental care has been formulated (e.g., Jamieson & Craig 1987; Jamieson 1989) and serves the important function of questioning the fundamental assumptions in our approach to the study of cooperative breeding and alloparental care. This alternative hypothesis has also forced workers in the field to phrase research questions more clearly and to be more critical in evaluating the nature of evidence required to support an adaptive hypothesis.

Until recently, there were few approaches, other than the Hamiltonian one, to the study of cooperative breeding. As Tinbergen (1963) pointed out many years ago, and as others have recently reiterated (Sherman 1988; Dewsbury 1992; Alcock & Sherman 1994), there are multiple ways of asking questions about behavior. A revival of interest in the ways in which mechanism and function are related is reflected in recent publications by Stamps (1991) and Huntingford (1993). These authors have reminded us that information on mechanisms may provide insights into the functional significance of behavior and vice versa. Recent papers by Reyer and Westerterp (1985); Reyer, Dittami, and Hall (1986); Schoech, Mumme, and Moore (1991); Vleck et al. (1991); Wingfield, Hegner, and Lewis (1992); and Poiani & Fletcher (1994) have been the first to address the proximate basis of cooperative breeding among birds. These questions have been addressed by workers on mammalian cooperative breeders for a decade or more (e.g., Creel & Creel 1990), and many of the contributions in this volume reflect the sophistication of our appreciation of the mechanisms underlying the behavioral and social attributes of cooperative breeding in mammals (see reviews by French, Asa, Creel & Waser, Carter & Roberts, and Faulkes & Abbott in this volume).

The study of avian cooperative breeding systems has led to considerable theoretical and empirical advances in behavioral biology, comparative psychology, and ecology. In the past decade four major books have been published on the behavioral ecology of cooperative breeding in birds, including Woolfenden and Fitzpatrick's volume on the Florida scrub jay (1984), Brown's review volume (1987), Koenig and Mumme's monograph on the acorn woodpecker (1987), and Stacey and Koenig's edited volume (1990). Each of these books has been extremely influential and is cited extensively in the literature.

Although cooperative breeding has been reported in numerous vertebrate species, interest has been focused on avian cooperative breeders. It is surprising that considerably less attention has been paid to these issues in cooperatively breeding mammals. This cluster of reproductive and social patterns referred to as cooperative breeding is exhibited by a variety of mammalian taxa including primates, canids, viverrids, and rodents. In many cases these species have been studied in great detail in the laboratory or for long periods in the field. As pointed out by Jennions and Macdonald (1994) and Mumme (this volume), mammals differ from birds in the form of primary parental care (suckling in mammals versus provisioning in birds), which may have a major impact on the incidence and expression of cooperative breeding. Although a few reviews of mammalian cooperative care exist, they focus primarily on functional explanations of alloparental behavior (e.g., Gittleman 1985; Jennions & Macdonald 1994). Examination of the proximate as well as ultimate factors that contribute to the expression of cooperative breeding is needed to provide some insight into the similarities and differences between avian and mammalian cooperative breeding. The current volume is the first on cooperative breeding in mammals and brings together a variety of investigators to summarize the state of our empirical and theoretical knowledge for mammalian species, as well as serving as a source of new ideas and directions for further study.

1.2 Terminology

The characteristic social structure that is the focus of this book has been referred to by a dizzying array of labels in the literature, including communal breeding, communal care, and cooperative breeding. It is worthwhile to consider carefully the use and implications of each term. Webster's *New World Dictionary of the American Language* (1984) defines "communal" as "shared or participated in by all" (from the Latin root *communis* – that which is common) and "cooperate" as "joint effort or operation" (from the Latin root *cooperatus*, meaning to work together). There is no implication of mutual benefits in the first definition of cooperation in Webster's dictionary. Another sense of the word "cooperate," however, and one closer to the ecological context of the word, is "to work or act together or jointly for a common purpose or benefit" (Webster's *Encyclopedic Unabridged Dictionary* 1989).

   Brown (1978) reviewed the history of the words "communal" and "cooperative," which have often been used synonomously in the avian literature. Prior to 1968, both "communal" and "cooperative" were applied to nesting or breeding. However, the terms were variously used to refer to systems with plural breeding (two or more breeding females per breeding unit), joint nesting (where there is a single nest for all the pairs in a breeding unit and all pairs feed young), and singular breeders (one breeding pair per unit). VanTyne and

Berger (1959) in their text, *Fundamentals of Ornithology*, restricted "communal" nesting to plural, joint nesters and "cooperative" to colonial breeders
lacking helpers at the nest. Subsequently in the avian literature, Lack (1968)
used both "communal" and "cooperative" synonomously. Since then, various
authors have redefined these two terms in various ways, sometimes using
"communal" to refer to species without helpers but with plurally breeding
females. Therefore, it is no surprise that in the avian literature, the two terms
have been used interchangeably. For example, in the introduction, summary,
and first six chapters of Stacey and Koenig's (1990) book, there are 49 references to cooperative breeders and 27 to communal.

In the mammalian literature, one of the earlier reviews of the evolution of
altruistic behavior in social carnivores uses both "communal" care (of young),
hunting, and territory defense as well as "socially cooperative" (Bulger 1975).
Even though Riedman (1982) focuses on alloparental care, she briefly
reviews the avian studies on cooperative rearing of young but used the terms
"communal" and "cooperative" interchangeably. More recently, Gittleman
(1985) has used communal care, which he contrasts as a form of parental care,
with maternal and biparental care. In a 1994 review article, Jennions and
Macdonald formalized the use of the term "cooperative" breeding in mammals. They argued that, like its use in the avian literature, cooperative breeding refers to parentlike behavior directed toward offspring by individuals
other than the parents. However, they suggested that the term be limited to situations in which parentlike behaviors, such as feeding, assistance in thermoregulation, or allosuckling, are costly to the actors (i.e., likely to reduce
direct fitness).

Brown (1978) takes a fairly strong stand against the use of the term "cooperative," even though it was used in the title of his 1970 paper. He prefers
"communal" to "cooperative" because cooperative implies (in the sense of
Hamilton 1964) certain fitness benefits that accrue to participants.
Furthermore, cooperation is also seen in some species that are not cooperative
breeders. The use of "communal" also has disadvantages because Emlen
(1984) and McGrew (1986) used it to describe a particular subset of cooperative breeders – those species characterized by shared parentage of offspring.
With shared maternity, young are then communally nursed (mammals) or
incubated and provisioned (birds). Therefore, we agree with Price and Evans
(1991) that communal breeding implies the strong likelihood of shared
parentage. Thus, we would argue that "communal" not be used to refer to systems in which breeders are monogamous (at least socially) and nonbreeding
group members assist in care of young. Unlike Price and Evans (1991), we do
not choose to refer to these social systems as communal rearing systems, even

those in which the major hallmark is shared care of young. We feel that this term ignores the other major characteristics of this system (at least in singular breeders): namely, delayed dispersal and delayed reproduction or reproductive suppression. For the previously mentioned reasons, we choose to use the term "cooperative breeding," although keeping in mind that this is not meant to imply benefits to participants a priori.

Multiple terms have also been used to refer to effects on reproduction of younger group members. "Reproductive suppression" is most comonly found in the literature, but some investigators have questioned the use of this term because it implies that an inability to reproduce is being imposed upon the younger individual by a parent or more dominant group member (in the sense of *Webster's* (1984) first definition of suppress: "to put down by force"). If the term "suppress" is used in this way, it presupposes that control lies in the hands of the dominant individual. (The implication of control by a dominant breeder may have arisen from the eusocial insect literature (see Michener [1974], Wilson [1971], and from Vehrencamp's [1983] influential paper on the evolution of dominance effects on reproduction.) It is possible, though, that the younger individual may be assessing its options and refraining from reproduction when it detects the presence of a certain individual, age class of individuals, or cues from these individuals (see chapters in this volume by French, Creel & Waser, and Solomon & Getz). Some have suggested (e.g., Snowdon pers. comm. 1994) the use of reproductive "inhibition," which from the Latin *inhibitus* means to hold back, restrain, or curb. This term seems more ambiguous because holding back implies that control is in the hands of the alloparent, but an inhibitor is also defined as a person or thing that prohibits or restrains (Webster 1984). Thus, the use of "inhibit" does not eliminate implications of having limitations on reproduction imposed upon an individual by another group member.

Other authors have used "reproductive delay," which carries with it no suggestion of the source or stimulus for the reduced likelihood of successful reproduction. Delay, however, suggests that reproduction eventually occurs, and that is not necessarily true in most instances. In addition, the term "delay" also carries with it an ontogenetic implication, namely, that reproduction may be prevented as a consequence of a time lag in the neuroendocrine maturation of the hypothalamic–pituitary–gonadal system (cf. Vandenbergh 1989). It is clear, however, from an examination of the endocrinology that underlies delayed reproduction in philopatric nonbreeders across a wide variety of mammalian taxa (see chapters by French, Carter & Roberts, Creel & Waser, Faulkes & Abbott, and Asa in this volume) that puberty delay is only one potential manifestation of altered endocrine status that can affect reproductive potential in these conditions. We have chosen to use the term "reproductive suppression" in our discussion but want to emphasize that this word choice does not imply anything about which individual controls reproduction by the alloparents. We use suppress as in Webster's (1984) third definition "to keep back; restrain; check."

Finally, there has been considerable discussion of the implications of the term "helper." The term, as used in the literature on cooperative breeding, is derived from its original usage by Skutch as "helper-at-the-nest" (Skutch 1935). In 1961, Skutch clarified his usage: "A 'helper' is a bird which assists in the nesting of an individual other than its mate, or feeds or otherwise attends a bird of whatever age which is neither its mate nor its dependent offspring" (Skutch 1961). Although some may believe that this term implies fitness benefits to the breeders or their offspring, Skutch's 1961 definition is clearly devoid of these implications and serves simply to describe and label the behavioral phenomenon. "Auxiliary" has been used in an attempt to find a better term, but on the basis of the Latin *auxiliaries* (meaning helpful) or *auxilium* (meaning aid), there does not appear to be a sufficient distinction between helper and auxiliary to justify the use of the latter. We prefer the term "alloparent" (Wilson 1975), which is defined as engaging in parentlike care of young that have been produced by individuals other than the caregiver. This term seems free of fitness implications and does not suggest that alloparental interactions produce either a benefit or cost to any of the parties.

1.3 The Book

Our primary goal in this book is to review the existing literature on mammalian cooperative breeding. Our approach is particularly broad and has been guided by the breakdown typically given for questions on cooperative breeding: Why delay dispersal? Why delay breeding? Why help? Additionally, the authors in this book illustrate the existent diversity among mammalian cooperative breeders. The next 10 chapters are grouped taxonomically and provide a discussion of proximate and functional explanations for one or more of the questions posed earlier. We initially asked authors to limit discussion to either proximate or ultimate levels of analysis but quickly recognized the fertile intellectual ground in the complex intertwining of these two broad perspectives. The chapters contained in this book admirably reflect a broad approach to the unique attributes of cooperative breeding, and readers will recognize each of Tinbergen's (1963) four questions – causation, ontogeny, function, and phylogeny – guiding research in this area.

The first two species-oriented chapters address the issues of alloparental care (Tardif) and reproductive suppression (French) in the marmosets and tamarins (order Primates, family Callitrichidae). The 40 to 50 species in this family exhibit considerable variation in mating systems, life history, and behavioral ecology. Hence, the group is a useful test case for examining the critical role these factors play in shaping the nature of the dynamics of cooperative breeding. The two chapters that follow concentrate on similar issues in the family Canidae. Moehlman and Hofer focus on body size, reproductive energetics, and suppression of subordinates, and Asa reviews potential proximate mechanisms that underlie suppression and the expression of parental care by breeders and alloparents. Dwarf mongooses (family Viverridae) occupy a central role in the chapters by Creel and Waser and Lucas, Creel, and Waser. The long-term field-research projects on this species, initiated by Jon Rood, provide a rich source of data for testing both qualitative and quantitative predictions regarding the origins and maintenance of cooperative breeding. In the chapter by Creel and Waser, the roles of reproductive energetics and alternative routes to fitness in shaping reproductive skew in mongoose social groups is highlighted, and in the chapter by Lucas, Creel, and Waser, the utility of dynamic programming to assess dispersal decisions by alloparents is presented. The nature of cooperative breeding among species in the order Rodentia is addressed in chapters by Solomon and Getz and Carter and Roberts. Important issues such as the costs and benefits of cooperative breeding are addressed in the former, whereas the latter presents detailed empirical analyses of the ontogenetic and proximate causation of delayed breeding and alloparental care, particularly in *Microtus*. The next two chapters are unusual in that they deal almost exclusively with a single species – the naked mole-rat, *Heterocephalus glaber*. The disadvantages inherent in taxonomic "tunnel vision" are, in our opinion, far outweighed by the contribution made by the study of this truly unique mammal. The naked mole-rat is a cooperatively breeding rodent found at the extreme end of the continuum of mammalian sociality (Sherman et al. 1995). The species is eusocial (i.e., displays overlap of generations, reproductive division of labor, and cooperative care of young), but eusociality occurs in many other mammals and birds (Jarvis & Bennett 1993; Solomon 1994; Sherman et al. 1995; Lacey & Sherman this volume). Lacey and Sherman assess the behavioral ecology of this species in the broad context of the evolution of both vertebrate and invertebrate sociality, whereas Faulkes and Abbott explore the behavioral and physiological mechanisms by which dominant individuals maintain their reproductive exclusivity.

Two final chapters provide important links between the previous contributions and other manifestations of cooperative breeding. Whereas all the other chapters deal with species that exhibit a singular breeding profile, Lewis and

Pusey explore commonalties and differences in the expression of communal care in species that exhibit plural breeding (i.e., multiple males and especially females in a group produce offspring). In a summary chapter, Mumme highlights the distinctive features of mammalian cooperative breeders as viewed from the perspective of a student of avian cooperative breeding. This contribution emphasizes some of the important similarities and differences between the two taxa and also identifies some of the intellectual quagmires that future students of cooperative breeding in any taxa would be wise to avoid.

1.4 Conclusion

The study of cooperative breeding draws together questions from the realms of evolution, behavioral ecology, behavioral endocrinology, population ecology, and genetics. Specifically, our understanding of issues such as alloparental care, philopatry, delayed dispersal, inclusive fitness, and reproductive suppression and maturation have been enhanced by the study of cooperative breeding. Furthermore, application of some of the concepts discerned from study of cooperative breeding can provide insight into problems stemming from conservation biology and wildlife and pest management.

The study of cooperative breeding is an active and expanding field. Although some of the early studies focused on alloparental care, recent reviews by Koenig et al. (1992) and Keller and Reeve (1994) have restimulated interest in philopatry and reproductive suppression. We feel that examination of mammalian cooperative breeders, especially in light of the fundamental differences in avian and mammalian reproductive biology, should lead to fruitful and heuristic outcomes for investigators interested in avian or mammalian social systems. We hope that this book as a whole will introduce this subject conceptually as well as make connections between the diversity of investigators studying various organisms by using diverse approaches. Finally, we sincerely hope that this book will introduce avian biologists interested in sociality and cooperative breeding to the diversity seen in mammalian cooperative breeders.

References

Alcock, J., & Sherman, P. (1994). The utility of the proximate-ultimate dichotomy in ethology. *Ethology* 86:58–62.
Brown, J. L. (1970). Cooperative breeding and altruistic behavior in the Mexican jay, *Aphelocoma ultramarina. Anim. Behav.* 18:366–378.
Brown, J. L. (1978). Avian communal breeding systems. *Ann. Rev. Ecol. Syst.* 9:123–155.

Brown, J. L. (1987). *Helping and communal breeding in birds*. Princeton: Princeton University Press.

Bulger, A. J. (1975). The evolution of altruistic behavior in social carnivores. *Biologist* 57:41–50.

Clutton-Brock, T. H. (1991). *The evolution of parental care*. Princeton: Princeton University Press.

Creel, S. R., & Creel, N. M. (1990). Energetics, reproductive suppression and obligate communal breeding in carnivores. *Behav. Ecol. Sociobiol.* 28:263–270.

Dewsbury, D. A. (1992). On the problems studied in ethology, comparative psychology, and animal behavior. *Ethology* 92:89–107.

Emlen, S. T. (1984). Cooperative breeding in birds and mammals. In *Behavioural ecology*, 2nd ed., ed. J. R. Krebs & N. B. Davies, pp. 305–339. Sunderland, Mass.: Sinauer Association.

Emlen, S. T., Reeve, H. K., Sherman, P. W., & Wrege, P. H. (1991). Adaptive versus nonadaptive explanations of behavior: the case of alloparental helping. *Am. Nat.* 138:259–270.

Gittleman, J. L. (1985). Functions of communal care in mammals. In *Evolution – Essays in honour of John Maynard Smith*, ed. P. J. Greenwood, P. H. Harvey, & M. Slatkin, pp. 187–205. Cambridge: Cambridge University Press.

Hamilton, W. D. (1964). The evolution of social behavior. *J. Theo. Biol.* 7:1–52.

Huntingford, F. A. (1993). Behavioral mechanisms in evolutionary perspective. *Trends Ecol. Evol.* 8:81–84.

Jamieson, I. G. (1989). Behavioral heterochrony and the evolution of birds' helping at the nest: an unselected consequence of communal breeding. *Am. Nat.* 133:394–406.

Jamieson, I. G., & Craig, J. L. (1987). Critique of helping behavior in birds: a departure from functional explanations. In *Perspectives in ethology*, vol. 7, ed. P. P. G. Bateson & P. H. Klopfer, pp. 79–98. New York: Plenum Press.

Jarvis, J. U. M., & Bennett, N. C. (1993). Eusociality has evolved independently in two genera of bathyergid mole-rats – but occurs in no other subterranean mammal. *Behav. Ecol. Sociobiol.* 33:253–260.

Jennions, M. D., & Macdonald, D. W. (1994). Cooperative breeding in mammals. *Trends Ecol. Evol.* 9:89–93.

Keller, L., & Reeve, H. K. (1994). Partitioning of reproduction in animal societies. *Trends Ecol. Evol.* 9:98–102.

Koenig, W. D., & Mumme, R. L. (1987). *Population ecology of the cooperatively breeding acorn woodpecker*. Princeton: Princeton University Press.

Koenig, W. D., Pitelka, F. A., Carmen, W. J., Mumme, R. L., & Stanback, M. T. (1992). The evolution of delayed dispersal in cooperative breeders. *Q. Rev. Biol.* 67:111–150.

Lack, D. (1968). *Ecological adaptations for breeding in birds*. Bristol, U.K.: Western Printing Services.

McGrew, W. C. (1986). Kinship terms and callitrichid mating patterns: a discussion note. *Primate Eye* 30:25–26.

Michener, C. D. (1974). *The social behavior of the bees*. Cambridge, Mass.: Belknap Press.

Poiani, A., & Fletcher, T. (1994). Plasma levels of androgens and gonadal development of breeders and helpers in the bell miner (*Manorina melanophrys*). *Behav. Ecol. Sociobiol.* 34:31–41.

Price, E. C., & Evans, S. (1991). Terminology in the study of callitrichid reproductive strategies. *Anim. Behav.* 42:1025–1027.

Reyer, H. U., & Westerterp, K. (1985). Parental energy expenditure: a proximate cause of helper recruitment in the pied kingfisher (*Cervle rudis*). *Behav. Ecol. Sociobiol.* 17:363–369.

Reyer, H. U., Dittami, J. P., & Hall, M. R. (1986). Avian helpers at the nest: are they psychologically castrated? *Ethology* 71:216–228.

Riedman, M. L. (1982). The evolution of alloparental care and adoption in mammals and birds. *Q. Rev. Biol.* 57:405–435.

Schoech, S. J., Mumme, R. L., & Moore, M. C. (1991). Reproductive endocrinology and mechanisms of breeding inhibition in cooperatively breeding Florida scrub jays (*Aphelocoma c. coerulescens*). *Condor* 93:354–364.

Sherman, P. W. (1988). The levels of analysis. *Anim. Behav.* 36:616–619.

Sherman, P. W., Lacey, E. A., Reeve, H. K., & Keller, L. (1995). The eusociality continuum. *Behav. Ecol.* 6:102–108.

Skutch, A. F. (1935). Helpers at the nest. *Auk* 52:257–273.

Skutch, A. F. (1961). Helpers among birds. *Condor* 63:198–226.

Solomon, N. G. (1994). Eusociality in a microtine rodent. *Trends Ecol. Evol.* 9:264.

Stacey, P. B., & Koenig, W. D. (1990). *Cooperative breeding in birds*. Cambridge: Cambridge University Press.

Stamps, J. A. (1991). Why evolutionary issues are reviving interest in proximate behavioral mechanisms. *Am. Zool.* 31:338–348.

Tinbergen, N. (1963). On the aims and methods of ethology. *Z. Tierpsychol.* 20:410–433.

Vandenbergh, J. G. (1989). Coordination of social signals and ovarian function during sexual development. *J. Anim. Sci.* 67:1841–1847.

VanTyne, J., & Berger, A. J. (1959). *Fundamentals of ornithology*. New York: Wiley.

Vehrencamp, S. L. (1983). A model for the evolution of despotic versus egalitarian societies. *Anim. Behav.* 31:667–682.

Vleck, C. M., Mays, N. A., Dawson, J. W., & Goldsmith, A. R. (1991). Hormonal correlates of parental and helping behavior in cooperatively breeding Harris' hawks (*Parabuteo unicinctus*). *Auk* 108:638–648.

Webster's *New world dictionary of the American language*. (1984). New York: Simon & Shuster.

Webster's *Encyclopedic unabridged dictionary*. (1989). New York: Portland House.

Wilson, E. O. (1971). *The insect societies*. Cambridge, Mass.: Harvard University Press.

Wilson, E. O. (1975). *Sociobiology*. Cambridge, Mass.: Harvard University Press.

Wingfield, J. C., Hegner, R. E., & Lewis, D. M. (1992). Hormonal responses to removal of a breeding male in the cooperatively breeding white-browed sparrow weaver, *Plocepasser mahali*. *Horm. Behav.* 26:145–155.

Woolfenden, G. E., & Fitzpatrick, J. W. (1984). *The Florida scrub jay: demography of a cooperative-breeding bird*. Princeton: Princeton University Press.

2

The Bioenergetics of Parental Behavior and the Evolution of Alloparental Care in Marmosets and Tamarins

SUZETTE D. TARDIF

2.1 Alloparental Behavior

Alloparental behavior, defined here as care of infants by individuals other than the mother, is relatively common in primates (Nicolson 1987). In primates, it reaches its most extreme levels in the South American marmosets and tamarins (family Callitrichidae). Popular literature on these small monkeys frequently describes the mothers as carrying the infants only during nursing, with fathers and other group members caring for infants the rest of the time. Though these popular reports exaggerate the limited role of the mother in infant care, they emphasize the long-standing recognition of non-maternal infant care in callitrichids. For this reason, they are a valuable group in which to examine the factors controlling mammalian alloparenting, or cooperative infant care.

The central question addressed in this review is, "Why do individuals other than mothers care for infants?" Clearly this question can be addressed on a number of levels from the proximate or physiological factors that affect interactions with infants (e.g., Dixson 1982; Pryce, Döbeli, & Martin 1993; for review see Pryce 1993) to the functional role that social and environmental interactions and constraints may play (e.g., Epple 1975; Price 1991, 1992a, 1992b; Tardif, Carson, & Gangaware 1992; Tardif, Harrison, & Simek 1993) to the possible adaptive advantages of such behavior (e.g., Rylands 1982; Goldizen 1987; Price 1990a; Baker 1991). This review will focus on the functional and adaptive explanations for alloparental behavior in callitrichids. Although these two forms of explanation may be formulated and evaluated separately, I believe that a joint examination of both strengthens our understanding of each.

Studies of alloparental behavior in both birds and mammals have generally assumed that such behaviors are adaptive; that is, they assume that such behavior involves benefits as well as costs (however, see Jamieson 1989 for an opposing view). Long-term studies in a few species of birds (e.g., Ligon & Ligon 1988; Emlen & Wrege 1989) and mammals (e.g., Lucas, Waser, & Creel this volume) provide persuasive evidence of the ways in which selection pressures may have shaped the identities of alloparents. However, for most species, including marmosets and tamarins, such comprehensive analyses are lacking. This review will therefore draw on a wide variety of information to assess the relative costs and benefits of alloparental behavior in callitrichids. Although the species being studied will be identified, the conclusions will be applied to callitrichids as a group. Although such conclusions are certainly oversimplifications that mask important differences within the family (e.g., see Ferrari & Lopes-Ferrari 1989; Tardif et al. 1993), there is not enough information at this point to meaningfully contrast alloparental behavior among callitrichid species.

The costs and benefits of cooperative breeding in callitrichids involves those of staying in the natal group (i.e., philopatry) as well as those of alloparenting. Studies of cooperative breeding frequently assess benefits/costs more in terms of those associated with philopatry and less with those associated with alloparenting. This is particularly true for the assessment of cost (Mumme this volume). Is there a meaningful cost associated with alloparenting, not just with staying? Immediate factors may be examined. Then the question may be asked: Do these costs translate into a fitness cost, such as decreased life span or decreased future reproduction. This review will center on the potential costs and benefits of alloparental behavior, per se, rather than asking, "Why are callitrichids philopatric?" An excellent discussion of ecological factors that may have shaped dispersal choices in callitrichids is provided by Rylands (1996). Additional discussion of the ways in which dispersal choices may relate to the phyletic dwarfism of callitrichids is found in Ford (1980).

2.2 Social Organization and Infant Care in Callitrichids

Marmosets and tamarins are small primates, with adult weights of 100 to 700 grams. They are the most fecund of anthropoid primates, typically producing fraternal twins and frequently displaying a fertile postpartum estrus. At birth, this twin litter weighs around 16 to 20 percent of adult body weight (Tardif et al. 1993; French this volume). As is true for all anthropoid primates, callitrichids physically transport their infants, typically carrying them on their backs. Nipples are located axially, and infants usually nurse while on the mother's back. The clinging ability of infants is crucial; marmosets and tamarins offer little help to infants in this respect (Rothe 1974; Price 1990b), although they will assist infants in the transfer from one carrier to another.

In a callitrichid social group, all individuals, including both male and female nonreproductive individuals, may participate in carrying infants. The extent to which any given individual contributes to infant care is highly variable (Epple 1975; Pryce 1988; Tardif et al. 1992). In some groups, breeding males are the primary carriers, whereas in others, nonbreeding individuals may carry infants more often. There is no consistent effect of sex or reproductive status on the extent of involvement in carrying. However, age is important; younger individuals (less than 12 to 14 months) participate less frequently (Ingram 1977; Cleveland & Snowdon 1984; Tardif et al. 1992).

Marmosets and tamarins provision infants with solid food (Brown & Mack 1978; Wolters 1978; Ferrari 1987; Feistner & Price 1990, 1991; Price & Feistner 1993; Tardif et al. 1993). Eisenberg (1981) suggests that this provisioning may be a critical factor in the survival of postweaning young. Unfortunately, there is insufficient information on the distribution of provisioning across potential alloparents to evaluate its possible adaptive significance as an alloparental behavior. This review will center on the better-studied alloparental behavior of infant transport.

Callitrichids were historically described as monogamous. As such, the mother's "helpers" were thought to consist of the father and older offspring. However, a number of recent field studies have revealed that in the wild callitrichids live in a wide array of social groups. Those that appear to be either polyandrous or monogamous (as indicated by mating patterns) are the most common, although polygynous groups have been reported (Goldizen 1987; Sussman & Garber 1987; Baker, Dietz, & Kleiman 1993; Digby & Ferrari 1994). French (this volume) provides a detailed discussion of the factors regulating the number of reproductive individuals in a callitrichid group. Regardless of reproductive pattern, all group members that are old enough are likely to participate in infant care.

There is debate over the extent to which adult males in multimale groups are likely to share paternity (see Goldizen 1987; Baker et al. 1993). However, the fact that multiple males copulate with the female has to reduce certainty of paternity from that of known monogamous groups. Frequently, then, the only parent that is unambiguously related to the infants is the mother. I will therefore include discussion of infant care by potential breeding males as an example of alloparental behavior.

There is very limited information available on infant care in callitrichids in the wild; only three or four field studies provide such information. However, there is a fairly broad base of information from studies of captive callitrichids. For this reason I will include both captive and field studies to address the questions of benefits and costs, while realizing that there are limits on the extent to which behavior in captivity can be used to assess such issues.

2.3 Potential Benefits of Alloparental Behavior

2.3.1 Hypotheses

Numerous hypotheses have been formulated regarding the potential direct and indirect benefits of alloparental behavior to the alloparent. Emlen & Wrege (1989), for example, formulated hypotheses regarding nine potential benefits of such behavior in the white-fronted bee-eater (*Merops bullockoides*). Insufficient data are available to assess the complete array of potential benefits of alloparental behavior in callitrichids. Four will be discussed here. They are not mutually exclusive and are not an exhaustive list of possible benefits; rather, they are those for which some empirical information is available from more than one species. For callitrichids, the following potential benefits derived from alloparenting have been proposed:

1. Improvement in infant care skill. Subsequent reproduction may be improved through experience gained through alloparenting.
2. Inheritance of a breeding slot. Alloparents may stay in the group and inherit a breeding position. This benefit can explain alloparental behavior only if such behavior increases the likelihood of inheriting a breeding position.
3. Increased access to the breeding female (i.e., infant carrying as courtship). This benefit would be limited to potential breeding males.
4. Inclusive fitness gains through caring for and subsequently improving the survival of relatives.

Evidence regarding each of these potential benefits follows.

2.3.1.1 Benefits from experience

It has frequently been noted in captive callitrichids that individuals lacking experience with infant siblings are poor parents. A number of studies have recorded higher survival of the infants of primiparous mothers if the mothers had previous experience with infant siblings than if they did not (for an exception, see Baker & Woods 1992), poor parental care being the primary cause of infant death (Table 2.1). These data argue for a potential benefit of experience with infants that is also suggested for other primates, such as

Table 2.1. *Relationship between previous allomaternal experience and subsequent infant survival*

Species	Inexperienced[a]		Experienced[b]		Reference
	n	Percentage[c] survival	*n*	Percentage survival	
Leontopithecus rosalia	2	0.0	2	100.0	Hoage 1978
" "	6	15.0	6	64.0	Baker & Woods 1992
Saguinus oedipus	3	0.0	6	54.5	Tardif et al. 1984
" "	3	0.0	11	17.0	Baker & Woods 1992
" "	3	0.0	7	18.0	Snowdon et al. 1985
Saguinus fuscicollis	16	23.5	5	80.0	Epple 1978
Saguinus labiatus	10	27.0	14	38.0	Pryce 1988
Saguinus imperator	8	38.0	5	43.0	Baker & Woods 1992
Callithrix jacchus	3	0.0	9	100.0	Pryce 1993
" "	6	54.5	3	100.0	Tardif et al. 1984

[a]Inexperienced: never in a group with infants prior to first delivery.
[b]Experienced: present in a group with infants (usually younger siblings) prior to first delivery
[c]Percentage of survival: percentage of first-delivery young surviving at least 3 months for all females in the sample (*n*).

vervet monkeys (Lancaster 1971; Fairbanks 1990). However, the data on captive callitrichids must be interpreted cautiously. Potentially important confounding factors, such as age and degree of social experience, usually have not been controlled, because observations of such experience effects are generally made in the course of captive management. In most cases, age and experience effects are confounded (Baker & Woods 1992). In one study of cotton-top tamarins (*Saguinus oedipus*), there was no difference in time spent carrying infants for experienced versus inexperienced animals of the same age (see Figure 2.1); experience with infant siblings was not a factor affecting subsequent alloparenting (Tardif et al. 1992).

Other attempts have been made to assess the role of experience as a benefit accrued by alloparents by testing predictions regarding the relative likelihood of alloparental behavior in animals of differing ages and sex. Baker (1991) proposed that, in free-ranging golden lion tamarins (*Leontopithecus rosalia*), experience was more important for females, given that most mothers in this species carry infants virtually all the time during the period of greatest risk for infant mortality, that is, the first week. Therefore, if experience is a primary benefit of alloparenting, inexperienced females should carry more than expe-

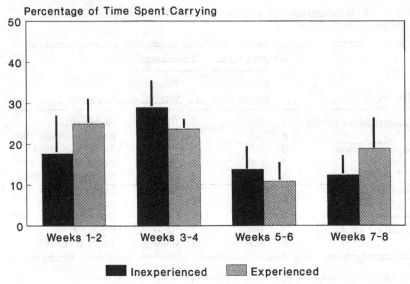

Figure 2.1. Comparison of the mean percentage of time (+S.E.) that experienced versus inexperienced subadult alloparents spent carrying infants in captive groups of cotton-top tamarins. Carrying percentages are calculated for two-week intervals from birth to Week 8. Samples sizes are (1) experienced and inexperienced, Weeks 1–4 = 9 each and (2) experienced and inexperienced, Weeks 5–8 = 7 each. There were no significant differences in percentage of time spent carrying relative to experience for any infant age. Figure modified from Tardif et al. 1992.

rienced females in order to gain such experience. Baker found that inexperienced females did carry more than experienced females and that they also carried more than males, regardless of the male's experience. He interpreted these results as supporting the hypothesis that gains in experience are a benefit accrued from alloparenting.

   This argument assumes that time spent carrying infants translates into gains in future maternal performance. However, none of the studies described so far have dealt specifically with what exactly might be gained by experience. Are experienced carriers more adept at retrieving from other carriers? Do they locomote more efficiently with infants than do inexperienced carriers? To what extent is actual experience in infant care (as opposed to exposure to infant siblings) correlated with improved reproductive performance? Juvenile primates have frequently been described as being inept at carrying, displaying poor carrying positions and frequently being intolerant of the infants on their backs. Experience seems to improve infant handling in other primates, such as vervet monkeys (Meaney, Lozos, & Stewart 1980). Whereas juveniles, which generally carry infants infrequently, may be inept, we have found that older (subadult and adult), inexperienced alloparents were as likely to retrieve infants adeptly and carry them as frequently as experienced alloparents (Tardif et al. 1992). Likewise, Goldizen (1987) observed that inexperienced year-old alloparents seemed relatively adept at carrying infants in free-ranging tamarin groups.

The differences in reproductive performance between experienced and inexperienced animals may reflect the lessening of a neophobic response in females that have previously been exposed to their infant siblings (Pryce 1992). Such benefits might be gained by simple exposure to infants or through observational learning (Tardif et al. 1992), as opposed to actual participation in infant care. If so, this benefit is not directly linked to alloparental behavior.

There is some evidence, then, that improved future reproductive success might result from experience gained through alloparental behavior. However, many questions remain regarding the extent to which actual infant care interactions are necessary. In addition, such benefits do not explain the alloparental behavior routinely displayed by older alloparents that already have extensive experience in infant care.

2.3.1.2 Benefit through inheritance of a breeding position

Callitrichids might use alloparental behavior as a passport, by "trading" their help for inheritance of a breeding slot. In order to assess this possibility, two questions must be asked. First, why do animals stay in their natal groups, and second, why do they care for infants when they stay?

Callitrichids are generally territorial; though there may be two to four adults of each sex in the social group, there is generally one breeding female and one to two breeding males. This pattern suggests that breeding opportunities in many populations in relatively saturated areas may be limited. Goldizen and Terborgh (1989), Baker, Dietz, and Kleiman (1993), and Rylands (1996) have dealt with the issue of philopatry in some detail. For the present discussion, I will deal primarily with the question of inheritance of a breeding position. Three field studies provide data on the incidence of individuals inheriting the breeding position in a group after the death or disappearance of the breeder of the same sex (Goldizen & Terborgh 1989; Ferrari & Diego 1992; Baker et al. 1993). The sample sizes were generally small, making it difficult to assess the relative impact of staying versus dispersing. Inheritance of a breeding position was found to be rare. However, in many populations, the ultimate chances for breeding may be highest for those individuals that stay in their natal groups rather than dispersing, given that achievement of a breeding position outside of the natal group was even more uncommon (Baker et al. 1993). There is some evidence that callitrichids may "prospect" by leaving their natal territory but return at a later date (Dawson 1978; Neyman 1978). Such prospecting has been reported for canids displaying alloparental care (e.g., jackals, Moehlman & Hoefer this volume). However, Ramirez (1984) reports that in free-ranging moustached tamarins (*Saguinus mystax*) animals that disperse are forced to carve out a territory among already established groups and frequently do not reproduce successfully.

As is true for other cooperatively breeding birds and mammals, then, remaining in the natal group may be the most successful route to future reproduction in some circumstances. However, in order for alloparental behavior to be related to such benefits, it must be demonstrated that infant-care behavior affects an individual's likelihood of remaining in a social group and inheriting; that is, if nondispersers can remain in the group and subsequently inherit a breeding position without alloparenting, then the behaviors that are adaptive are those related to dispersal decisions and not to alloparental behavior.

There is no information on free-ranging or captive animals that suggests that infant care by nonreproductive individuals improves their likelihood of remaining in the natal territory or inheriting a breeding position. Figure 2.2 presents the percentage of time that members of 10 sets of twin siblings carried infants in captive cotton-top tamarin groups. Notice that for many sets, one twin carried substantially more than did the other. Because there was such variation within twin litters, there was the potential to assess the impact of alloparenting upon group tenure. For 60 percent of these groups (6/10), at least one of the twins had to be removed from the family group as a result of aggression from the same-sex parent. In 4/6 cases the sibling that carried most frequently was expelled from the group; in 2/6 cases the sibling that carried least frequently was expelled. There appeared to be no advantage for those that carried infants frequently. It has been reported that breeding females, in both field and captive studies, are more likely to expel female alloparents two to three months after the birth of infants, again suggesting that alloparental behavior does not serve as a "passport" to group membership (Price & McGrew 1991).

Although remaining in a nonreproductive position may be a superior strategy to dispersing in some environments, this does not mean that subsequent participation in alloparental behavior is also a superior strategy. There appears to be little evidence that benefits derived from alloparenting allow a nonreproductive individual to remain in a group or breed in that group in the future.

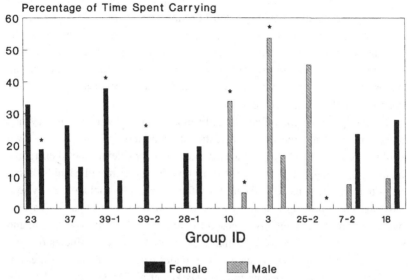

Figure 2.2. Comparison of the percentage of time infants were carried (from birth to Week 4) by individuals in 10 sets of twin alloparents versus their subsequent expulsion from the group. Figure modified from Tardif et al. 1992. Some groups have a double designation because they were observed across more than one litter of infants. For example, 39–1 is the first litter observed in group 39, and 7–2 is the second litter observed in group 7.

2.3.1.3 Benefit of infant care as courtship

Infant care by potential breeding males might be related to breeding in two ways: (1) if infant care affects mate selection by the female and (2) if the male's motivation toward infant care is influenced by access to mating.

Rylands (1982) and Price (1990a) have proposed that infant carrying by potential breeding males may function as a form of courtship. Price proposed that it would "pay females to choose as mates males that are competent caretakers." In support of this hypothesis, Price (1990a) observed that captive male cotton-top tamarins were more likely to successfully copulate when they were carrying infants than when they were not. However, the interpretation of these results is confounded by the fact that data were pooled over a period during which the infants were maturing; in such a situation, the frequency with which adult males carried infants during the female's estrus might be confounded somewhat with the relative maturity of infants. It is difficult to determine cause and effect in such a scenario. Baker et al. (1993) compared

the extent of infant care of dominant and subordinate males in free-ranging
two-male–one-female trios of golden lion tamarins (*Leontopithecus rosalia*)
and found no difference in the percentage of time that dominant and subordinate males carried infants, although dominant males appeared able to sexually
monopolize the female during probable estrus periods. Further studies on the
role of female choice in mate selection would improve our assessment of this
potential benefit of infant care by males.

Males might care for infants of only those females with which they had
mated. In this way, the male's investment in infants that are not his own
would be reduced. Both empirical evidence and a consideration of their reproductive physiology suggest that this is not the case in callitrichids. First,
males that have not mated will care for infants. Examples include care by sexually mature, nonbreeding males within established social groups, and care by
males that are experimentally exposed to novel, unrelated infants (Cleveland
& Snowdon 1984; Wamboldt, Gelhard, & Insel 1988). Also, studies of free-
ranging golden lion tamarins suggest that males care for infants regardless of
their breeding access to estrus females (Baker et al. 1993). Finally, because
callitrichid males experience a 130 to 180 day delay between mating and
infant care (i.e., the length of gestation), there is little opportunity for a direct
causal link between mating and subsequent willingness to care for infants.
Callitrichid males do not appear, therefore, to apportion infant care relative to
mating, unlike some polyandrous birds in which care of nestlings follows
mating by 10 to 20 days (Davies et al. 1992).

2.3.1.4 Benefit through improved indirect fitness

The strength of this hypothesis is dependent upon two testable assumptions:
First, that individuals are more likely to care for related than unrelated infants,
and second, that such help in infant care improves infant survival. The first
point was considered almost axiomatic as long as callitrichids were assumed
to live in groups consisting of a monogamous pair and their offspring.
However, field studies over the past decade have revealed a wide array of
social systems in free-ranging callitrichids. Ferrari & Digby (1996) suggest
that marmosets (genus *Callithrix*) have a group structure that is consistent
with a family (i.e., a breeding pair and their offspring, both adult and immature). If this is the case, then alloparental behavior may be directed almost
entirely to siblings. However, field studies of tamarins (particularly *Saguinus*)
reveal that both male and female subadults and adults frequently migrate from
one group to another; there is therefore potential for individuals to be in social
groups in which the infants are not related to them, or for more variation in

levels of relatedness between potential alloparents and infants. Unfortunately for most field studies of callitrichids, the degree of genetic relatedness between infants and potential alloparents is unknown, unless they are known to be the siblings. Only one study of free-ranging callitrichids has examined the relationship between alloparents and infants. Baker (1991) examined the percentage of time that golden lion tamarins spent carrying infants to which they were more or less closely related. Males carried more closely related infants more often. For females, there was no difference. If inclusive fitness gains were the sole explanation for alloparental care, there would be no obvious reason for the benefits of carrying more closely related infants to be greater for males than for females, suggesting that other factors need to be examined in order to explain this difference. Baker (1991) proposed that females might be less likely to differentiate among levels of relatedness because of the importance of gaining experience. He also proposed that males accrued additional benefits from care of younger brothers, in terms of providing themselves with a potential dispersal partner in the future (see also Rylands 1996). Because females did not tend to disperse with their siblings, this benefit was assumed to be gender-specific.

Two cautionary points should be made in interpreting these results. First, the level of relatedness between alloparents and most infants was relatively high (estimated between 0.25 and 0.5). Second, the data used in this analysis involved only polygynous groups in which there was more than one litter of infants in the group. In five of seven cases, the female whose infants were more closely related to the potential alloparents was the dominant female, whereas the female with less closely related infants was subordinate. Subordinate female common marmosets (*Callithrix jacchus*) are less likely to relinquish their infants to other carriers than are dominant females (Digby & Ferrari 1994). The effects of these two factors are difficult to separate, making results difficult to interpret.

In captivity, individuals will care for infants that are not their own (Cleveland & Snowdon 1984). Wamboldt et al. (1988) presented related and unrelated infants to family groups of captive pygmy marmosets (*Cebuella pygmaea*). They found that breeding males and alloparents did not differ in their responsiveness to the two types of infants, whereas mothers were significantly more responsive to their own infants. These observations suggest that there is not a clear recognition of related infants by nonmothers and that breeding males and alloparents will provide care to socially unfamiliar infants. However, these observations do not reveal whether the type or amount of care afforded to related versus nonrelated infants might differ.

The second testable assumption behind the proposed adaptive advantage of caring for relatives is that such alloparental behavior increases the chances for survival of infants. The extent to which this may be so is unclear (see Solomon & Getz, this volume, for a discussion of similar issues in cooperatively breeding rodents). There have been some attempts to gauge the effects of alloparents by comparing group size with reproductive success. Garber, Moya, and Malaga (1984) found a relationship between the number of surviving infants and the number of adult males in the group (but not with total group size) in moustached tamarins. However, no relationship between the number of alloparents in the group and the number of surviving infants was found for free-ranging golden lion tamarins (Dietz & Baker 1993) and captive common marmosets (Rothe, Koenig, Darms 1993). It has been pointed out in the study of cooperatively breeding birds that many factors may be correlated with group size or number of alloparents (territory size and quality, for example) and that these factors may also explain differences seen in reproductive success (Koenig & Mumme 1987; Ligon & Ligon 1988). Dawson (1978) found such a relationship between habitat quality and number of surviving offspring in free-ranging cotton-top tamarins, although there was no relationship between infant survival and group size.

There is evidence in other mammals that alloparents may improve the future reproductive success of their siblings, not through increased survival of those siblings but through improved condition. Specifically, prairie vole (*Microtus ochrogaster*) pups reared with older siblings are significantly larger at separation from the parents than those without alloparents present (Solomon 1991). This increased size translates into improved reproductive performance as adults (Solomon 1994). No published data are available on the relationship between the presence of alloparents and the condition of offspring in free-ranging callitrichids. However, in one study of captive common marmosets, the presence of alloparents did not significantly affect infant growth (Jaquish 1993).

The impact of number of alloparents may be particularly difficult to assess if it is an important factor only during group formation. Studies of both captive and free-ranging callitrichids suggest that three or four animals usually account for 90 percent or more of infant carrying (Rylands 1986; Goldizen 1987; Koenig & Rothe 1991a; Rothe & Darms 1993). This means that the number of "necessary" carriers may be below the mean group size for most callitrichids, which is 5.5 to 11.5 individuals, depending upon species (Ferrari & Lopes-Ferrari 1989). Therefore, in areas of stable territory structure and group size, there may be no evidence of the effects of group size. Only if group formation results in frequent groupings of fewer than four animals would group size be related to infant survival.

2.4 Costs of Alloparental Behavior

2.4.1 Levels of Cost

The costs associated with a given behavior can be assessed at two levels: immediate costs in terms such as energy expenditure, increased predation risks, or reduced foraging efficiency; and fitness costs, in terms of the effects of the behavior on present or future reproductive success. These two types of costs may or may not be related to each other; for example, the energetic expense of alloparental behavior may or may not have an impact upon fitness (Stearns 1992). Likewise, the immediate reproductive effort that alloparents forfeit may or may not be reflected in the immediate costs of such behavior. In most studies of birds and mammals, costs to alloparents are ascribed to the costs of philopatry rather than to those of alloparental behavior per se. The immediate and long-term costs of alloparental behaviors themselves are often assumed although rarely examined. However, just as knowledge of the fitness consequences of a behavior are important in determining its adaptive function, assessment of potential cost also is important in determining the likelihood that the behavior represents a product of selection (Stearns 1992; Mumme this volume). What are the costs of caring for infants, and how may they be expressed? If, for example, philopatry reduces fitness, but alloparental behaviors do not, then the strength of some adaptive arguments may be affected. Callitrichids are an excellent group in which to examine this question, given that infant care has long been assumed to be the primary feature shaping callitrichid social behavior (Kleiman 1977; Leutenegger 1980).

2.4.1.1 Immediate costs of alloparental behavior

Three immediate costs of alloparental behavior in callitrichids have been hypothesized: the energetic cost of infant transport, increased risk of predation, and the reduced foraging time associated with infant transport. The energetic cost has been assumed but not measured, but data on effects of infant transport on other activities are available.

The caloric cost of transporting infants can be estimated with information on travel speeds from field data and data on relative adult and infant weights (see Taylor 1980). Figure 2.3 provides a comparison of the estimated daily cost specific to infant transport versus the cost of maintenance, lactation, and adult travel without an infant load. The figures are provided as daily expenditure of 1,000 calories relative to body mass. Lactation represents a significant

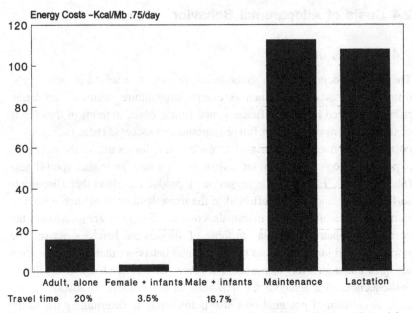

Figure 2.3. Comparison of the energy costs (in kcal/metabolic mass/day) of travel for adult tamarin alone (traveling 20% of time), female transporting infants (3.5% travel time), male transporting infants (16.7% travel time), maintenance, and lactation. The travel times used to estimate costs are from Goldizen 1987 for free-ranging saddle-backed tamarins. The costs of maintenance and lactation taken from Kirkwood and Underwood (1984) on captive cotton-top tamarins.

energetic cost almost equal to maintenance cost (Kirkwood & Underwood 1984). Carrying two 30-day-old infants would result in a minimum 21 percent increase in the caloric cost of traveling per minute over the cost of traveling without infants. However, free-ranging and captive callitrichids spend less time in travel when they are carrying infants than when they are not (Table 2.2). For example, in one free-ranging saddle-back tamarin group (Goldizen 1987), males reduced travel time (as percentage of total observation time) from around 20 to 16.7 percent, whereas carrying infants and breeding females reduced carrying time to 3.5 percent while transporting infants. So, although traveling with infants results in a 21 percent increase per minute carried over the cost of not carrying, individuals may reduce the actual cost by traveling less often when they are carrying (Tardif & Harrison 1990).

Empirical data are available from both field and captive studies that indicate that individuals transporting infants are limited in their other activities (see Table 2.2). Goldizen (1987) found that free-ranging saddle-backed

Table 2.2. *Comparison of time spent in travel or foraging versus resting relative to infant transport*

Subject description	Time in travel/ time resting		Time in feeding/ time resting		Reference
	While carrying	While not carrying	While carrying	While not carrying	
Free-ranging saddle-back tamarins; breeding females	0.03	0.26	0.03	0.53	Goldizen 1987a
Free-ranging saddle-back tamarins; breeding males	0.23	0.33	0.08	0.37	Goldizen 1987a
Captive cotton-top tamarins; breeding & nonbreeding males	0.03	0.11	0.02	0.35	Price 1992b

tamarins were less likely to forage while carrying infants. She proposed that a single pair of individuals (i.e., a mother and father alone) would be unable to meet their foraging needs while sharing the care of infants. This argument assumes that the relationship between infant carrying and foraging is invariant, which may or may not be true. However, it does provide a basis for predictions regarding the cost of infant care in general and alloparenting in particular.

Price (1992b) found a similar relationship between foraging and infant transport in captive cotton-top tamarins; animals that were carrying infants were less likely to forage. This lack of activity in infant carriers might be related to the energetic or mechanical limitations placed upon the carrier. However, Price raised another interesting possibility. She noted that carriers of infants were more likely to be hidden than individuals not carrying infants. Because predation is likely to be an important pressure on callitrichids (Caine 1993), Price proposed that infant carriers may be forced to use concealment, given the possible limitations in their mobility. Comparisons across all small-bodied neotropical primates suggest that antipredation strategies have shaped infant-care patterns in these species as much if not more than energetic costs of reproduction (Caine 1993; Tardif 1994).

There is evidence, therefore, suggesting a cost associated with infant care in callitrichids. What is unknown is whether this cost translates into a meaningful force, in terms of selection. How much cost is too much cost? At what point does the cost associated with a behavior imply the requirement of benefits to balance such cost? Do the costs described here translate into a potential effect on fitness? There are presently no answers to these questions.

2.5 Evolution of Alloparental Behavior in Marmosets and Tamarins

The occurrence of alloparental behavior in primates is controlled by attraction to infants as well as by access to infants. Attraction to infants is widespread and evolutionarily conservative in primates. In the platyrrhines (South American primates), infant transport by females other than mothers is reported in *Aotus* and *Callicebus* (Wright 1984), *Saimiri* (Williams et al. 1994), and *Cebus* (O'Brien & Robinson 1991), as well as the Callitrichidae; in two of these genera (*Saimiri* and *Cebus*) females also are reported to nurse other females' infants. Infant transport by males is routinely observed in a smaller number of species, including *Aotus, Callicebus,* possibly *Pithecia,* and the Callitrichidae (Wright 1984; Vogt 1984). Given a lack of clear agreement on what species are likely to represent an ancestral form in platyrrhines, it is difficult to assess whether care by males arose separately for these genera. However, it is clear that care by males is not unique to callitrichids.

Although attraction to infants and motivation to carry are common across the platyrrhines, access to infants is highly variable across species. Because the primary infant-care activity of primates (i.e., physical transport of infants) can be accomplished by only one individual at a time, opportunities for care may be limited. Among the Old World monkeys, such differences in access to infants have been used to explain interspecific variation in allomaternal behavior. McKenna (1979), for example, suggests that alloparental behavior in langurs is common because mothers readily allow access to their infants in this relatively loose social arrangement. In macaques, however, the risk associated with loss of infant contact in a society with strong dominance hierarchies may be too great (Nicolson 1991). Evidence from common marmoset groups with more than one reproductive female also suggests that subordinate females in this callitrichid species may be less willing to relinquish their infants to others than are dominant females (Koenig & Rothe 1991b; Digby & Ferrari 1994).

The distribution of alloparental behavior in both Old World and New World primates suggests that the general frequency of its occurrence is better explained by variation in maternal tolerance for other caregivers rather than by variation in the willingness or motivation of potential caregivers. Virtually all primates are social, frequently living in groups of 10 to 100 individuals. In

Table 2.3. *Comparison across types of caregivers of ability to retrieve infants in captive cotton-top tamarins*

Carrier type[a]	n	Percentage with failed retrieval[b]	Percentage with no failed retrieval
Mother	36	13.9	86.1
Father	36	33.3	67.1
Nonreproductive alloparents	46	43.5	56.5

[a]Differences between carrier types are significant ($\chi^2 = 8.30$; $p < 0.025$).
[b]Failed retrieval: individual unsuccessfully attempts to take infant from another carrier; individuals with at least one observed failed retrieval are included in the percentage with failed retrieval.

such groups, mothers will relinquish their infants if such tolerance provides a benefit to the mother without excessive costs. Many callitrichid species offer such a setting because the breeding female is most likely to be dominant to every other member of the group. In such a social setting, mothers may be relatively sure of regaining their infants. For example, Table 2.3 compares the relative percentage of infant-retrieval attempts that were successfully resisted in captive cotton-top tamarins relative to the identity of the retriever. Mothers were more likely to successfully retrieve infants than were either fathers or alloparents.

Jamieson (1989) has proposed that alloparental behavior in birds results from the fact that care-taking responses are highly conserved, in general, as a result of the obvious impacts that adequate parental care has on fitness in species with altricial young. In conjunction, he proposes that alloparental behavior may not have been selected for but rather be a byproduct of strong selection for care behavior directed toward offspring. Likewise, Quaitt (1979) has proposed that alloparental behavior in primates is the result of strong selection for maternal care. Comparisons of the occurrence of alloparental behavior across primates support this argument (McKenna 1979). However, numerous researchers studying cooperative breeding in birds have taken exception to Jamieson's position. Emlen et al. (1991) propose that strong selection for parental behavior may form the basis for the origin of alloparental behavior but that it does not explain apparent adaptive modifications of such behaviors. If alloparental behavior is solely the byproduct of strong selection for parental behavior, the distribution of alloparental behavior across different categories of individuals (age, gender, etc.) should simply reflect population demography. If, however, alloparental behavior has been subjected to selection pressures, one can hypothesize specific effects on the type of individual that displays such behavior. Emlen et al. (1991) cite numerous examples of such modifications from social insects, a variety of cooperatively breeding birds, naked mole-rats, and dwarf monogooses.

The hypothesized benefits proposed in this review might be expected to shape the identity of those displaying alloparental behavior relative to age, sex, reproductive state, environment, and relatedness. Variation among individuals in infant-care behaviors has frequently been noted in callitrichids (Epple 1975; Wolters 1978; Tardif et al. 1986, 1990; Pryce 1988). This variation suggests that callitrichids may display adaptive modifications of alloparental behavior, because not everyone participates equally. However, there is no particularly strong support for any of the four hypotheses presented in this review. Specifically, there is some support for the hypothesis that alloparental behavior may improve future reproductive performance by providing experience in infant care. However, the evidence fails to provide specific mechanisms through which such experience might improve future reproductive performance and does not eliminate the possibility that simple exposure to infants, rather than infant-care experience per se, is the factor that differentiates the reproductive performance of experienced and inexperienced animals. The hypothesis that alloparental behavior enhances indirect fitness is supported by observation of higher degrees of infant care directed toward more closely related infants in at least one callitrichid species, the golden lion tamarin. However, evidence regarding the effects of such care by alloparents on the survival, growth, and future reproduction of relatives is either negative or lacking. Although philopatry may be the most successful strategy for acquiring a breeding position, there is no evidence that alloparental behaviors affect the success or failure of such acquisitions. Finally, there is some evidence that males that are potential breeders may improve their likelihood of mating by participating in infant care.

2.6 Conclusions

The best and most conservative conclusion regarding callitrichid alloparental behavior is that the available data are inadequate for assessing its adaptive value. Further data are essential on the following aspects:

1. Evidence regarding the effect of alloparents upon infant survival.
2. Evidence regarding the relative involvement of related versus nonrelated alloparents. Of specific interest is whether individuals that are differentially

related to a given infant differ in their alloparental behavior toward that infant. In this way, the behavior of the mother and her relationship to potential alloparents is not confused with dominance or parity.

3. Evidence regarding the effect of alloparenting upon retaining group membership and inheriting a breeding position. Here the benefits of philopatry must be separated from the benefits of alloparental behavior.

4. More evidence of the relationship between male infant care and breeding opportunities.

5. More detailed studies on the distribution of food provisioning among potential alloparents and its effects on infant survival.

Acknowledgments

I thank the editors of this volume (Nancy Solomon and Jeffrey French) for both the opportunity to participate in this endeavor and their helpful comments on earlier versions of this chapter. I also thank Anthony Rylands and Andrew Baker for their many thoughtful comments, which substantially improved this work. My research, which is reported upon in this chapter, is supported by National Institutes of Health grant R01 RR02022.

References

Baker, A. J. (1991). *Evolution of the social system of the golden lion tamarin (Leontopithecus rosalia): Mating system, group dynamics and cooperative breeding.* PhD dissertation, University of Maryland, College Park, Md.

Baker, A. J., & Woods, F. (1992). Reproduction of the emperor tamarin (*Saguinus imperator*) in captivity, with comparison to the cotton-top and golden lion tamarins. *Am. J. Primatol.* 26:1–10.

Baker, A. J., Dietz, J. M., & Kleiman, D. G. (1993). Behavioural evidence for monopolization of paternity in multi-male groups of golden lion tamarins. *Anim. Behav.* 46:1091–1103.

Brown, K., & Mack, D. S. (1978). Food sharing among captive *Leontopithecus rosalia. Folia primatol.* 29:268–290.

Caine, N. (1993). Flexibility and co-operation as unifying themes in *Saguinus* social organization and behaviour: the role of predation pressures. In *Marmosets and tamarins: Systematics, behaviour and ecology,* ed. A. Rylands, pp. 200–219. Oxford: Oxford University Press.

Cleveland, J., & Snowdon, C. T. (1984). Social development during the first twenty weeks in the cotton-top tamarin (*Saguinus o. oedipus*). *Anim. Behav.* 32:432–444.

Davies, N. B., Hatchwell, B. J., Robson, T., & Burke, T. (1992). Paternity and parental effort in dunnocks *Prunella modularis:* how good are male chick-feeding rules? *Anim. Behav.* 43:729–746.

Dawson, G. A. 1978. Composition and stability of social groups of the tamarin, *Saguinus oedipus geoffroyi,* in Panama: ecological and behavioral implications.

In *The biology and conservation of the Callitrichidae,* ed. D. G. Kleiman, pp. 23–37. Washington, D.C.: Smithson. Institution Press.

Dietz, J. M., & Baker, A. J. (1993). Polygyny and female reproductive success in golden lion tamarins, *Leontopithecus rosalia. Anim. Behav.* 46:1067–1078.

Digby, L., & Ferrari, S. F. (1994). Multiple breeding females in free-ranging groups of *Callithrix jacchus. Int. J. Primatol.* 15:389–397.

Dixson, A. F. (1982). Prolactin and parental behaviour in a male New World primate. *Nature, Lond.,* 299:551–553.

Eisenberg, J. F. (1981). *The mammalian radiations,* pp. 163–164. Chicago: University of Chicago Press.

Emlen, S. T., & Wrege, P. H. (1989). A test of alternate hypotheses for helping behavior in white-fronted bee-eaters of Kenya. *Behav. Ecol. Sociobiol.* 25:303–319.

Emlen, S. T., Reeve, H. K., Sherman, P. W., Wrege, P. H., Ratnikes, F. L. W., & Shellman-Reeve, J. (1991). Adaptive versus nonadaptive explanations of behavior: the case of alloparental helping. *Am. Nat.* 138:259–270.

Epple, G. (1975). Parental behavior in *Saguinus fuscicollis* ssp. (Callitrichidae). *Folia primatol.* 24:221–238.

Epple, G. (1978). Reproductive and social behavior of marmosets with special reference to captive breeding. In *Marmosets in experimental medicine,* ed. N. Gengozian & F. W. Deinhardt, pp. 50–62. Basel: S. Karger.

Fairbanks, L. A. (1990). Reciprocal benefits of allomothering for female vervet monkeys. *Anim. Behav.* 40:553–562.

Feistner, A. T. C., & Price, E. C. (1990). Food-sharing in cotton-top tamarins *(Saguinus oedipus). Folia primatol.* 54:34–45.

Feistner, A. T. C., & Price, E. C. (1991). Food offering in New World primates: two species added. *Folia primatol.* 57:165–168.

Ferrari, S. F. (1987). Food transfer in a wild marmoset group. *Folia primatol.* 48:203–206.

Ferrari, S. F., & Diego, V. H. (1992). Long-term changes in a wild marmoset group. *Folia primatol.* 58:215–218.

Ferrari, S. F., & Digby, L. J. (1996). Wild *Callithrix* groups: stable extended families? *Am. J. Primatol.* 38:19–28.

Ferrari, S. F., & Lopes-Ferrari, M. (1989). A re-evaluation of the social organization of the Callitrichidae, with reference to the ecological differences between genera. *Folia primatol.* 52:132–147.

Ford, S. M. (1980). Callitrichids as phyletic dwarfs and the place of the Callitrichidae in Platyrrhini. *Primates* 21:31–43.

Garber, P. A., Moya, L., & Malaga, C. (1984). A preliminary field study of the moustached tamarin monkey *(Saguinus mystax)* in northeastern Peru: questions concerned with the evolution of a communal breeding system. *Folia primatol.* 42:17–32.

Goldizen, A. W. (1987). Facultative polyandry and the role of infant-carrying in wild saddle-back tamarins *(Saguinus fuscicollis). Behav. Ecol. Sociobiol.* 20:99–109.

Goldizen, A. W., & Terborgh, J. (1989). Demography and dispersal patterns of a tamarin population: possible causes of delayed breeding. *Am. Nat.* 134:208–224.

Hoage, R. J. (1978). Parental care in *Leontopithecus rosalia rosalia:* sex and age difference in carrying behavior and the role of prior experience. In *The biology and conservation of the Callitrichidae,* ed. D. Kleiman, pp. 293–305. Washington, D.C.: Smithsonian Institution Press.

Ingram, J. C. (1977). Interactions between parents and infants, and the development of independence in the common marmoset *(Callithrix jacchus). Anim. Behav.* 25:811–827.

Jamieson, I. G. (1989). Behavioral heterochrony and the evolution of birds' helping at the nest: an unselected consequence of communal breeding. *Am. Nat.* 133:394–406.

Jaquish, C. E. (1993). *Genetic, behavioral and social effects on fitness components in marmosets and tamarins (family: Callitrichidae).* PhD dissertation, Washington University, St. Louis, Mo.

Kirkwood, J. K., & Underwood, S. J. (1984). Energy requirements of captive cotton-top tamarins, *Saguinus oedipus oedipus. Folia primatol.* 42:180–187.

Kleiman, D. G. (1977). Monogamy in mammals. *Q. Rev. Biol.* 52:39–69.

Koenig, A., & Rothe, H. (1991a). Social relationships and individual contribution to cooperative behaviour in captive common marmosets (*Callithrix jacchus*). *Primates* 32:183–195.

Koenig, A., & Rothe, H. (1991b). Infant carrying in a polygynous group of the common marmoset (*Callithrix jacchus*). *Am. J. Primatol.* 25:185–190.

Koenig, W. D., & Mumme, R. L. (1987). *Population ecology of the cooperative breeding acorn woodpecker.* Princeton: Princeton University Press.

Lancaster, J. (1971). Play-mothering: the relations between juvenile females and young infants among free-ranging vervet monkeys (*Cercopithecus aethiops*). *Folia primatol.* 15:161–182.

Leutenegger, W. (1980). Monogamy in callitrichids: a consequence of phyletic dwarfism. *Intl. J. Primatol.* 1:95–98.

Ligon, J. D., & Ligon, S. H. (1988). Territory quality: key determinant of fitness in the group-living green woodhoopoe. In *The ecology of social behavior,* ed. C. N. Slobodchikoff, pp. 229–253. New York: Academic Press.

McKenna, J. J. (1979). The evolution of allomothering behavior among colobine monkeys: function and opportunism in evolution. *Am. Anthropol.* 81:818–840.

Meaney, M. J., Lozos, E., & Stewart, J. (1980). Infant carrying by nulliparous females vervet monkeys (*Cercopithecus aethiops*). *J. Comp. Psychol.* 104: 377–381.

Neyman, P. F. (1978). Aspects of the ecology and social organization of the free-ranging cotton-top tamarins (*Saguinus oedipus*) and the conservation status of the species. In *The biology and conservation of the Callitrichidae,* ed. D. G. Kleiman, pp. 39–71. Washington, D.C.: Smithson Institution Press.

Nicolson, N. A. (1987). Infants, mothers and other female. In *Primate societies,* eds. B. Smuts, D. Cheney, R. Seyfarth, R. Wrangham, & T. Struhsaker, pp. 330–342. Chicago: University of Chicago Press.

Nicolson, N. A. (1991). Maternal behavior in human and nonhuman primates. In *Understanding behavior,* ed. J. D. Loy & C. B. Peters, pp. 17–50. Oxford: Oxford University Press.

O'Brien, T. G., & Robinson, J. G. (1991). Allomaternal care by female wedge-capped capuchin monkeys: effects of age, rank and relatedness. *Behaviour* 119:30–50.

Price, E. C. (1990a). Infant carrying as a courtship strategy of breeding male cotton-top tamarins. *Anim. Behav.* 40:784–786.

Price, E. C. (1990b). Parturition and perinatal behaviour in captive cotton-top tamarins (*Saguinus oedipus*). *Primates* 31:523–535.

Price, E. C. (1991). Competition to carry infants in captive families of cotton-top tamarins (*Saguinus oedipus*). *Behaviour* 118:66–88.

Price, E. C. (1992a). Contributions to infant care in captive cotton-top tamarins (*Saguinus oedipus*): the influence of age, sex and reproductive status. *Intl. J. Primatol.* 13:125–141.

Price, E. C. (1992b). The costs of infant carrying in captive cotton-top tamarins. *Am. J. Primatol.* 26:23–33.

Price, E. C., & Feistner, A. T. C. (1993). Food sharing in lion tamarins: tests of three hypotheses. *Am. J. Primatol.* 31:211–221.

Price, E. C., & McGrew, W. C. (1991). Departures from monogamy in colonies of captive cotton-top tamarins. *Folia primatol.* 57:16–27.

Pryce, C. R. (1988). Individual and group effects on early caregiver–infant relationships in red-bellied tamarin monkeys. *Anim. Behav.* 36:1455–1464.

Pryce, C. R. (1992). A comparative systems model of the regulation of maternal motivation in mammals. *Anim. Behav.* 43:417–441.

Pryce, C. R. (1993). The regulation of maternal behaviour in marmosets and tamarins. *Behav. Processes* 30:201–224.

Pryce, C. R., Döbeli, M., & Martin, R. D. (1993). Effects of sex steroids on maternal motivation in the common marmoset (*Callithrix jacchus*): development and application of an operant system with maternal reinforcement. *J. Comp. Psychol.* 107:99–115.

Quiatt, D. (1979). Aunts and mothers: adaptive implications of allomaternal behavior of nonhuman primates. *Am. Anthropol.* 81:310–319.

Ramirez, M. (1984). Population recovery in the moustached tamarin (*Saguinus mystax*): management strategies and mechanisms of recovery. *Am. J. Primatol.* 7:245–259.

Rothe, H. (1974). Further observations on the delivery behaviour of the common marmoset (*Callithrix jacchus*). *Z. Saugetierk.* 39:135–142.

Rothe, H., & Darms, K. (1993). The social organization of marmosets: a critical evaluation of recent concepts, In *Marmosets and tamarins: Systematics, behaviour and ecology,* ed. A. Rylands, pp. 176–199. Oxford: Oxford University Press.

Rothe, H., Koenig, A., & Darms, K. 1993. Infant survival and number of helpers in captive groups of common marmoset (*Callithrix jacchus*). *Am. J. Primatol.* 30:131–137.

Rylands, A. B. (1982). *Behaviour and ecology of three species of marmosets and tamarins (Callitrichidae, Primates) in Brazil.* PhD dissertation, University of Cambridge, Cambridge.

Rylands, A. B. (1986). Infant-carrying in a wild marmoset group, *Callithrix humeralifer:* evidence for a polyandrous mating system. In *Primatologia No. Brasil-2,* ed., M. T. deMello, pp. 131–144. Brasilia: Sociedade Brasileira de Primatologia.

Rylands, A. B. (1996). Habitat and the evolution of social and reproductive behaviour in Callitrichidae. *Am. J. Primatol.* 38:5–18.

Snowdon, C. T., Savage, A., & McConnell, P. B. (1985). A breeding colony of cotton-top tamarins (*Saguinus oedipus*). *Lab. Anim. Sci.* 35:477–480.

Solomon, N. G. (1991). Current indirect fitness benefits associated with philopatry in juvenile prairie voles. *Behav. Ecol. Sociobiol.* 29:277–282.

Solomon, N. G. (1994). Effect of the pre-weaning environment on subsequent reproduction in prairie voles. *Microtus ochrogaster. Anim. Behav.* 48:331–341.

Stearns, S. C. (1992). *The evolution of life histories,* pp. 72–90. Oxford: Oxford University Press.

Sussman, R. W., & Garber, P. A. (1987). A new interpretation of the social organization and mating system of the Callitrichidae. *Intl. J. Primatol.* 8:73–92.

Tardif, S. D. (1994). Relative energetic cost of infant care in small-bodied neotropical primates and its relation to infant-care patterns. *Am. J. Primatol* 34:133–144.

Tardif, S. D., Carson, R. L., & Gangaware, B. L. (1986). Comparison of infant-care in family groups of the common marmoset (*Callithrix jacchus*) and the cotton-top tamarin (*Saguinus oedipus*). *Am. J. Primatol* 11:103–110.

Tardif, S. D., Carson, R. L., & Gangaware, B. L. (1990). Infant-care behavior of mothers and fathers in a communal-care primate, the cotton-top tamarin (*Saguinus oedipus*). *Am. J. Primatol.* 22:73–85.

Tardif, S. D., Carson, R. L., & Gangaware, B. L. (1992). Infant-care behavior of non-reproductive helpers in a communal-care primate, the cotton-top tamarin (*Saguinus oedipus*). *Ethology* 92:155–167.

Tardif, S. D., & Harrison, M. L. (1990). Estimates of the energetic cost of infant transport in tamarins. *Am. J. Phys. Anthropol.* 81:306 (abstract).

Tardif, S. D., Harrison, M. L., & Simek, M. A. (1993). Communal infant care in marmosets and tamarins: relation to energetics, ecology and social organization. In *Marmosets and tamarins: Systematics, behaviour and ecology,* ed. A. Rylands, pp. 220–234. Oxford: Oxford University Press.

Tardif, S. D., Richter, C. B., & Carson, R. L. (1984). Effects of sibling-rearing experience on future reproductive success in two species in Callitrichidae. *Am. J. Primatol.* 6:377–380.

Taylor, C. R. (1980). Energetic cost of generating muscular force during running. *J. Exp. Biol.* 86:9–18.

Vogt, J. L. (1984). Interactions between adult males and infants in prosimians and New World monkeys. In *Primate paternalism,* ed. D. M. Taub, pp. 346–376. New York: Van Nostrand.

Wamboldt, M. Z., Gelhard, R. E., & Insel, T. R. (1988). Gender differences in caring for infant *Cebuella pygmaea:* the role of infant age and relatedness. *Dev. Psychobiol.* 21:187–202.

Williams, L., Gibson, S., McDaniel, M., Bazzel, J., Barnes, S., & Abee, C. (1994). Allomaternal interactions in the Bolivian squirrel monkey (*Saimiri boliviensis boliviensis*). *Am. J. Primatol.* 34:145–156.

Wolters, H.-J. (1978). Some aspects of role taking behaviour in captive family groups of the cotton-top tamarian. In *Biology and behaviour of marmosets,* ed. H. Rothe, H. J. Wolters, & J. P. Hearn, pp. 259–278. Gottingen: Eigenverlag Rothe.

Wright, P. C. (1984). Biparental care in *Aotus trivirgatus* and *Callicebus moloch.* In *Female primates: Studies by women primatologists,* ed. M. Small, pp. 59–75. New York: Liss.

3

Proximate Regulation of Singular Breeding in Callitrichid Primates

JEFFREY A. FRENCH

3.1 Introduction

The fundamental issues that confront students of cooperative breeding are delayed dispersal or philopatry, helping behavior, and delayed breeding. These are central questions, regardless of the taxonomic group under study or the level of analysis (mechanistic, developmental, functional, or evolutionary) that guides the formation of questions about cooperative breeding (Brown 1987; Emlen 1991; Mumme this volume). The first set of questions concern philopatry and delayed dispersal. What are the ecological factors that promote retention of individuals, often offspring of the breeding adults, in the social group? Does the pattern of intragroup social interactions differ in species that exhibit delayed dispersal from those in which offspring disperse at sexual maturation (Chepko-Sade & Halpin 1987)? What are the relative roles of the benefits for delayed dispersers that may accrue from philopatry with respect to potential constraints on dispersal (Koenig et al. 1992; Mumme this volume)?

The second question concerns the expression of helping or alloparental behavior. The primary issues here concern, on a proximate level, the underlying physiological (e.g., Carter & Roberts this volume), motivational (Pryce 1992; Pryce, Döbeli, & Martin 1993), and behavioral (Tardif, Harrison, & Simek 1993; Tardif, this volume) regulation of care-giving behavior in alloparents. Are these mechanisms controlling alloparental care similar to those that control maternal and paternal behavior, or have the control systems been modified and specially adapted for alloparental care (e.g., Jamieson 1989)? On an ultimate level, the major questions concern the functional consequences of alloparental behavior for the recipients of this care and the fitness consequences of engaging in (or failing to engage in) these activities, both for the nonbreeding caregiver and for the breeding adults in the group.

The third central question of cooperative breeding concerns the identity and number of individuals that engage in reproductive activity within each group. One of the most distinctive variations in systems of cooperative breeding is with respect to the number of adult females per group that typically produce offspring. In one form, reproduction in groups is monogynous: Breeding is typically limited to a single adult female. The avian literature has historically referred to this subtype as "singular breeding." For only a single female to produce offspring, in spite of multiple females of breeding age in groups, requires either a delay in the onset of reproductive activity in individuals that mature within a group or a suppression of reproduction in otherwise reproductively competent females who may join a group. The alternative expression of cooperative breeding occurs in groups in which more than one adult female is actively breeding and producing offspring. Tradition in the avian literature has referred to this polygynous form as "plural breeding." Among mammals, the expression of cooperative breeding is more likely to take the singular than the plural form (Emlen 1991; Creel & Waser this volume; but see Solomon & Getz this volume). Cooperative breeding systems present behavioral biologists with this dilemma: *How and why is reproduction typically limited to a single individual?* The nature of the stimuli that induce and maintain reduced reproductive success in subordinates, and the transduction of these signals into physiological or reproductive consequences, can provide insights into the proximate regulation of singular breeding. It has been argued (e.g., Abbott 1987) that the maintenance of singular breeding in cooperatively breeding groups is one of the most dramatic demonstrations of rank-related variance in reproductive success. On an ultimate level, delayed breeding (reproductive suppression) potentially represents a substantial cost in terms of lifetime direct fitness, compared to individuals that do not delay breeding.

The evolution and maintenance of reproductive suppression in subordinates has typically been approached from the perspective of dominance interactions and reproductive hegemony (e.g., Vehrencamp 1983; Abbott 1987; French & Inglett 1991). Recent work, however, has emphasized an evaluation of delayed breeding and reproductive suppression in a life-history context (Wasser & Barash 1983), particularly with reference to the energetics of reproduction (see Creel & Creel 1991; Creel & Waser 1991; Creel & Waser this volume). Recent attention has been directed toward the issue of tolerance of reproductive suppression by subordinates. Growing evidence suggests that there are conditions under which dominant breeders cannot (or do not) inhibit reproduction in subordinates, or under which subordinates will not tolerate inhibition (Keller & Reeve 1994). The nature of the social and physical environments that contribute to the occurrence of breeding by subordinates in typically singular breeding systems have recently come to the fore.

Approximately 40 species in four genera comprise the marmosets and tamarins (order Primates, family Callitrichidae). They include the monophyletic pygmy marmoset (genus *Cebuella*), the marmosets (genus *Callithrix*), the tamarins (genus *Saguinus*), and the lion tamarins (genus *Leontopithecus*). Some recent taxonomies of New World primates (e.g., Kay 1990) include Goeldi's monkey (*Callimico goeldi*) with the marmosets and tamarins at the family level. As we shall see later, this species provides some useful comparative leverage for understanding the maintenance of singular breeding. Callitrichid primates are distributed widely throughout the primary and secondary forests of the neotropics. As a whole, these species are characterized by small, highly territorial groups of 4 to 15 individuals.

The callitrichid primates exhibit the three classic diagnostic characteristics associated with cooperative breeding, including delayed dispersal (Goldizen & Terborgh 1989), alloparental care (Tardif this volume), and delayed breeding by subordinates (Abbott 1987; French & Inglett 1991). The focus of this chapter will be on the latter phenomenon: delayed or suppressed reproduction. I will examine the complex interplay between behavioral interactions, social demography, reproductive life history, and reproductive physiology in the regulation of singular breeding in primates of the family Callitrichidae. I will review the proximal mechanisms – olfactory, behavioral, social, and physiological – by which reproductive activity is typically limited to a single female in callitrichid social groups. Next will follow a discussion and analysis of the causes and contexts of departures from singular breeding in callitrichid groups. As in other mammalian taxa (e.g., the families Canidae and Viverridae) (Moehlman & Hofer this volume; Creel & Waser this volume), cases of exceptions to species-typical profiles pose interesting challenges for theories of reproductive biology in these species. Understanding and accounting for these variations in the degree of reproductive skew in group members (Vehrencamp 1983; Keller & Reeve 1994), both within and among species, can shed light on the evolution of cooperation. I will conclude with a discussion of the importance of variations in reproductive life histories in the interpretation of taxonomic differences in the mechanisms that maintain singular breeding. The chapter will therefore provide a preliminary framework for understanding intra- and interspecific variation in the regulation of reproductive function in subordinate females in the family Callitrichidae.

A preliminary comment is required on the taxonomic scope of this review. I have included descriptions of mechanisms underlying reproductive skew in females for species representing all four formally recognized genera of callitrichid primates, *Cebuella, Callithrix, Saguinus,* and *Leontopithecus.* I have also included a discussion of these issues in Goeldi's marmoset, *Callimico goeldi.* The taxonomic relationship of this species to representatives of the other four taxa of callitrichids is contentious (see reviews in Ford & Davis 1992; Rylands, Coimbra-Filho, & Mittermeier 1993). However, its inclusion in this review is warranted on the basis of several arguments. First, *Callimico* and the callitrichid primates are considered to be monophyletic, at least at the family level, in several recent taxonomies (Rosenberger 1981; Ford 1986; Kay 1990). Second, a comparison of the two groups reveals widespread cross-taxa similarity on some of the major diagnostic characters used to differentiate neotropical primates, including the possession of clawed digits and second molar reduction. Finally, one of the areas in which the two groups differ substantially is in reproductive biology, with *Callimico* typically producing singleton infants whereas modal litter size among the callitrichid genera is twin infants. This difference becomes particularly relevant in the discussion to follow later in the chapter on taxonomic variation in mechanisms of suppression.

3.2 Proximate Mechanisms Maintaining Singular Breeding in Callitrichid Primates

Kleiman (1980) outlined a conceptual orientation of the potential mechanisms that could produce reproductive failure or reduced reproductive success in subordinate animals. The mechanisms that would produce reproductive failure would vary along two dimensions, one temporal and one mechanistic. With respect to the temporal dimension, Kleiman divided the suppression continuum into a preconception and a postconception period. In other words, reproduction in subordinates could be inhibited or suppressed either prior to the conception of offspring by subordinates or at some point after conception, including the time of gestation, the perinatal period, and the postnatal period of infant dependence. On the mechanistic dimension, Kleiman suggested that reproductive failures could be either physiological or behavioral in origin. Figure 3.1 portrays the two-by-two matrix that is produced by these two dimensions, along with examples of how the potential combination of these factors might affect reproduction. The obvious conclusion is that the same functional outcome – the reduction or suppression of reproductive performance in subordinate individuals – can be produced by a tremendous diversity of underlying mechanisms. We turn now to a discussion of the specific mechanisms underlying reproductive inhibition in subordinate females in callitrichid primates.

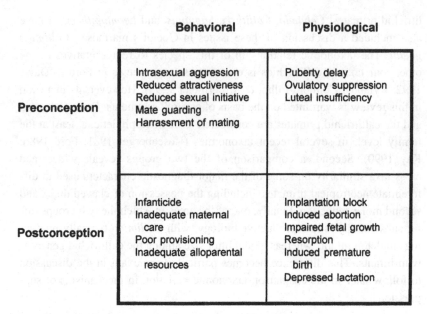

	Behavioral	**Physiological**
Preconception	Intrasexual aggression Reduced attractiveness Reduced sexual initiative Mate guarding Harrassment of mating	Puberty delay Ovulatory suppression Luteal insufficiency
Postconception	Infanticide Inadequate maternal care Poor provisioning Inadequate alloparental resources	Implantation block Induced abortion Impaired fetal growth Resorption Induced premature birth Depressed lactation

Figure 3.1. Two-by-two matrix of potential sources of skew in reproduction output among females in cooperatively breeding social groups. The sources are divided according to mechanistic (physiological vs. behavioral) and temporal (preconception vs. postconception) variables. Original categories highlighted by Kleiman (1980).

3.2.1 Physiological Mechanisms

3.2.1.1 Puberty delay

Strictly speaking, puberty delay refers to a temporal delay in the maturation of the neuroendocrine processes that underlie normal reproductive function, and it is conceptually distinct from other forms of ovulatory or reproductive failure. Acceleration or stimulation of puberty refers to the advancement of the onset of reproductive capacity. Several reports on callitrichid primates suggest that the timing of the ontogeny of reproductive competence in females may be regulated by social factors, but it is not clear whether these examples represent clear cases of delay in pubertal processes instead of some more generalized reproductive disruption. In the cotton-top tamarin (*Saguinus oedipus*), Tardif (1984) demonstrated that the mean age at the first sign of reproductive maturation (elevated levels of plasma progesterone) occurred at less than 500 days of age in six females housed away from their natal group, but maturation occurred in only one of five females housed in their natal group prior to 550 days of age. Thus, if one defines puberty as the process that culminates in the first ovulation, puberty is delayed in subordinate daughters. Studies by French, Abbott, and Snowdown (1984), Ziegler et al. (1987), and Savage, Ziegler, and Snowdon (1988) on the cotton-top tamarin; those of Evans and Hodges (1984) on common marmosets (*Callithrix jacchus*); Epple and Katz (1984) on saddleback tamarins (*S. fuscicollis*); and Küderling, Evans, and Abbott (1992) on red-bellied tamarins (*S. labiatus*), demonstrate that first ovulation in daughters either does not occur in the natal family group or is delayed while in the natal family group. However, there is some variability among studies in the proportion of daughters that experience anovulation in natal family groups (Abbott 1984; Tardif 1984; Saltzman et al. 1994). Thus, the available evidence on reproductive development in many species of female callitrichid primates is consistent with puberty delay, but these effects may be difficult to dissociate from more generalized effects on reproductive function (see later discussion of ovulatory suppression).

Olfactory cues from dominant females may serve an important role as part of the cluster of proximate stimuli that may produce delays in first ovulation in subordinate daughters. Urinary estrogen cycles in females that are removed from their family groups but that continue to receive olfactory stimuli associated with the dominant adult female or the natal family group exhibit atypical and probably nonovulatory ovarian cycles (cotton-top tamarin: Savage et al. 1988; saddleback tamarins: Epple & Katz 1984). Thus, there may be strong inhibitory influences on puberty as a result of olfactory cues produced by the dominant breeding female. However, as will be seen later in this review, these olfactory cues are not sufficient to maintain the anovulatory status of subordinate females.

Several lines of evidence suggest that the maturational delay in first ovulation observed in females housed in their natal family group may be a consequence of the absence of appropriate stimuli that induce or stimulate reproductive maturation, rather than, or in addition to, the presence of stimuli that may inhibit ovulation. First, older daughters in natal family groups show higher basal levels of estrogen excretion than do younger ones, suggesting that some degree of maturation may occur even while daughters are in the presence of breeding females (French et al. 1984; Ziegler et al. 1987; Savage et al. 1988). In addition, young female cotton-top tamarins housed away from the natal group but in the absence of stimulation from unfamiliar animals (especially males) exhibit increases in levels of urinary estrogens and gonadotropins, but fail to show organized ovulatory cycles (Ziegler et al. 1987; Savage et al. 1988). However, several studies have reported the onset of normal ovarian function in older daughters in family groups accompanied by the death or removal of the breeding adult female (French et al. 1984;

Heistermann et al. 1989) or in the absence of stimulation by unfamiliar males (Tardif 1984). Clearly, the ontogeny of reproductive competency in female callitrichid primates is a complex and multiply determined process.

Studies over the past few years by Ziegler and Snowdon and colleagues have revealed some of the complexity of reproductive development in the cotton-top tamarin. It appears that at least for young females (around 1 year of age), removal from the natal group and being housed in isolation is associated with increased ovarian activity (higher levels of urinary estrogen), but the qualitative nature of the hormonal profiles suggests that normal ovulatory cycles are not produced under these conditions (Ziegler et al. 1987; Savage et al. 1988). These findings confirm that natal family groups are the source of stimuli that limit ovarian function but further suggest that other critical stimuli may be required to initiate functional ovulatory cycles (see Carter & Roberts, this volume, for similar processes in prairie voles, *Microtus ochrogaster*).

Widowski et al. (1990) refined the class of stimuli that potentially stimulate reproductive development in female cotton-top tamarins by monitoring reproductive development in females under one of four conditions: in the natal family group, housed with familiar males (siblings), housed singly but exposed to urinary cues from unfamiliar males, and paired with an unfamiliar male. The endocrine status of females housed with familiar males did not differ from females housed in natal family groups, indicating that familiar males lack the capacity to stimulate the onset of reproductive function. Isolated females had significantly higher urinary estrogen levels than natal group females, although the patterns did not appear to be organized into ovulatory cycles. Urinary cues from unfamiliar males alone were not sufficient to influence any measure of reproductive or behavioral maturity. Once paired with unfamiliar males, however, five of six females showed organized ovulatory cycles or became pregnant shortly after pairing. Thus, it appears that a more complex set of stimuli emanating from unfamiliar adult males, including auditory, visual, and olfactory contact (Widowski et al. 1992), is necessary to produce the full expression of ovulatory competency in cotton-top tamarins. Such male stimulation may be more important for younger subordinate females than for older subordinates, because the onset of reproductive cyclicity in older females can occur in the absence of stimulation from unfamiliar males (French et al. 1984; Heistermann et al. 1989). These findings, nonetheless, reveal the importance of male stimulation, in addition to the potential role of female-mediated inhibition, in the regulation of singular breeding in callitrichids.

3.2.1.2 Suppression of ovulation

There is considerable evidence in a number of species that ovulation is inhibited or suppressed in all but the breeding female in the social group. To date, ovulatory failure in subordinates has been clearly demonstrated in representatives of two of the four genera of callitrichid primates (*Saguinus oedipus:* French et al. 1984; Ziegler et al. 1987; Savage et al. 1988; Heistermann et al. 1989; *S. fuscicollis:* Epple & Katz 1984; *S. labiatus:* Küderling et al. 1992; *Callithrix jacchus:* Abbott & Hearn 1978; Abbott 1984; Evans & Hodges 1984; Abbott, Hodges, & George 1988). Preliminary data from our laboratory suggest a similar phenomenon in the black tufted-ear marmoset (*Callithrix kuhli*) as well (Smith, Schaffner, & French 1995). Figure 3.2 presents examples of the distinctive ovulatory failure associated with social subordination in representative species of *Saguinus* and *Callithrix*.

In the common marmoset, there may be some distinction between the level or intensity of ovulatory suppression as it is expressed in intact family groups versus experimentally constituted "peer groups" of unrelated adult males and females. In groups of unrelated adults established in captivity, there is a high rate of ovulatory failure in all but the dominant female (Abbott 1984, 1986, 1993; Abbott & George 1991). In family groups, however, there is evidence that a proportion of daughters may exhibit normal ovulatory cycles (Abbott 1984; Hubrecht 1989; Saltzman et al. 1994a; but see Evans & Hodges 1984). Interestingly, variations in ovulatory status do not translate into variations in sociosexual behavior, at least within the intact family group. Daughters that exhibit ovulatory patterns of plasma progesterone fail to show sexual behavior and do not become pregnant while in the family group (Abbott 1984). However, they readily engage in sexual solicitations and copulatory patterns when presented with unrelated males (Hubrecht 1989).

Olfactory cues have also been implicated in the mediation of ovulatory suppression, at least in the common marmoset. However, the role of olfaction in the inhibition of ovulation is complex. The transfer of odors collected from the breeding female to a subordinate and acyclic female delays the time until first ovulation from 10 days (control scent transfer) to 31 days after the subordinate has been removed from the cage containing the dominant female (Barrett, Abbott, & George 1990). This effect appears to be produced by individual (and not status-related) cues in the scent marks, since the transfer of marks associated with an unfamiliar dominant female do not delay ovulation in subordinates who have been removed from their social groups (Smith, 1994). Olfactory cues appear to be important in the *establishment* but not the

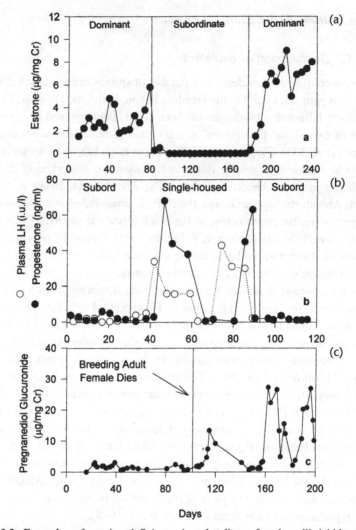

Figure 3.2. Examples of ovarian deficiency in subordinate female callitrichids. (a) Urinary estrogen profiles in cotton-top tamarin (*Saguinus oedipus*) female show adult patterns of hormone excretion while housed alone with a male, then suppressed ovarian function while housed in presence of dominant female, then a return to cyclicity while housed alone with another, unfamiliar male (modified and reprinted with permission from French et al. 1984); (b) subordinate female common marmoset exhibits lack of ovarian cyclicity while subordinate in social group, then displays two normal ovulatory cycles while housed singly, and finally shows ovulatory failure when placed back as subordinate in social group (from Smith 1994, with permission); (c) profile of urinary pregnanediol-glucuronide excretion in year-old female black tufted-ear marmoset (*Callithrix kuhli*) immediately before and following death of dominant breeding female in group. Death of the adult female was associated with onset of ovarian cycles. Vertical lines indicate ±1 s.c.m. From Smith, Schaffner, & French, unpublished results.

maintenance of singular breeding in the common marmoset, since destruction of the primary and accessory olfactory systems eliminates dominance-related differences in reproductive function among females in newly established groups but not in well-established groups (Barrett, Abbott, & George 1993). Cues arising from dominance-related behavioral interactions may be sufficient to maintain singular breeding after hierarchies have been firmly established.

A detailed discussion of the physiological mechanisms underlying reproductive inhibition in female callitrichids is beyond the scope of this chapter. Abbott et al. (1990) provide an excellent review of this topic in the common marmoset, and the locus of reproductive failure in subordinates appears to be hypothalamic and/or pituitary insufficiency. It is relatively clear, however, that in *Callithrix* and *Saguinus* reproductive failure or delay in first ovulation is *not* produced by stress-induced pathology or by high levels of adrenal corticosteroids. In acyclic subordinate female common marmosets, levels of plasma corticosteroids are lower than those in dominant cycling females (Abbott et al. 1981). In cotton-top tamarins, as in the common marmoset, noncycling daughters in natal family groups exhibit lower levels of corticosteroid excretion than do recently paired females (Ziegler, Scheffler, & Snowdon 1995). In fact, adrenal activity in the common marmoset is positively correlated with social status and hence with ovarian status. Upon group formation and the establishment of clear dominance hierarchies among females, marmosets that became cyclic showed increases in plasma cortisol, whereas those that were cyclic prior to group formation but became noncyclic (and subordinate) showed decreases in plasma cortisol (Saltzman et al. 1994b). These data are therefore inconsistent with the interpretation that reproductive inhibition in subordinates is mediated through adrenocortical activity. Rather, they suggest that the link between dominance status and adrenocortical activity is mediated by reproductive, and not social, status.

There has been one recent report on the endocrinology of ovarian cycles and pregnancy in the pygmy marmoset, *Cebuella* (Ziegler et al. 1990). However, to date nothing is known about the reproductive status of subordinate females and daughters in this species. On the basis of reproductive demographics from wild populations of pygmy marmosets, the pygmy marmoset resembles *Callithrix* and *Saguinus* in that breeding and sexual activity is limited to a single breeding female in each group (Soini 1987, 1988, 1993).

The fourth genus of callitrichid primates is the lion tamarin, genus *Leontopithecus*. Of the four known species (or subspecies; see Rylands et al. 1993) of lion tamarins, information on female reproductive biology has been published for only the golden lion tamarin, *L. rosalia* (see summary in French

& Inglett in press). Again, on the basis of the reproductive demographics of
captive breeding, this species resembles other tamarins and marmosets –
namely, production of offspring is limited to a single female (Kleiman et al.
1982). However, the endocrine profile of reproductively inactive subordinate
females is markedly different in this species.

We have monitored reproductive function in a number of daughters and
subordinate females over the past decade (French & Stribley 1985; 1987;
French, Inglett, & Dethlefs 1989; French & Inglett 1991; French et al. 1992).
There is a clear maturational effect on hormonal profiles. Daughters under 12
months of age in family groups exhibit extremely low concentrations of estrogen and gonadotropin excretion, and no signs of coordinated cycles are
observed in young females. However, all females older than 16 months of age
living in their natal family groups exhibit high concentrations of estrogen
excretion, levels that are statistically indistinguishable from those of dominant breeding females (Figure 3.3). Most surprisingly, all daughters and/or
behaviorally subordinate females exhibit normal ovarian cycles. Even subtle
details of ovarian cycle dynamics, such as cycle length and luteal phase
length, do not differ between subordinate daughters and females housed apart
from their natal family groups. Figure 3.4 presents summary statistics on
cycle length from females housed in the two social contexts (Inglett 1993).
Qualitatively, the cycles in subordinate daughters are indistinguishable from
those of dominant breeding females, exhibiting the sinusoidal pattern of estrogen elevation that is consistent with ovulatory cycles (French et al. 1992). Our
results also suggest that cycles in subordinates represent viable ovulatory
cycles, because overall cycle length and luteal phase length do not differ
among subordinate and dominant females (Figure 3.4). Thus, the luteal insufficiency seen in subordinate female cycles in other species (e.g., *C. jacchus*,
Abbott & George 1991) does not occur in the cycles of subordinate female
golden lion tamarins. In addition, there is no evidence of sporadic or "oligo-
cyclic" profiles of ovulation (Saltzman et al. 1994) in subordinate female lion
tamarins.

Overall, we have monitored urinary estrogen cycles and the potential
occurrence of sexual behavior in eight subordinate females (seven daughters
and one subordinate female in a two-female "peer group"). These observations covered a total of 119 female-months. In all cases, subordinate females
exhibited normal urinary estrogen cycles, and in no cases were there even
short periods of acyclicity (French et al. 1989; unpublished observations). In
spite of this apparently normal reproductive physiology, subordinate females
do not receive or solicit sexual activity from males in the group that, in the
case of the seven daughters, were either male siblings or fathers. The behav-

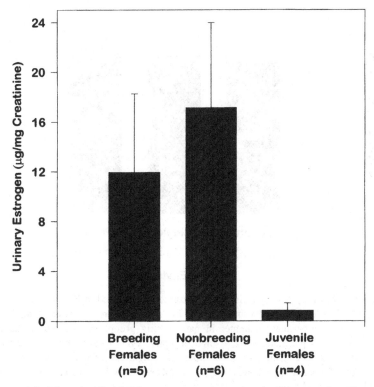

Figure 3.3. Mean levels of urinary estrogen excretion in lion tamarins. *Breeding females* = adult females that occupy the dominant and breeding position in their social group; *nonbreeding females* = adult-aged daughters (>2 years old) living in natal family groups; *juvenile females* = daughters less than 14 months of age in natal family groups. Data from French and Stribley 1985 and French et al. 1989.

ioral potential for sexual activity, however, is clearly present in these subordinate females. Inglett (1993) arranged encounters in neutral cages between daughters housed in family groups and unfamiliar, unrelated male partners. During these encounters, all daughters engaged in sexual activity. During the 30-minute tests, females received high levels of anogenital investigation from males, adopted species-typical solicitation patterns, and engaged in relatively high levels of copulatory behavior. Upon return to their home cage, these females again adopted a subordinate role and neither solicited nor received sexual interactions with potential male partners (Inglett 1993).

In golden lion tamarins, therefore, a different picture emerges with respect to the role of physiological mechanisms in the production and maintenance of

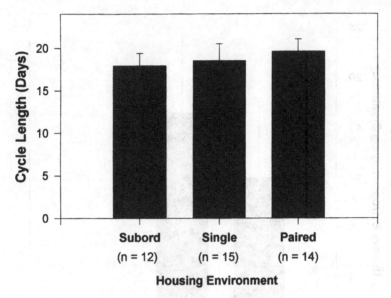

Figure 3.4. Mean cycle length (in days, measured as described in French et al. 1992) in adult female lion tamarins housed in three different social conditions. *Subordinate* = subordinate daughter in natal family group; *single* = housed alone without family or pairmate; *paired* = nonconceptive cycle length while housed with unrelated adult male. Data for *Subordinate* and *Single* from Inglett 1993; data for *Paired* from French and Stribley, 1985; *n* = number of complete ovarian cycles represented in each condition.

singular breeding. Reproductive physiology appears to be unaffected in subordinate females. Social context exerts a dramatic influence on the quality of sociosexual interactions between subordinate females and males. In the context of the natal family group, daughters do not appear to be attractive to males (or at least males do not engage in sexually oriented anogenital sniffing), daughters do not actively solicit sexual attention from males, and copulatory behavior has never been observed with a subordinate female. Outside the context of the natal family group, subordinate females are highly attractive to unrelated males, actively solicit sexual activity, and fully express the behavioral potential for copulatory interactions.

One study has addressed the endocrinology of reproductive function in subordinate females in Goeldi's monkey, *Callimico goeldii* (Carroll et al. 1990). Animals were housed in trios consisting of a single adult male and two adult females. In two of the three trios, both females exhibited ovarian cycles and became pregnant. Both also carried the infants to term. In the third trio, the two females exhibited ovarian cyclicity prior to the introduction of the male. Once the male was introduced, both females ceased cycling, and only one resumed ovarian cyclicity. Thus, in *Callimico,* physiological suppression of ovulation does not appear to be common, and multiple reproductively active females can co-exist in social groups. If reproductive success differs between dominant and subordinate female in this species, then, the mechanisms producing this outcome may more closely resemble those in lion tamarins than in the other callitrichid primates.

3.2.1.3 Postconception physiological regulation of fertility in subordinates

Although Kleiman's categories suggest a variety of postconception physiological mechanisms that could potentially disrupt reproduction in subordinate female callitrichids, there is as yet no convincing demonstration that these mechanisms may operate. There is indirect evidence of stress- or aggression-induced abortions or miscarriages in subordinate females in two captive groups of pygmy marmosets (Snowdon, pers. comm. 1994). Furthermore, several field studies have reported multiple pregnant females at capture and census, but only one set of offspring has subsequently been observed in the group (Savage 1990; Garber et al. 1993). Although these rare observations may be examples of fetal loss, behavioral explanations (e.g., infanticide, infant abandonment), as opposed to physiological ones, may also account for these cases of reduced reproductive success in subordinate females that attempt to breed.

3.2.2 Behavioral Mechanisms Maintaining Singular Breeding

3.2.2.1 Preconception mechanisms

Research on captive groups of callitrichid primates has revealed a host of potential behavioral mechanisms that may minimize the likelihood that multiple females produce offspring in a social group. First, there tend to be high levels of targeted aggression among females within family groups that, under field conditions, may result in the peripheralization or dispersal of subordinate females. High levels of intragroup aggression among females has been documented in lion tamarins (Kleiman 1979; Inglett et al. 1989), cotton-top tamarins (McGrew & McLuckie 1986; Tardif 1988), and common marmosets (Rothe 1975). Aggression is exhibited both between mothers and daughters and between same-sex pairs of siblings. In the cotton-top tamarin, removal of females from social groups because of targeted aggression is more common than the removal of males for the same reasons (McGrew & McLuckie 1986; Tardif 1988). In the lion tamarin, intrafamily fights among males are equally as common as fights among females. However, the consequence of agonistic interactions are much more severe among females. In a survey of intrafamily aggressive episodes, Inglett et al. (1989) noted that only 10 percent of the attacks led to the death of the target of the attacks in male–male interactions, whereas one third of the attacks targeted by females to other females (i.e, toward daughters or sibling sisters) resulted in the death of the recipient of the attacks in spite of veterinary intervention. Thus, the high rates and high intensity of intragroup female–female agonism represent important behavioral mechanisms that may reduce the likelihood that multiple females per group engage in breeding activity.

Laboratory studies have also revealed important aggressive dispositions among breeding females that may reduce the likelihood that a second breeding female could successfully enter an existing group. In lion tamarins, arranged encounters between breeding pairs and unfamiliar strangers vary dramatically as a function of the sex of the stranger (French & Inglett 1989). Encounters between pairs and out-group males are characterized by low levels of agonistic displays and an absence of overt physical aggression. In contrast, encounters between established pairs and unfamiliar females produce extremely high levels of female–female display and active aggression. Although the unfamiliar intruders were protected from physical harm by placing them in a small cage, the interactions between females appeared sufficiently intense that serious injuries and possibly death would have resulted had no precautions been taken. Resident females, interestingly, discriminated between intruding females who occupied the breeding role in their social groups from adult-age female intruders who were subordinate in their social group. Because as has been discussed, adult-age subordinates in lion tamarins are reproductively cyclic, the discrimination is made on the basis of social, and not endocrine, status of the intruding female. This aggression among females observed in captivity is also present in wild populations: in the eight years of observations of free-ranging lion tamarins at Poço das Antas in Brazil, no dispersing female was *ever* successful in joining a group that already contained a breeding female (Dietz & Baker 1993).

In other callitrichid species, sex differences in the pattern of aggression toward strangers is less pronounced than in the lion tamarin (e.g., see French & Snowdon 1981), but female-biased targeting of agonistic displays and attacks by female residents is the norm (see reviews in French & Inglett 1991; Anzenberger 1992). Other features of group demography may also modulate aggressive tendencies among females. For instance, in small groups (single pairs), aggression by female residents toward female intruders is low in intensity and incidence (French & Inglett 1989; French et al. 1995). However, as group size increases, levels of aggression by resident females increases and becomes more severe (French & Inglett 1989; French & Schaffner unpublished results). Breeding females in groups with few alloparental resources may thus modulate aggression to recruit additional group members, whereas breeding females in large groups with many potential alloparents may be intolerant of other potential reproductive competitors. In any event, however, it seems likely that intolerance of unrelated females and aggressive exclusion of them from social groups constitutes an important force in the maintenance of singular breeding.

3.2.2.2 Postconception mechanisms

Of the several possible ways in which subordinates might experience reduced reproductive success as a consequence of behavioral mechanisms after conception, none has been widely documented in either captive or wild groups of callitrichid primates. Cases of intentional infanticide (as opposed to infant neglect) by individuals other than the breeding adult female and male have not, to my knowledge, ever been directly observed in the wild. In a wild population of common marmosets, Digby (1994) recently documented the death of an infant born to a subordinate female in the group. Although the attack was not directly observed, the 25-day-old infant was found fatally injured on the ground, with several animals, including a subordinate male caretaker and the group's dominant female, leaving the area. In a captive group of common marmosets, Alonso (1986) observed a group in which an adult female caused the death of two infants of a second breeding female. That infanticide has *not* been more widely observed in either the wild or in captivity may represent the relative paucity of cases in which subordinate females carry infants to term in social groups (but see exceptions discussed later). Atypical or maladaptive parental behavior has been commonly interpreted to be a consequence of inappropriate or insufficient experience with infants (Snowdon, Savage, & McConnell 1985; Johnson, Petto, & Sehgal 1991). It appears that all subadult and adult callitrichids express considerable motivation to carry and provision infants, even under conditions in which they might encounter unrelated or unfamiliar infants or infant-related stimuli (Koenig & Rothe 1991; Pryce et al. 1993; Tardif this volume). Thus, infanticide as a common mechanism for reducing subordinate female fertility and maintaining singular breeding within a social group seems unlikely.

Reduced levels of important alloparental resources to subordinate females is a second postconception behavioral outcome that could contribute to reduced reproductive success in these females and hence to the maintenance of singular breeding. Again, there are few observations to substantiate that this mechanism plays an important role in callitrichid primates. In two cases in the common marmoset in which carrying effort was noted in groups with two concurrent breeding females, the subordinate female carried her infants more than was usual in this species, at least during the first two weeks of life (Koenig & Rothe 1991; Digby 1994). However, this may reflect a greater tendency for the subordinate female to retain possession of her infants, rather than a lack of interest by potential caregivers (Digby 1994).

3.3 Plural Breeding in Social Groups of Callitrichid Primates

Callitrichid social biology has been characterized as highly flexible, and a number of variations in social and mating structure have been observed in both the wild and in captive settings. Although callitrichids have historically been considered monogamous in both social and mating systems (Epple 1975; Kleiman 1977), there have been increasing numbers of reports of variations on this theme. Most notable among these are observations of cooperative polyandry in a handful of species (*Saguinus fuscicollis,* Goldizen 1987b; *Callithrix humeralifer,* Rylands 1986; *Saguinus mystax,* Garber, Moya, & Malaga 1984). The evidence for shared paternity in groups at this point is inferential, since there is no case that has assessed paternity in offspring. The evidence for cooperative polyandry consists of the following (from strongest to weakest evidence): (1) there have been scattered observations of multiple males engaged in copulation with a receptive female within a single group (Terborgh & Goldizen 1985; Rylands 1986; Baker, Dietz, & Kleiman 1993); (2) relatively low levels of agonistic behavior have been reported between adult males, even during times of heightened sexual activity (Goldizen 1987a; Garber et al. 1993); (3) high levels of infant care are expressed by more than one adult male per group (Garber et al. 1984; Goldizen, 1987b; Baker et al. 1993), and (4) group census data that reveals multiple adult males in groups (e.g., Garber et al. 1984). The relative social stability of captive groups that contain multiple adult males (Epple 1975), especially males that are related (Kleiman 1978), is consistent with observations of group demography from the field. However, no information regarding genetic relatedness is available to confirm multiple or mixed paternity in these groups, so the precision of the label "cooperative polyandry" has not as yet been conclusively evaluated (see Baker et al. 1993 for a discussion of alternative models).

Groups in which multiple females are reproductively active are relatively rare. However, in the past few years increasing numbers of reports from both captive and wild settings have indicated that in some species and under some conditions groups may contain multiple reproductively active females. Analysis of the context in which subordinates engage in breeding attempts, and the ultimate success of these attempts, may provide clues to the function of reproductive suppression in a broader life history context. The following sections provide a review of "failures" of reproductive suppression in callitrichid primates. The summaries are divided into observations in the wild and those reported from social groups in captivity.

A preparatory comment for this section is required prior to evaluating the literature. There is considerable variability among investigators in criteria for plural breeding among females. Some studies regard the presence of multiple adult females that have previously produced offspring as de facto evidence of plural breeding (e.g., Dietz & Baker 1993). Others use behavioral data (multiple females copulating, Terborgh & Goldizen 1985), morphological measures (nipple elongation and/or uterine palpation, e.g., Savage 1990; Garber et al. 1993), and the presence of multiple litters (e.g., Roda 1989; Digby & Ferrari 1994) as indexes of plural breeding. Obviously, there is variance among measures in the confidence with which we can interpret various findings. The systematic application of molecular techniques to populations of callitrichids is just now underway (e.g., Cheverud et al. 1994), and I expect that answers to many previously intractable problems in callitrichid primate biology, including the origins of multiple breeding females, will be provided by these techniques.

3.3.1 Departures From Singular Breeding in Wild Populations

Table 3.1 presents a summary of the findings on departures from singular breeding in wild callitrichid primate populations, and a brief narrative on each case follows.

3.3.1.1 Pygmy marmosets (genus Cebuella)

Plural breeding has never been documented in wild groups of pygmy marmosets. Soini (1988, 1993) has trapped over 80 intact groups, and the majority of these groups consisted of the adult male and female and offspring from

Table 3.1. *Field studies reporting multiple reproductively active females in free-ranging groups of callitrichid and callimiconid primates*

Genus	Reports	Measure of reproduction	Outcome	Sources
Cebuella	No groups observed with multiple breeding females.	Manual palpation, observation of nipple extension, presence of multiple litters.		Soini 1988, 1993
Callithrix	Three of 5 groups contained multiparous females. In one group, three infants present and two females lactating.	Morphological evaluation of nipples, of multiple litters.	In one of 5 groups, likely that offspring of multiple females reared.	Scanlon et al. 1988
	Three groups in which two females produced offspring.	Presence of offspring, observations of infant nursing and survival.	Infants of subordinates survived in 2 of 3 groups.	Digby & Ferrari 1994
	Two groups in which two females gave birth to offspring.	Presence of multiple litters.	Outcome unknown.	Roda 1989
Saguinus	Three of 13 monitored groups contained two pregnant females (or one pregnant and one lactating female).	Morphological evaluation of nipples and breasts, manual palpation.	Only one litter per group was reared, only one female nursed offspring.	Garber et al. 1993

	One group contained two reproductive females.	Unknown measure.	Outcome unknown.	Raimirez 1984
	Two polygynous groups: (1) adult male mated with mate and daughter; (2) two parous females produced offspring.	Mating behavior and presence of multiple litters.	(1) No reports of multiple pregnancies. (2) One female left group shortly after birth of litter, left infants behind.	Terborgh & Goldizen 1985
	One group contained two pregnant females.	Manual palpation.	Infants of only one female observed after parturition.	Savage 1990
Leontopithecus	In nine of 11 cases of potentially 2 polygynous groups (= 2 parous females) both females produced offspring ($n = 8$) or pregnant ($n = 1$).	Observation of multiple litters, nursing and carrying, manual palpation.	Infants survived in 6 of 8 groups, infants of subordinate females did not survive to weaning in 2 of 8 groups.	Dietz & Baker 1993
Callimico	Two groups contained multiple reproductive females.	Direct observation of parturition and group census indicating more than one infant.	Infants of both females survived in each of the groups.	Masataka 1981a, 1981b
	One study group had multiple reproductive females.	Presence of multiple infants of same age.	Infants of both females survived.	Pook & Pook 1981

one to four litters. A small proportion of the groups contained two adult females, but in no case was there evidence of more than one reproductively active female, either on the basis of the presence of multiple dependent litters of offspring or on the basis of morphological evidence (e.g., nipple elongation). In some socially polyandrous groups, Soini (1987) noted multiple males copulating with the breeding female, but the dominant male maintained exclusive social and sexual access to the female at the suspected time of maximal fertility.

3.3.1.2 Marmosets (genus Callithrix)

A variety of field studies on species in the genus *Callithrix* provide evidence consistent with the notion that breeding activity is typically limited to a single female. A group of buffy headed marmosets (*Callithrix flaviceps*) was monitored more or less continuously for six years by Ferrari and his colleagues (Ferrari & Lopes-Ferrari 1989; Ferrari & Diego 1992). Group size varied widely during this time, from 5 to 16 individuals, and the group contained from one to six females of potential breeding age. At all times, however, only one female was reproductively active, because only twin litters were present. The identity of the breeding female changed during the course of the study, with the reproductive role transferring from the dominant female to her daughter. A similar pattern was documented for *Callithrix humeralifer,* with a single breeding female in each group and a transfer of reproductive status from mother to daughter (Rylands 1986; Stevenson & Rylands 1988).

Several field studies on free-ranging common marmosets suggest considerable variability among populations in the presence or absence of strict singular breeding among females. Studies of two populations (Hubrecht 1984; Stevenson & Rylands 1988; Alonso & Langguth 1989, cited in Digby & Ferrari 1994) indicated that breeding was limited to a single adult female. These assessments were based on group demography, morphological evaluation upon trapping, and observation of the number of litters in each group. Three field reports indicate the presence of multiple breeding females in common marmoset groups. Scanlon, Chalmers, & Monteiro da Cruz (1988) found indirect evidence of multiple parous females in some of the groups trapped at the Tapacurá Ecological Station. In three of the five groups, more than one female possessed elongate nipples, suggesting a previous pregnancy. Simultaneous reproduction by multiple females was suggested in only one of the groups, which contained more than two dependent infants at one time and two females with swollen breasts. Roda (1989, cited in Digby & Ferrari 1994) reported simultaneous parturitions by two females in two study groups.

Digby & Ferrari (1994) present the most compelling evidence of simultaneous breeding by more than one female in free-ranging *Callithrix*. During the course of an 18-month field study, breeding attempts by two females occurred in each of the three study groups. In two groups, the two breeding females produced two to three litters each and copulated with a single adult male from the group during the postpartum period. Reproductive outcomes were generally successful for both females; offspring survived to weaning in the case of five of the six breeding females. In the case of the sixth female, her singleton infant disappeared several days after birth, and she subsequently emigrated from her group less than two months following the death of her infant. Several characteristics of these groups may be associated with multiple breeding females. First, all three groups were large (maximum group sizes of 8, 14, and 15 individuals), at or above mean group size of approximately 8 individuals for this species (Stevenson & Rylands 1988). Second, all groups had small home ranges and thus a high relative population density. Finally, the groups were located on home ranges adjoining cultivated plantations, which might have been a source of a superabundance of gum-producing trees, an important resource for *Callithrix* (Ferrari & Lopes-Ferrari 1989; Ferrari 1993).

Thus, in the genus *Callithrix,* there is mixed evidence with regard to the presence of multiple breeding females, and the incidence appears to vary across populations and study sites. Plural breeding by females does occur with some frequency, and at least in the case of one study site, offspring of both dominant and subordinate females survived. Finally, there is a hint that ecological and demographic variables may play an important role in shaping breeding decisions, since plural breeding appears to be associated with high population density, large group size, and high food availability. Singular breeding, in contrast, is associated with low population density and low food availability. Further field studies of groups that differ on these variables will be necessary to outline the relative importance of these factors.

3.3.1.3 Tamarins (genus Saguinus)

There are three well-documented examples of plural breeding females in wild tamarins. At least two of the groups of saddleback tamarins under observation by Terborgh and Goldizen (1985) contained two reproductively (or sexually) active females. In one group, a breeding adult male copulated with the breeding adult female and one daughter, but only a single pregnancy in the dominant female ensued. In the second group, two parous females produced viable offspring, but one of the females disappeared several days after the birth of the infants, leaving the litter behind in the group. In cotton-top tamarins in

Colombia, a single group contained two pregnant females upon capture, as determined by manual palpation. Subsequently, however, offspring of only one female were noted in the group. The fate of the offspring from the second female is not known (Savage 1990). Finally, multiple breeding females (either pregnant or lactating) were observed in three of 13 groups of moustached tamarins (*S. mystax*). As in cotton-top tamarins, however, only one litter per group was successfully reared, and only one female was observed to nurse dependent offspring (Garber et al. 1993). Thus, in the genus *Saguinus,* the presence of multiple breeding females is a rare event. Furthermore, in the event that two or more females with signs of reproductive activity are present in a group, the reproductive outcome for the second female is poor, because only one litter appears to have been reared successfully.

3.3.1.4 Lion tamarins (genus Leontopithecus)

Information on the one well-studied population of golden lion tamarins suggests that plural breeding is a relatively common occurrence in this taxon, at least when compared to *Cebuella* and *Saguinus*. Dietz & Baker (1993) reviewed eight years of group demography and reproduction in the Poço das Antas population of free-ranging lion tamarins. Of the 211 semiannual group samples, 20 (10.6%) consisted of polygynous groups (defined as groups that contained more than one female that showed morphological evidence of being parous, whether or not both females produced offspring or were pregnant during that particular group sample). The polygynous group samples were derived from 11 pairings of known parous females. Polygynous groups tended to consist of related females, since 6 of 11 duos consisted of known mothers and daughters, two more were suspected to be mothers and daughters, and two were sisters. There was a clear dominance relationship between the two females, with the older female usually dominant over the younger. Both females produced offspring in 8 of the 11 groups, but fewer offspring of the subordinate female survived to weaning than in the dominant female.

3.3.1.5 Goeldi's monkey (genus Callimico)

Two field studies of moderate length have been conducted on wild populations of *Callimico goeldi*. The 5-month-long study by Pook & Pook (1981) provides indirect evidence of multiple breeding females, since one group contained two subadult yearlings. Surviving twins in *Callimico* are rare, even in captivity (Beck et al. 1982); thus, it is likely that the two subadults were offspring of separate females. Masataka (1981a) provides both indirect and direct evidence of multiple breeding females. He noted a group that contained two infants, again an unlikely event if a single female is reproductively active. In addition, he closely monitored for six months a group that contained a single adult male and two adult females. Each of the females produced a single offspring within two weeks of each other, and both infants were successfully reared. The male copulated with each of the females during the postpartum estrus. There were pronounced changes in the social dynamics of the group after the birth of the offspring (Masataka, 1981b). The lactating females showed much more affilitative behavior (e.g., grooming, proximity) with each other shortly after the birth of their infants than they did prior to birth or following infant independence. There was also considerable interest in and care of each other's infants. The postpartum period was also accompanied by increased aggressive targeting of a third female in the group that was of breeding age, and eventually she became peripheralized. One other neighboring group near Masataka's study site also contained two infants, again suggesting that multiple females were reproductively active.

3.3.2 Departures From Singular Breeding in Captive Populations

Three recent reviews have collected information on plural breeding groups in a number of species, including tamarins (*Saguinus:* Price & McGrew 1991), common marmosets (*Callithrix:* Rothe & Koenig 1991), and one review that covers a variety of callitrichids (Carroll, 1986). For the purposes of this review, I include only "spontaneously" plural breeding (or potentially plurally breeding) groups. Groups with plural breeding that arose through experimental manipulation or radical recombination of animals (such as the constitution of "peer groups" of multiple unrelated males and females; see Abbott 1984, 1987) were specifically excluded. As was the case for studies on wild populations, there is considerable variability in definitions of plural breeding in captive situations. Unlike studies on wild populations, however, reports on captive groups provide more information on the endocrine status of subordinate females (e.g., French et al. 1984; Tardif 1984; Heistermann et al. 1989) and on the fate of offspring produced by these subordinates (e.g., Price & McGrew 1991). Estimating the relative incidence of plural breeding among captive populations, either within or among taxa, is difficult. When departures from singular breeding are reported from a captive colony, it is rare to see an accompanying statement of the size of the population of groups in which these exceptions have occurred.

A summary of the reports of plural breeding is presented in Table 3.2. These findings are categorized according to one of three major demographic

Table 3.2. *Incidence of multiple reproductively active females in captive groups of callitrichid and callimiconid primates, separated by social context*

Social Context	Species	Finding	n	Citation
Adult female removed	*Callithrix sp.*	No sexual activity among daughters and fathers/sons for 10 months.	2	Carroll 1986
	S. oedipus	Daughter became cyclic, but mating with father rare, no pregnancy.	1	French et al. 1984
	S. oedipus	Eldest daughter showed ovarian cyclicity, no pregnancies.	4	Heistermann et al. 1989
	Callimico	No sign of sexual activity in group.	1	Carroll 1986
	S. fuscicollis	No mating among daughter and brothers for 7.5 years after death of adult male.	1	Carroll 1986
Adult male removed	*Callimico*	No signs of sexual activity during period in which heterosexual partners were limited to siblings.	3	Carroll 1986
	C. jacchus	One daughter in each group was cyclic, but no mating or sexual behavior.	2	Hubrecht 1989
	S. oedipus	Two cases (F1 and F3) in which removal of father was followed by endocrine signs of ovulation.	2	Tardif 1984

	Species	Description	N	Reference
Adult male removed and replaced	*S. oedipus*	Half of reported cases of potentially polygynous groups were father removal and replacement.	4	Price & McGrew 1991
	C. jacchus	Mother and daughter pregnant following replacement of male.	3	Abbott 1984
	C. jacchus	Of 10 polygynous groups, 7 involved father removal and replacement.	7	Rothe & Koenig 1991
	C. argentata	Mother and daughter became pregnant after male replacement.	1	Carroll 1986
	Callimico	Mother and daughter became pregnant and offspring of both females survived.	1	Carroll 1986
No social changes in group	*S. oedipus*	Two of 8 polygynous groups occurred with no removal of breeding adults.	2	Price & McGrew 1991
	C. jacchus	One of 10 cases of polygynous mating occurred in absence of removal of breeding adults.	1	Rothe & Koenig 1991
	Cebuella	Mother and daughter pregnant, only mother delivers offspring. Daughters in both cases had opportunity to mate with neighboring males.	2	Snowdon pers. comm. 1994

changes in social groups: removal of the breeding adult female, removal of the breeding adult male (with no replacement), and removal of the breeding adult male and replacement of the male with an unrelated adult male. Several extremely important themes regarding the potential for plural breeding in callitrichid social groups emerge from these summaries. First, removal of the dominant female (in most cases, the mother of the subordinate female) has been linked with the attainment of ovarian cyclicity in the subordinate female (especially in *Saguinus*). However, removal of the mother alone, in the absence of any other social change, does not typically lead to pregnancy in adult daughters. Second, removal of the male with no subsequent replacement by an unrelated male does not appear to be associated with a high incidence of multiply breeding females in social groups. However, this social change may be associated with modifications in the ovarian status of subordinates. In the common marmoset, for instance, daughters in intact and stable natal groups do not show ovarian cycling (Abbott 1984; Evans & Hodges 1984), whereas a higher proportion of daughters and subordinates in groups without the adult male or in groups for which the dominance hierarchies are just being established tend to exhibit ovulatory cycles (Abbott 1984; Hubrecht 1989).

One of the themes to emerge from these summaries is the role played by the relatedness of the dominant and breeding adult males to subordinate females in the social group. A high proportion of the cases in which multiple females are reproductively active in a social group (either ovulating or producing offspring) is associated with the removal or death of the resident breeding male accompanied by replacement with an unrelated adult male. Thus, under social conditions in which potential adult male partners are related, subordinate females are likely to exhibit ovulatory quiescence and fail to become pregnant. However, under conditions in which potential male partners are not related, the incidence of reproductive cyclicity by subordinates and the production of offspring by these individuals is higher. These findings mirror endocrine profiles in cooperatively breeding birds, in which subordinate helpers that are related to breeders tend to be physiologically suppressed, whereas helpers unrelated to the breeders are reproductively mature (e.g., Reyer, Dittami, & Hall 1986; Mays, Vleck, & Dawson 1991).

This pattern of results suggests that inbreeding avoidance has been an important selective factor shaping callitrichid breeding strategies, at least among subordinates. First, inbreeding in these species, as in others, is associated with reduced reproductive success. Analysis of captive breeding records indicate that inbreeding is associated with markedly reduced reproductive success, even under the relatively benign conditions of captivity. Ralls and Ballou (1982) compared rates of offspring survival to six months of age in three separate captive populations of tamarins (*S. fuscicollis illigeri, S. f. fuscicollis,* and *L. rosalia*) for pairs that were noninbred (estimated coefficient of relatedness = zero) versus inbred pairs whose estimated coefficient of relatedness was greater than zero. In all three populations, inbred pairs had significantly higher infant mortality than did noninbred pairs. In the case of *S. f. fuscicollis,* the mortality rate for inbred pairings (60.0% mortality by six months) was almost double that of noninbred pairs (32.3% mortality; Ralls & Ballou 1982). The tremendous cost associated with inbreeding is further suggested by preliminary evidence from a wild population of lion tamarins. Dietz and Baker (1993) recorded the fate of the first reproductive attempt in a total of 19 females. Of these 19, five females had potential sexual partners that were either distant relatives or unrelated (= noninbred matings), whereas for the remaining 14 females the only potential in-group sexual partner was either a father or male sibling (= inbred matings). *None* of the offspring from the presumed inbred matings survived to weaning. In contrast, however, 86 percent of the matings that were presumed to be noninbred resulted in offspring surviving to six months of age. Thus, for subordinates, breeding attempts with close relatives in the natal family group may compound the already reduced reproductive success associated with lower social status.

3.4 Species Differences in Mechanisms of Reproductive Suppression: The Cost of Reproduction

In *Saguinus* and *Callithrix,* singular breeding is maintained primarily through endocrine mechanisms, whereas in *Leontopithecus* the mechanisms that limit reproduction to a single female appear to be exclusively behavioral. How, then, can we account for these remarkable species differences in the mechanisms underlying reproductive inhibition among otherwise strikingly similar species of callitrichid primates?

Reproductive energetics have proven to be extremely useful in analyzing trends in social and mating systems across a wide taxonomic span in ways that are relevant to the present discussion. Gittleman (1985; Gittleman & Oftedal 1987) first drew attention to the association between reproductive energetics and parental systems in carnivores. Across the order Carnivora, there is tremendous variation in the cost of reproduction, both in prenatal investment (as measured by the cost of gestation) and in postnatal investment (as measured by the growth rates of pups and litters). Gittleman's analyses reveal a distinctive link between these estimates of reproductive costs and the expression of parental care. Uniparental (i.e., maternal) care predominates for those species with low initial litter weights or slow litter growth rates. In other words, exclusively maternal care is associated with low reproductive costs in carnivores. In contrast, biparental care or communal care is typical for those species with high initial litter weights and rapid litter growth rates. Thus, reproductive costs are predictive of broad taxonomic differences in the expression of parental care in the carnivores.

Creel and his colleagues (Creel & Creel 1991; Creel & Waser this volume) have recently extended Gittleman's analysis to an examination of the relationship between reproductive energetics and mechanisms of reproductive suppression in carnivores that exhibit communal breeding. Traditional hypotheses from the literature on cooperative breeding suggest that high reproductive skew in a social group may be a consequence of strong ecological or demographic constraints on successful reproduction (e.g., Brown 1987; Keller & Reeve 1994). One of these constraints might be related to the cost of engaging in a reproductive attempt. Creel & Creel (1991) therefore predicted that reproductive suppression of subordinates might vary across taxa as a function of the cost of reproduction. Evaluation of this hypothesis was accomplished by comparing estimates of reproductive costs in singular breeding species of carnivores that typically exhibit physiological or obligate subordinate suppression (e.g., dwarf mongooses, *Helogale parvula;* African wild dogs, *Lycaon pictus*) with costs in species that typically display plural breeding with no ovulatory suppression of socially subordinate females (e.g., lions, *Panthera leo;* spotted hyenas, *Crocuta crocuta*). In keeping with the hypothesis, the analyses revealed that the costs of both gestation and postnatal investment in offspring were greater in species with suppression than in species without suppression. Creel and Creel (1991) suggest, therefore, that reproduction by subordinates in singularly breeding species may be more risky than reproduction by subordinates in plurally breeding carnivores, and that the evolution of mechanisms, such as suppression of ovulation that minimize wasteful reproduction in subordinates, might be favored.

What is the relationship between reproductive costs and mechanisms of suppression in the callitrichid primates? Good measures of the cost of reproduction have only recently become available for the callitrichid primates. Tardif, Harrison, and Simek (1993) provided estimates of *post*natal investment in offspring by comparing growth curves for infants throughout the period of dependence on nursing. Sufficient data were available to estimate rates of weight gain for *Saguinus, Leontopithecus,* and *Callithrix.* Variations in absolute weight gain by infants among species were apparent, with the larger-bodied *Saguinus* infants gaining more weight than the smaller-bodied *Callithrix.* However, when growth rates were adjusted for adult body weight, the slopes of the regression lines describing weight gain trajectories did not

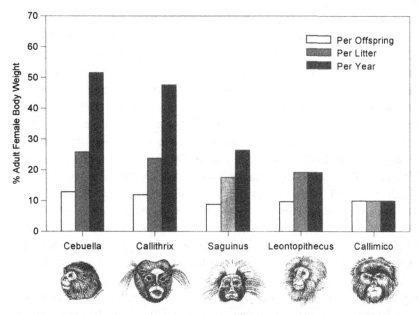

Figure 3.5. Relative costs of gestation in callitrichid and callimiconid primates, estimated by the ratio (in percentages) between neonate weight and adult female body weight. Estimates are provided per offspring, per litter (= offspring weight × modal litter size) and per annum (litter weight × litters per year).

differ significantly across species. Thus, there appears to be insufficient variation in postnatal maternal investment in offspring among taxonomic groups in the callitrichids that could be used to support differential predictions about the presence versus absence of reproductive suppression.

Measures of taxonomic variation in prenatal investment in offspring in callitrichids have been difficult to assess. Although published information on offspring and maternal weight is available in several sources (Leutteneger 1973; Hershkovitz, 1977), sample sizes tend to be small and other problems plague the data sets. Recently, Ford and Davis (1992) provided a useful summary of genus-level patterns in reproductive life histories among the callitrichids, including estimates of litter size, reproductive rate, and maternal and neonate weights. Figure 3.5 summarizes the data for the four genera in the family Callitrichidae and for the closely related Goeldi's marmoset, *Callimico goeldi.*

On the basis of individual neonate-to-maternal ratios, the differences among taxa are unremarkable, varying by less than 3 percent. These subtle differences among taxa, however, are amplified when the gestation costs are estimated by litter-to-maternal-weight ratios, because the modal litter size in callitrichids is two. (The modal litter size in *Callimico* is one infant; Beck et al. 1982.) There is also conspicuous differences across taxa in the likelihood of a postpartum conception and hence the typical number of litters produced each year. *Cebuella* and *Callithrix* regularly produce two litters each year in both captive and field settings (Soini 1988; Stevenson & Rylands 1988). *Saguinus* tends to be slightly more variable, but most species have interbirth intervals of less than one year and exhibit the potential to produce 1.5 litters each year (Snowdon & Soini 1988; Ford & Davis 1992). For instance, inter-birth intervals in cotton-top tamarins is approximately 200 days (French 1984; Ziegler et al. 1987), allowing for an average of at least 1.5 litters each year. In contrast, the modal number of litters each year in *Leontopithecus* is one (Kleiman, Ballou, & Evans 1982; Dietz, Baker, & Maglioretti 1994; French, Pissinatti, & Coimbra-Filho in press). Only a small proportion of females (approximately 20%) produce multiple litters per year at any point during their lifetime. Like lion tamarins, Goeldi's marmoset engages in only one reproductive attempt each year. The widespread occurrence of two litters per year in the smaller-bodied marmosets and tamarins, and the limitation of breeding to one attempt per year in the larger-bodied lion tamarins and Goeldi's monkey, further magnify differences in the prenatal costs of repro-duction among the taxa when calculated on an annualized basis for a breeding female. For *Cebuella* and *Callithrix,* breeding females produce the offspring equivalent at birth the equivalent of 50 percent or more of their body weight per year, whereas breeding females in *Leontopithecus* and *Callimico* produce the offspring equivalent at birth of only 10 to 20 percent of their body weight per year (Figure 3.5). Thus, there is considerable variation among taxa in the gestational costs of reproduction.

Is the variance in reproductive energetics consistent with Creel and Creel's (1991) hypothesis regarding the mechanisms of suppression in callitrichids? Table 3.3 summarizes the information presented in this chapter regarding sub-ordinate reproduction. These measures include species-typical mechanisms of subordinate suppression (presence or absence of physiological suppression of ovulation), the relative incidence of plural breeding by females, and the likeli-hood that offspring produced by subordinates will be reared successfully. Although our knowledge is limited for many of these features in a number of species, the results on subordinate reproduction tend to be consistent with the energetics hypothesis. Suppression is more likely to be physiological in species for which reproduction is more costly than in species in which the prenatal maternal investment is lower. The pygmy marmoset, for which we

Table 3.3. *Summary of findings on reproductive energetics and suppression*

Measure	Cebuella	Callithrix	Saguinus	Leontopithecus	Callimico
Annual gestation costs[a]	51.6%	47.6%	26.4%	19.2%	9.9%
Ovulation suppressed?	Unknown	Usually	Yes	No	No
Subordinate pregnancy common?	No	Rare	Rare	Often	Common
Survival of offspring born to subordinates?	No	Sometimes	No	Common	Common

[a]Expressed as percentage of sum of annual neonatal weight divided by maternal weight; see text for additional details.

have no published information on mechanisms of reproductive suppression, constitutes a critical test case for this scenario, because the gestational costs of reproduction for this species are the highest among the callitrichids.

The incidence of pregnancy among subordinates also appears to vary consistently with reproductive costs. Although subordinate pregnancy appears to be a common occurrence in *Callimico* and at least in wild populations of golden lion tamarins, it is a rare occurrence in tamarins of the genus *Saguinus*. Subordinate female pregnancy is rare in some populations of *Callithrix,* but it may be common in others. Subordinate pregnancy has never been documented in wild populations of pygmy marmosets, and only twice in captivity.

Finally, subordinate offspring survival appears to vary with reproductive energetics. High rates of subordinate offspring survival are known in taxa in which costs of reproduction are lowest, whereas subordinate offspring survival in the remaining taxa is lower. *Callithrix* again appears to be a taxa in which considerable variability is expressed, with at least one recent study suggesting that subordinates experience some success in rearing offspring (Digby & Ferrari 1994).

Although a preliminary evaluation of the observations in Table 3.3 yields a fit between prediction and observation, a variety of cautionary statements are necessary at this point. First, as already pointed out, our knowledge base for many species of callitrichids is exceedingly sparse. If the case of *Callithrix* marmosets is a guide, then what appears at first blush to be a simple relationship between social status and reproduction may in fact be much more complicated and variable than initially envisioned. Additional field and laboratory studies will surely provide additional evidence with which to evaluate the schema that have been presented. Second, much of our information on variation in the mechanistic aspects of suppression is based upon the study of three species: the cotton-top tamarin, the common marmoset, and the lion tamarin. We need to know the degree to which data on these three species is representative of their genera. Within each genus there is considerable variation in a host of ecological and life-history variables (see reviews in Rylands 1993) that may prove useful in further refinements of the relationships among energetics, social structure, reproductive suppression, and cooperative breeding in the callitrichids. Third, it is important to appreciate individual differences and the role they may play in reproductive suppression. Individual differences in diet and weight are associated with variations in reproductive potential in common marmosets, with heavier females tending to have more ova per ovulatory event and less reduction in litter size during gestation (Tardif 1992; Tardif & Jaquish 1994). These variables may be important in shaping the relationships among breeding, reproductive energetics, and offspring production in subordinates. Finally, it bears repeating that the pygmy marmoset represents a key link in the arguments presented here. The scenario predicts that pygmy marmosets should exhibit pronounced physiological suppression of ovulation among subordinates. The existing field data on group structure and the paucity of observations on plural breeding in captive groups are consistent with the prediction.

3.5 Conclusions

Tamarins and marmosets exhibit a remarkable diversity of mechanisms that promote the maintenance of singular breeding, including both prevention of conception and reduced postnatal survivorship in the offspring of subordinates, produced by either behavioral or physiological means. There is growing recognition, however, that departures from singular breeding may be a much more common occurrence in both captive and wild groups than previously suspected. Although there are many gaps in our knowledge of how and why singular breeding is maintained in callitrichid groups, the following tentative conclusions based on the available evidence seem warranted.

First, within-species variation in singular versus plural breeding appears to be critically dependent upon group demography. The absence, removal, or disappearance of the breeding female is associated with the onset of reproductive potential (e.g., onset of reproductivity cyclicity or increased levels of scent marking and sexual solicitation behavior). However, removal of the dominant female is not always associated with conception and the production of offspring in subordinate females. The variable most conspicuously associated with the production of offspring by subordinate females is the presence of an unrelated male. Species-typical mechanisms of fertility suppression appear to fail with greater regularity when subordinates have the opportunity to interact with and potentially mate with an unrelated male. Therefore, reproductive inhibition in callitrichids may have as much to do with inbreeding avoidance and reproductive restraint by subordinates as with the maintenance of reproductive dominance and reduction of potential reproductive competitors by dominant females (Abbott 1987, 1993; French & Inglett 1991).

It will be of interest in the future to evaluate the role of a host of other complex ecological variables in the expression of plural breeding in callitrichids. These variables include, but are not limited to, the effects of (1) variations in habitat quantity and quality on female breeding profiles (Dietz & Baker 1993), (2) encounter rates with neighboring groups and the opportunity to interact with out-group males (Hubrecht 1989), (3) group size and hence number of available infant caregivers (Savage, 1990), (4) opportunities for dispersal (Goldizen & Terborgh 1989), and (5) habitat saturation or range restriction (Dietz & Baker 1993). Certainly much within-taxa variation in many of the varibles associated with delayed dispersal and reproduction is likely to be accounted for by these demographic and ecological features, as has been demonstrated in avian cooperative breeders (Komdeur 1992).

Second, genus-level differences in mechanisms maintaining singular breeding appear to be associated with taxonomic differences in the energetics of reproduction. This review has tentatively identified two variables that co-vary with genus-typical mechanisms of suppression: body size and annual gestation costs for females. Species-typical mechanisms of suppression tend to be physiological in nature for the smaller-bodied callitrichids and those with high gestation costs. Mechanisms maintaining singular breeding tend to be nonphysiological in the largest-bodied callitrichid primates, the lion tamarins, and in the closely related Goeldi's monkey, for which the annual costs of gestation are relatively lower. Thus, a link between reproductive energetics and the mechanisms of suppression is implicated in at least two distinct mammalian taxa (carnivores and callitrichid primates) that express cooperative breeding systems.

Third, the proximate regulation of singular breeding in female callitrichid primates is a complex melange of inhibitory cues, probably arising from the breeding adult female, and stimulatory cues, probably arising from unfamiliar or unrelated adult males. The precise nature of interactions among these processes and establishing the relative importance of each in determining the reproductive potential of subordinate females remain challenges for the future.

Acknowledgments

The studies on lion tamarins and black tufted-ear marmosets conducted in my laboratory at the Callitrichid Research Facility at the University of Nebraska at Omaha were supported in part by grants from the Public Health Service (HD 23,139), the Cattell Foundation, and the National Science Foundation (CRB 90–00094, IBN 92–09528, and OSR 92–55225 awarded to the University of Nebraska). I thank all my collaborators at UNO, especially my students Judy Stribley, Allen Salo, Betty Inglett, Colleen Schaffner, and John Schneider, and my colleagues Bill deGraw and Shel Hendricks, for a decade of stimulating collegiality and productivity. I especially thank Denise Hightower for her excellence and dedication in animal care. The following people provided detailed critiques of earlier versions of this chapter: Dan Leger, Jo Manning, Colleen Schaffner, Tessa Smith, John Schneider, Suzette Tardif. Jacque Franzen kindly provided copy-editing services on the penultimate draft. I especially thank Dave Abbott for his input to and feedback on the ideas presented in this paper, both historically and in recent times. I finally wish to acknowledge the contributions of Mary Lincoln and Aaron and Anna French to my capacity to prepare this work and for their understanding as I spent long weekends and evenings looking at, thinking and reading about, and sometimes even writing about callitrichid primates.

References

Abbott, D. H. (1984). Behavioral and physiological suppression of fertility in subordinate marmoset monkeys. *Am. J. Primatol.* 6:169–186.

Abbott, D. H. (1986). Social suppression of reproduction in subordinate marmoset monkeys (*Callithrix jacchus jacchus*). In *A Primatologia no Brasil*, ed. M. T. deMello, pp. 1–16. Brasilia: Sociedade Brasileira de Primatologia.

Abbott, D. H. (1987). Behaviorally mediated suppression of reproduction in female primates. *J. Zool., Lond.* 213:455–470.

Abbott, D. H. (1993). Social conflict and reproductive suppression in marmoset and tamarin monkeys. In *Primate social conflict*, ed. W. A. Mason & S. P. Mendoza, pp. 331–372. Albany: State University of New York Press.

Abbott, D. H., & George, L. M. (1991). Reproductive consequences of changing social status in female common marmosets. In *Primate responses to environmental change*, ed. H. O. Box, pp. 295–310. London: Chapman & Hall.

Abbott, D. H., George, L. M., Barrett, J., Hodges, K. T., O'Byrne, K. T., Sheffield, J. W., Sutherland, I. A., Chambers, G. R., Lunn, S. F., & Ruiz de Elvira, M.-C. (1990). Social control of ovulation in marmoset monkeys: a neuroendocrine basis for the study of infertility. In *Socioendocrinology of Primate Reproduction*, ed. T. E. Ziegler & F. B. Bercovitch, pp. 135–158. New York: Wiley-Liss.

Abbott, D. H., & Hearn, J. P. (1978). Physical, hormonal and behavioural aspects of sexual development in the marmoset monkey, *Callithrix jacchus*. *J. Reprod. Fert.* 53:155–166.

Abbott, D. H., Hodges, J. K., & George, L. M. (1988). Social status controls LH secretion and ovulation in female marmoset monkeys (*Callithrix jacchus*). *J. Endocrinol.* 117:329–339.

Abbott, D. H., McNeilly, A. S., Lunn, S. F., Hulme, M. J., & Burden, F. J. (1981). Inhibition of ovarian function in subordinate female marmoset monkeys (*Callithrix jacchus*). *J. Reprod. Fert.* 63:335–345.

Alonso, C. (1986). Fracasso na inibição da reprodução de uma fêmea subordinada e troca de hierarquia em um grupo familiar de *Callithrix jacchus jacchus*. In *A Primatologia no Brasil,* ed. M. Thiago de Mello, p. 203. Brasilia: Sociedade Brasileira de Primatologia.

Alonso, C., & Langguth, A. (1989). Ecologia e comportamento de *Callithrix jacchus* (Callitrichidae, Primates) numa ilha de floresta atlántica. *Rev. Nordestina Biol.* 6:105–137.

Anzenberger, G. (1992). Monogamous social systems and paternity in primates. In *Paternity in primates: genetic tests and theories,* ed. R. D. Martin, A. F. Dixson, & E. J. Wickings, pp. 203–224. Basel, Switzerland: S. Karger.

Baker, A. J., Dietz, J. M., & Kleiman, D. G. (1993). Behavioural evidence for monopolization of paternity in multi-male groups of golden lion tamarins. *Anim. Behav.* 46:1091–1103.

Barrett, J., Abbott, D. H., & George, L. M. (1990). Extension of reproductive suppression by pheromonal cues in subordinate female marmoset monkeys, *Callithrix jacchus*. *J. Reprod. Fert.* 90:411–418.

Barrett, J., Abbott, D. H., & George, L. M. (1993). Sensory cues and the suppression of reproduction in subordinate female marmoset monkeys, *Callithrix jacchus*. *J. Reprod. Fert.* 97:301–310.

Beck, B. B., Anderson, D., Odgen, J., Rettberg, B., Brejla, C., Scola, R., & Warneke, M. (1982). Breeding the Goeldi's monkey *Callimico goeldii* at Brookfield Zoo, Chicago. *Int. Zoo Yearbk.* 22:106–114.

Brown, J. L. (1987). *Helping and communal breeding in birds: ecology and evolution.* Princeton: Princeton University Press.

Carroll, J. B. (1986). Social correlates of reproductive suppression in captive callitrichid family groups. *Dodo* 23:80–85.

Carroll, J. B., Abbott, D. H., George, L. M., Hindle, J. E., & Martin, R. D. (1990). Urinary endocrine monitoring of the ovarian cycle and pregnancy in Goeldi's monkey (*Callmico goeldii*). *J. Reprod. Fert.* 89:149–161.

Chepko-Sade, D., & Halpin, Z. T., eds. (1987). *Mammalian dispersal patterns.* Chicago: University of Chicago Press.

Cheverud, J., Routman, E., Jaquish, C., Tardif, S., Peterson, G., Belfiore, N., & Forman, L. (1994). Quantitative and molecular genetic variation in captive cotton-top tamarins (*Saguinus oedipus*). *Cons. Biol.* 8:95–105.

Creel, S. R., & Creel, N. M. (1991). Energetics, reproductive suppression and obligate communal breeding in carnivores. *Behav. Ecol. Sociobiol.* 28:263–270.

Creel, S. R., Creel, N. M., Wildt, D. E., & Monfort, S. L. (1992). Behavioural and endocrine mechanisms of reproductive suppression in Serengeti dwarf mongooses. *Anim. Behav.* 43:231–245.

Creel, S. R., & Waser, P. M. (1991). Failures of reproductive suppression in dwarf mongooses (*Helogale parvula*): accident or adaptation? *Behav. Ecol.* 2:7–13.

Dietz, J. M., & Baker, A. J. (1993). Polygyny and female reproductive success in golden lion tamarins, *Leontopithecus rosalia*. *Anim. Behav.* 46:1067–1078.

Dietz, J. M., Baker, A. J., & Maglioretti, D. (1994). Seasonal variation in reproduction, juvenile growth, and adult body mass in golden lion tamarins (*Leontopithecus rosalia*). *Amer. J. Primatol.* 34:115–132.

Digby, L. J. (1994). Infanticide, infant care, and female reproductive strategies in a wild population of common marmosets. *Amer. J. Phys. Anthropol.* Suppl. 18:80–81.

Digby, L. J., & Ferrari, S. F. (1994). Multiple breeding females in free-ranging groups of *Callithrix jacchus. Int. J. Primatol.* 15:389–398.

Emlen, S. T. (1991). Evolution of cooperative breeding in birds and mammals. In *Behavioural ecology: An evolutionary approach,* 3rd ed., ed. J. R. Krebs & N. B. Davies, pp. 301–335. London: Blackwell.

Epple, G. (1975). The behavior of marmoset monkeys (*Callitrichidae*). In *Primate behavior,* ed. L. Rosenblum, pp. 195–239. New York: Academic Press.

Epple, G., & Katz, Y. (1984). Social influences on estrogen excretion and ovarian cyclicity in saddle-back tamarins (*Saguinus fuscicollis*). *Amer. J. Primatol.* 6:215–227.

Evans, S., & Hodges, J. K. (1984). Reproductive status of adult daughters in family groups of common marmosets (*Callithrix jacchus jacchus*). *Folia primatol.* 42:127–133.

Ferrari, S. F. (1993). Ecological differentiation in the Callitrichidae. In *Marmosets and tamarins: Systematics, behaviour, and ecology,* ed. A. B. Rylands, pp. 314–328. Oxford: Oxford University Press.

Ferrari, S. F., & Diego, V. H. (1992). Long-term changes in a wild marmoset group *Folia primatol.* 58:215–218.

Ferrari, S. F., & Lopes-Ferrari, M. A. (1989). A re-evaluation of the social organisation of the Callitrichidae, with special reference to the ecological differences between genera. *Folia primatol* 52:132–147.

Ford, S. M. (1986). Systematics of the New World monkeys. In *Comparative primate biology, Vol. 1: Systematics, evolution, and anatomy,* ed. D. R. Swindler & J. Erwin, pp. 73–135. New York: Liss.

Ford, S. M., & Davis, L. C. (1992). Systematics and body size: implications for feeding adaptations in New World monkeys. *Am. J. Phys. Anthropol.* 88:415–468.

French, J. A. (1984). Lactation and fertility: an examination of nursing and interbirth intervals in tamarins (*Saguinus oedipus*). *Folia primatol.* 40:276–282.

French, J. A., Abbott, D. H., & Snowdon, C. T. (1984). The effects of social environment on estrogen excretion, scent marking, and sociosexual behavior in tamarins (*Saguinus oedipus*). *Am. J. Primatol.* 6:155–167.

French, J. A., deGraw, W. A., Hendricks, S. E., Wegner, F., & Bridson, W. E. (1992). Urinary and plasma gonadotropin concentrations in golden lion tamarins (*Leontopithecus rosalia*). *Amer. J. Primatol.* 26:53–59.

French, J.A., & Inglett, B. J. (1989). Female–female aggression and male indifference in response to unfamiliar intruders in lion tamarins. *Anim. Behav.* 37:487–497.

French, J. A., & Inglett, B. J. (1991). Responses to novel social stimuli in tamarins: A comparative perspective. In *Primate responses to environmental change,* ed. H. Box, pp. 275–294. London: Chapman and Hall.

French, J. A., & Inglett, B. J. (in press). Reproductive behavior and biology of female lion tamarins. In *Conservation biology of the golden lion tamarin,* ed. D. G. Kleiman. Washington, D.C.: Smithsonian Institution Press.

French, J. A., Inglett, B. J., & Dethlefs, T. M. (1989). The reproductive status of nonbreeding group members in captive golden lion tamarin social groups. *Am. J. Primatol.* 18:73–86.

French, J. A., Pissinatti, A., & Coimbra-Filho, A. F. (in press). Reproduction in captive lion tamarins (*Leontopithecus*): Seasonality, infant survival, and sex ratios. *Am. J. Primatol.*

French, J. A., Schaffner, C. M., Shepherd, R. E., & Miller, M. E. (1995). Familiarity with intruders modulates agonism toward outgroup conspecifics in Wied's black tufted-ear marmoset (*Callithrix kuhli*). *Ethology* 99:24–38.

French, J. A., & Snowdon, C. T. (1981). Sexual dimorphism in responses to unfamiliar intruders in the tamarin, *Saguinus oedipus*. *Anim. Behav.* 29:822–829.

French, J. A., & Stribley, J. A. (1985). Patterns of urinary oestrogen excretion in female golden lion tamarins (*Leontopithecus rosalia*). *J. Reprod. Fert.* 75:537–546.

French, J. A., & Stribley, J. A. (1987). Ovarian cycles are synchronized between and within social groups of lion tamarins (*Leontopithecus rosalia*). *Amer. J. Primatol.* 12:469–478.

Garber, P. A., Moya, L., & Malaga, C. (1984). A preliminary field study of the moustached tamarin monkey (*Saguinus mystax*) in northeastern Peru: questions concerned with the evolution of a communal breeding system. *Folia primatol.* 42:17–32.

Garber, P. A., Encamacion, F., Moya, L., & Pruetz, J. D. (1993). Demography and reproductive patterns in moustached tamarin monkeys (*Saguinus mystax*): implications for reconstructing platyrrhine mating systems. *Amer. J. Primatol.* 29:235–254.

Gittleman, J. L. (1985). Functions of communal care in mammals. In *Evolution – essays in honour of John Maynard Smith*, ed. P. J. Greenwood & M. Slatkin, pp. 187–205. Cambridge: Cambridge University Press.

Gittleman, J. L., & Oftedal, O. T. (1987). Comparative growth and lactation energetics in carnivores. *Symp. Zool. Soc. Lond.* 57:41–77.

Goldizen, A. W. (1987a). Facultative polyandry and the role of infant carrying in wild saddle-back tamarins (*Saguinus fuscicollis*). *Behav. Ecol. Sociobiol.* 20:99–109.

Goldizen, A. W. (1987b). Tamarins and marmosets: communal care of offspring. In *Primate societies*, ed. B. B. Smuts, D. L. Cheney, R. M. Seyfarth, R. W. Wrangham, & T. T. Struhsaker, pp. 34–43. Chicago: University of Chicago Press.

Goldizen, A.W., & Terborgh, J. (1989). Demography and dispersal patterns of a tamarin population: possible causes of delayed breeding. *Am. Nat.* 134:208–224.

Heistermann, M., Kleis, E., Pröve, E., & Wolters, H.-J. (1989). Fertility status, dominance, and scent marking behavior of family-housed female cotton-top tamarins (*Saguinus oedipus*) in absence of their mothers. *Am. J. Primatol.* 18:177–189.

Hershkovitz, P. (1977). *Living New World primates, Part 1*. Chicago: University of Chicago Press.

Hubrecht, R. C. (1984). Field observations on group size and composition of the common marmoset (*Callithrix jacchus jacchus*) at Tapacurá. *Primates* 25:13–21.

Hubrecht, R. C. (1989). The fertility of daughters in common marmoset (*Callithrix jacchus jacchus*) family groups. *Primates* 30:423–432.

Inglett, B. J. (1993). *The role of social bonds and the female reproductive cycle on the regulation of sociosexual behavior in the golden lion tamarin*. PhD dissertation, University of Nebraska, Omaha.

Inglett, B. J., French, J. A., Simmons, L. G., & Vires, K. W. (1989). Dynamics of intrafamily aggression and social reintegration in lion tamarins. *Zoo Biol.* 8:67–78.

Jamieson, I. (1989). Behavioral heterochrony and the evolution of birds' helping at the nest: an unselected consequence of communal breeding? *Am. Nat.* 133:394–406.

Johnson, L. D., Petto, A. J., & Sehgal, P. K. (1991). Factors in the rejection and survival of captive cotton-top tamarins (*Saguinus oedipus*). *Am. J. Primatol.* 25:91–102.

Kay, R. F. (1990). The phyletic relationships of extant and fossil Pitheciinae (Platyrrhini, Anthropoidae). *J. Hum. Evol.* 19:175–208.
Keller, L., & Reeve, H. K. (1994). Partitioning of reproduction in animal societies. *Trends Ecol. Evol.* 9:98–102.
Kleiman, D. G. (1977). Characteristics of reproduction and sociosexual interactions in pairs of lion tamarins (*Leontopithecus rosalia*) during the reproduction cycle. In *The biology and conservation of the Callitrichidae,* ed. D. G. Kleiman, pp. 181–190. Washington, D.C.: Smithsonian Press, 1977.
Kleiman, D. G. (1978). The development of pair preferences in the lion tamarin (*Leontopithecus rosalia*): male competition or female choice. In *Biology and behaviour of marmosets,* ed. H. Rothe, H.-J. Wolters, & J. P. Hearn, pp. 203–207. Göttingen, FRG:Eigenverlag-H. Rothe.
Kleiman, D. G. (1979). Parent-offspring conflict and sibling competition in a monogamous primate. *Amer. Nat.* 114:753–760.
Kleiman, D. G. (1980). The sociobiology of captive propagation. In *Conservation biology, an evolutionary-ecological perspective,* ed. M. E. Soule & B. A. Wilcox, pp. 243–261. Sunderland, Mass.: Sinauer Press.
Kleiman, D. G., Ballou, J. D., & Evans, R. F. (1982). An analysis of recent reproductive trends in captive golden lion tamarins *Leontopithecus r. rosalia* with comments on their future demographic management. *Int. Zoo Yrbk.* 22:94–101.
Koenig, A., & Rothe, H. (1991). Infant carrying in a polygynous group of common marmosets (*Callithrix jacchus*). *Amer. J. Primatol.* 25:185–190.
Koenig, W. D., Pitelka, F. A., Carmen, W. J., Mumme, R. L., & Stanback, M. T. (1992). The evolution of delayed dispersal in cooperative breeders. *Q. Rev. Biol.* 67:111–150.
Komdeur, J. (1992). Importance of habitat saturation and territory quality for evolution of cooperative breeding in the Seychelles warbler. *Nature* 358:493–495.
Küderling, I., Evans, C. S., & Abbott, D. H. (1992). Differential excretion of urinary estradiol in alpha-females and adult daughters of red bellied tamarins (*Saguinus labiatus,* Callitrichidae). Paper presented at XIVth Congress of International Primatological Society. Strasbourg, France, Abstract #409.
Leutenegger, W. (1973). Maternal-fetal weight relationships in primates. *Folia primatol.* 20:280–294.
Masataka, N. (1981a). A field study of the social behavior of Goeldi's monkeys (*Callimico goeldii*) in North Bolivia. I. Group composition, breeding cycle, and infant development. In *Kyoto Overseas Research Reports of New World Monkeys,* pp. 23–32. Kyoto, Japan: Kyoto University Primate Research Institute.
Masataka, N. (1981b). A field study of the social behavior of Goeldi's monkeys (*Callimico goeldii*) in North Bolivia. II. Grouping pattern and intragroups relationship. In *Kyoto Overseas Research Reports of New World Monkeys,* pp. 33–41. Kyoto, Japan: Kyoto University Primate Research Institute.
McGrew, W. C., & McLuckie, E. C. (1986). Philopatry and dispersion in the cottontop tamarin, *Saguinus o. oedipus:* An attempted laboratory simulation. *Int. J. Primatol.* 7:399–420.
Mays, N. A., Vleck, C. M., & Dawson, J. (1991). Plasma luteinizing hormone, steroid hormones, behavioral role, and nest stage in cooperatively breeding Harris' hawks (*Parabuteo unicinctus*). *Auk* 108:619–637.
Pook, A. G., & Pook, G. (1981). A field study of the socio-ecology of the Goeldi's monkey (*Callimico goeldii*) in Northern Bolivia. *Folia primatol* 35:288–312.
Price, E. C., & McGrew, W. C. (1991). Departures from monogamy in colonies of captive cotton-top tamarins. *Folia primatol.* 57:16–27.

Pryce, C. R. (1992). A comparative systems model of the regulation of maternal motivation in mammals. *Anim. Behav.* 43:417–441.

Pryce, C. R., Döbeli, M., & Martin, R. D. (1993). Effects of sex steroids on maternal motivation in the common marmoset (*Callithrix jacchus*): development and application of an operant system with maternal reinforcement. *J. Comp. Psychol.* 107:99–115.

Ralls, K., & Ballou, J. (1982). Effects of inbreeding on infant mortality in captive primates. *Int. J. Primatol.* 3:491–505.

Ramirez, M. (1984). Population recovery in the moustached tamarin (*Saguinus mystax*): management strategies and mechanisms of recovery. *Amer. J. Primatol.* 7:245–259.

Reyer, H. U., Dittami, J. P., & Hall, M. R. (1986). Avian helpers at the nest: are they psychologically castrated? *Ethology* 71:216–228.

Roda, S. A. (1989). Ocorrencia de duas femeas reprodutivas em grupos selvagens de *Callithrix jacchus* (Primates, Callitrichidae). In *Resumos do XVI congresso Brasileiro de zoologia,* p. 122. Joào Pessoa, Brasil: Universidade Federal da Paraiba.

Rosenberger, A. R. (1981). Systematics: The higher taxa. In *Ecology and behaviour of neotropical primates,* ed. A. F. Coimbra-Filho & R. A. Mittermeier, pp. 111–168. Rio de Janeiro: Academia Brasiliera de Ciencias.

Rothe, H. (1975). Some aspects of sexuality and reproduction in groups of captive marmosets (*Callithrix jacchus*). *Z. Tierpsychol.* 37:255–273.

Rothe, H., & Koenig, A. (1991). Variability of social organization in captive common marmosets (*Callithrix jacchus*). *Folia primatol.* 57:28–33.

Rylands, A. B. (1986). Infant-carrying in a wild marmoset group *Callithirx humeralifer:* evidence for a polyandrous mating system. In *A primatologia no Brasil – 2,* ed. M. T. de Mello, pp. 131–144. Brasilia, Brazil: Sociedade Brasileira de Primatologia.

Rylands, A. B. (1993). *Marmosets and tamarins: Systematics, ecology and behaviour.* Oxford: Oxford University Press.

Rylands, A. B., Coimbra-Filho, A. F., & Mittermeier, R. A. (1993). Systematics, geographic distribution, and some notes on the conservation status of the Callitrichidae. In *Marmosets and tamarins: Systematics, behaviour, and ecology,* ed. A. B. Rylands, pp. 11–77. Oxford: Oxford University Press.

Saltzman, W., Haker, M. W., Schultz-Darken, N. J., & Abbott, D. H. (1994a). Ovulatory activity, social behavior, and "dispersal" tendencies of post-pubertal female common marmosets (*Callithrix jacchus*) living in their natal families. *Am. J. Primatol.* 33:238.

Saltzman, W., Schultz-Darken, N. J., Scheffler, G., Wegner, F. H., & Abbott, D. H. (1994b). Social and reproductive influences on plasma cortisol in female marmoset monkeys. *Physiol. Behav.* 56:801–810.

Savage, A. (1990). *The reproductive biology of the cotton-top tamarin* (Saguinus oedipus oedipus) *in Colombia.* PhD dissertation, University of Wisconsin, Madison.

Savage, A., Ziegler, T. E., & Snowdon, C. T. (1988). Sociosexual development, pair bond formation, and mechanisms of fertility suppression in female cotton-top tamarins (*Saguinus oedipus oedipus*). *Am. J. Primatol.* 14:345–359.

Scanlon, C. E., Chalmers, N. R., & Monteiro da Cruz, M. A. O. (1988). Changes in the size, composition, and reproductive conditions of wild marmoset groups (*Callithrix jacchus jacchus*) in northeast Brazil. *Primates* 29:295–305.

Smith, T. E. (1994). *Role of odor in the suppression of reproduction in female naked mole-rats and common marmosets and the social organisation of these two species.* PhD dissertation, University of London.

Smith, T. E., Schaffner, C. M., & French, J. A. (1995). Regulation of reproductive function in subordinate female black tufted-ear marmosets (*Callithrix kuhli*). *Amer. J. Primatol.* 36:156–157.

Snowdon, C. T., Savage, A., & McConnell, P. B. (1985). A breeding colony of cotton-top tamarins (*Saguinus oedipus*). *Lab. Anim. Sci.* 35:477–480.

Snowdon, C. T., & Soini, P. (1988). The tamarins, genus *Saguinus*. In *Ecology and behavior of Neotropical primates, Vol. 2*, ed. R. A. Mittermeier, A. B. Rylands, A. F. Coimbra-Filho, & G. A. B. Fonseca, pp. 223–298. Washington, D. C.: World Wildlife Fund.

Snowdon, C. T., Ziegler, T. E., & Widowski, T. M. (1993). Further hormonal suppression of eldest daughter cotton-top tamarins following birth of infants. *Amer. J. Primatol.* 31:11–21.

Soini, P. (1987). Sociosexual behavior of a free-ranging *Cebuella pygmaea* (Callitrichidae, Platyrrhini) troop during postpartum estrus of its reproductive female. *Am. J. Primatol.* 13:223–230.

Soini, P. (1988). The pygmy marmoset, genus *Cebuella*. In *Ecology and behavior of Neotropical primates, Vol. 2*, ed. R. A. Mittermeier, A. B. Rylands, A. Coimbra-Filho, & G. A. B. Fonseca, pp. 79–129. Washington, D.C.: World Wildlife Fund.

Soini, P. (1993). The ecology of the pygmy marmoset, *Cebuella pygmaea*: some comparisons with two sympatric tamarins. In *Marmosets and tamarins: Systematics, behaviour, and ecology*, ed. A. B. Rylands, pp. 257–261. Oxford: Oxford University Press.

Stevenson, M. F., & Rylands, A. B. (1988). The marmosets, genus *Callithrix*. In *Ecology and behavior of Neotropical primates, Vol. 2*, ed. R. A. Mittermeier, A. B. Rylands, A. Coimbra-Filho, G. A. B. Fonseca, pp. 131–222. Washington, D.C.: World Wildlife Fund.

Tardif, S. D. (1984). Social influences on sexual maturation of female *Saguinus oedipus oedipus*. *Amer. J. Primatol.* 6:199–209.

Tardif, S. D. (1988). Intragroup aggression in cotton-top tamarins. Paper presented at International Primatological Society, Brasilia, Brazil.

Tardif, S. D. (1992). Relation between caloric content of diet, body weight, and ovulation number in the marmoset monkey. *Biol. Reprod.* 46 Suppl. 1:106.

Tardif, S. D., Harrison, M. L., & Simek, M. A. (1993). Communal infant care in marmosets and tamarins: relation to energetics, ecology, and social organization. In *Marmosets and tamarins: Systematics, behaviour, and ecology*, ed. A. B. Rylands, pp. 220–234. Oxford: Oxford University Press.

Tardif, S. D., & Jaquish, C. E. (1994). The common marmoset as a model for nutritional impacts upon reproduction. *Ann. N. Y. Acad. Sci.* 709:214–215.

Terborgh, J., & Goldizen, A. W. (1985). On the mating system of the cooperatively breeding saddle-backed tamarin (*Saguinus fuscicollis*). *Behav. Ecol. Sociobiol.* 16:293–299.

Vehrencamp, S. L. (1983). A model for the evolution of despotic versus egalitarian species. *Anim. Behav.* 31:667–682.

Wasser, S. K., & Barash, D. P. (1983). Reproductive suppression among female mammals: implications for biomedicine and sexual selection. *Q. Rev. Biol.*, 58:513–558.

Widowski, T. M., Ziegler, T. W., Elowson, A. M., & Snowdon, C. T. (1990). The role of males in stimulating reproductive function in female cotton-top tamarins, *Saguinus o. oedipus*. *Anim. Behav.* 40:731–741.

Widowski, T. M., Porter, T.A., Ziegler, T. E., & Snowdon, C. T. (1992). The stimulatory effect of males on the initiation but not the maintenance of ovarian cycling in cotton-top tamarins (*Saguinus o. oedipus*). *Am. J. Primatol.* 26:97–108.

Ziegler, T. E., Bridson, W. E., Snowdon, C. T., & Eman, S. (1987). The endocrinology of puberty and reproductive functioning in female cotton-top tamarins (*Saguinus oedipus*) under varying social conditions. *Biol. Reprod.* 37:618–627.

Ziegler, T. E., Scheffler, G., & Snowdon, C. T. (1995). The relationship of cortisol levels to social environment and reproductive functioning in female cotton-top tamarins, *Saguinus oedipus. Horm. Behav.* 29:407–424.

Ziegler, T. E., Snowdon, C. T., & Bridson, W. E. (1990). Reproductive performance and excretion of urinary estrogens and gonadotropins in the female pygmy marmoset (*Cebuella pygmaea*). *Am. J. Primatol.* 22:191–204.

4

Cooperative Breeding, Reproductive Suppression, and Body Mass in Canids

PATRICIA D. MOEHLMAN and HERIBERT HOFER

4.1 Introduction

The family Canidae is unusual among mammals in that its pervasive mating system is obligatory monogamy (Kleiman 1977). Canids have large litter sizes and a long period of infant dependency (Kleiman & Eisenberg 1973). They also characteristically have a high degree of (1) intraspecific flexibility in social organization, and (2) cooperative behavior within social groups, ranging from hunting and food sharing to the provisioning of sick adults and dependent pups (Macdonald & Moehlman 1983; Moehlman 1989). In this chapter, we use an up-to-date and expanded data set to conduct a review of canid life-history traits and ask how body mass, resources, and behavior interact to facilitate the development of cooperative breeding. We pursue two lines of inquiry. We first discuss the influence of body mass on life-history traits and test hypotheses that consider ecological and behavioral correlates of reproductive output. We then consider the costs and benefits of alloparental care in greater detail and ask how the evolution of reproductive suppression is related to individual reproductive tactics.

Previous analyses of the allometric scaling of life-history traits in canids revealed positive and significant relationships between female body mass, gestation, neonate mass, litter size, and litter mass, which appeared to be linked to breeding behavior in a systematic fashion. These relationships suggested that larger canids might require helpers for the successful rearing of young (Moehlman 1986, 1989). In carnivores, cooperatively breeding and biparental species have a higher litter mass than species with strictly maternal care (Gittleman 1985a, 1986). In a more recent analysis of order Carnivora, which incorporated an expanded data base and phylogenic correlations, Gittleman (1993) came to the conclusion that an allometric approach might not be useful for elucidating the evolution of carnivore life-history traits unless it provided new insights into the factors and mechanisms involved in evolutionary changes. This is because previous allometric analyses have been characterized by a search for patterns and correlations (and exceptions to them, Calder 1984) rather than having been guided by testing hypothesis using appropriate fitness measures. Lindstedt and Swain (1988) and Gordon (1989) illustrate that erroneous conclusions are likely if breeding success per season is chosen as the fitness measure rather than lifetime reproductive output (lifetime litter size and lifetime litter mass). In this chapter we extend previous analyses based on yearly production of offspring to examining lifetime output. We show that canids have significantly positive allometry for lifetime litter mass. Thus, do larger canids have a larger lifetime reproductive output and does this translate into increased fitness through larger numbers of surviving offspring? If larger canid species had relatively larger litters, they would presumably require more postpartum input to pups than smaller canid species. Do larger canid females anticipate that their offspring will receive a total postpartum alloparental and parental input that exceeds the input potentially available to offspring of smaller females? We test this proposition by investigating three hypotheses:

1. Larger canids produce larger litters because they can rely on recruiting a sufficient number of alloparental helpers by suppressing their reproduction. This predicts that (a) group size scales with body mass, (b) the incidence of alloparental care scales with body mass, and (c) the incidence of reproductive suppression scales with body mass and litter size.
2. Larger canids produce larger litters because the size and type of food they consume improves the economy of provisioning in that it increases the benefit–cost ratio of foraging for provisioning. This predicts that (d) diet scales with body mass in that larger canids are more likely to eat large (i.e., vertebrate) items, (e) the incidence of cooperative hunting scales with body mass if cooperative hunting improves the benefit–cost ratio of foraging, and (f) the benefit–cost ratio of foraging increases with body mass.
3. Larger canids produce larger litters because the resources on a breeding pair's home range do not constrain survival and recruitment of alloparental caretakers. This predicts that (g) the incidence of cooperative hunting scales with maximum prey size so that cooperative hunting facilitates access to prey sizes (and thereby the variability of the amount of resources per home range) otherwise unobtainable, (h) breeding individuals are less likely to reject additional group members in larger canids, (i) the occurrence of starvation of caretakers is less likely in larger canids, (j) the

amount of resources per home range increases with body mass, and (k) rates of resource depletion should decrease with body mass.

It is equally important to appreciate the behavioral and ecological components of social organization and breeding systems and the variability observed within and between species of similar size. The established method for examining the role of food resources in determining carnivore social systems has been to record distribution and abundance and to examine density-dependent relationships (Macdonald 1983; von Schantz 1984a,b). This approach has been useful, but it is basically one step removed from a rigorous approach that examines how the energetics of food acquisition (costs) affect parental input to pups (provisioning) and fitness benefits (pup survival). An important but little appreciated distinction is made between parental effort (the energy devoted to producing care) and parental input (the care received by the young). One purpose of our review is to point out that there are currently few data sets that combine breeding behavior, parental and alloparental input to individual young, and individual parent and alloparent energetic costs and benefits of food acquisition. Without such information, a thorough understanding of the evolution of cooperative breeding and reproductive suppression remains incomplete. We present some simple models of parental and alloparental input to demonstrate how variation in costs (energy expenditure for food) and benefits (pup survival) might affect individual reproductive tactics and what data field researchers ought to look for when they consider the impact of alloparental care.

4.2 Methods

Data on life-history traits were extracted from primary sources. Previous compilations of canid life-history traits can be found in Ewer (1973), Bekoff and Jamieson (1975), Bekoff, Diamond, and Mitton (1981), Eisenberg (1981), Gittleman (1984, 1986), and Moehlman (1986). We compiled a new data set from original sources, including the results of many recently published studies and resulting in an expanded data base.

The majority of the 33 species reviewed (for scientific names see Table 4.A1) have been subjected to short-term studies, and for some larger canids, such as coyotes or wolves, many studies concentrated on population aspects (population density, territory size, diet) rather than detailed studies of breeding behavior and success. Thus, the type and quality of data varied considerably from brief casual observations to detailed long-term behavior studies, limiting the resolution of comparative analysis.

4.2.1 Data Selection

Results from field studies were preferred whenever available. Data were used from captive studies for traits that were assumed to be relatively unaffected by conditions of captivity. These include gestation, neonate mass, age at eye opening, age when teeth erupt, age when eating first solids, age at weaning, age when reaching adult body mass, and age at sexual maturity. Age at sexual maturity was based on proven records of breeding. Few data on longevity were available from field studies, and thus records from captivity were used. Captive studies also provided information on the occurrence of behavioral traits (e.g., males provisioning pups), that due to the lack of studies had not been observed in the wild. The data on life-history traits and breeding biology of canids and their sources are listed in Tables 4.A2 and 4.A3.

We took particular care to minimize sampling errors for calculating species means of female body mass and litter size, the two traits that are most easily measured in the field. We attempted to use only the results of studies that measured both traits in the same population. Population means were often the only statistic reported by a study. If such data were available from more than one population, more intensively studied populations were preferred to short studies with small sample sizes. Results of more than one study were available for only a minority of species. In these cases, we calculated means (of means) across populations. Estimates of litter size were available as counts of placental scars, embryos, or pups at first emergence from the den. Placental scar counts may (1) underestimate litter size at birth if the period between parturition and examination of scars is long, because placental scars can fade and (2) overestimate litter size at birth due to intrauterine mortality or persistence of scars from previous litters (Lindström 1981). Validated placental scar counts were provided by Lindström (1981) for red foxes and by McEwen and Scott (1956) and Macpherson (1969) for arctic foxes. Counts of embryos may overestimate litter size at birth if embryos are not near full term and could be aborted. Comparison of counts of embryos and placental scars in arctic foxes showed no significant differences (Hersteinsson 1984). Counts of emerged pups may underestimate litter size at birth if (1) observations and/or emergence occur several weeks after birth as some mortality may go unnoticed, or (2) counts are based on trapping (Hersteinsson 1984). If more than one type of information was available, counts of embryos and placental scars were preferred to counts of pups at the den.

Group size was recorded as the mean or modal social group size of adults observed in the field. In some cases the sizes were based on known individuals in social groups, in other cases on random observations of adults seen together. Cooperative hunting was scored as yes or no; a negative record does not preclude the possibility that it may be observed in the future. Diet was scored as vertebrate eater (V) or invertebrate eater (I) if the key component of the diet exceeded 60 percent, and as omnivore (O) if none of the components exceeded 60 percent. The mass of the largest prey taken by a species was scored independently of its importance in the population's diet. The rearing of young in the wild has been observed in 20 species, but few studies were sufficiently extensive to record details of and variation in breeding behavior and rearing of young. Because of small sample sizes in most cases and the anecdotal nature of some of the information, it was difficult to assign species to narrowly defined categories of breeding or rearing behavior. We have opted instead to (1) report the variation observed and (2) produce two scores for the occurrence of reproductive suppression per species that recognize the quality of information available.

In a strict sense, reproductive suppression occurs when there are sexually mature individuals in a social unit that could reproduce but are prevented from doing so because other group members use behavioral or chemical means to suppress a breeding attempt. This has been demonstrated in captivity for gray wolves (Packard et al. 1985), bush dogs (Porton, Kleiman, & Rodden 1987), red foxes (Macdonald 1980), and arctic foxes (Kullberg & Angerbjörn 1992). In field studies, reproductive suppression may be recognized if (1) group composition is completely known and (2) the presence of nonreproductive adults (NRA) subordinate to breeding individual(s) is recorded. In all well-studied cases so far, the presence of NRA has meant that NRA assisted in rearing young by guarding and provisioning (i.e., were alloparents or helpers, Skutch 1935, 1961). More indirect evidence are observations of groups of more than two adults if only one breeding pair is known but group composition and social relationships are unknown. If the number of litters per group is unknown, a record of one male and two females in a group may indicate either (1) a NRA that is reproductively suppressed and acts as an alloparent (helper) or (2) a polygynous breeding system with multiple litters – the opposite of reproductive suppression. In recognition of these ambiguities, reproductive suppression was defined to occur (1) in the broad sense if NRA were recorded whether or not social organization and social relationships were known; (2) in the strict sense if NRA were observed to help by guarding and provisioning young. This is still less than optimal because it assumes that the extra adults were sexually mature. A species was considered to lack reproductive suppression if no evidence in favor of reproductive suppression was available (in the two senses just given), or if additional adults per group were always associated with mating systems other than monogamy, multiple litters, communal denning, or communal nursing. If data existed in favor of and against reproductive suppression, the species was scored as being reproductively suppressed. Thus, a positive score for reproductive suppression means that incidences of suppression have been recorded rather than that a species has obligatory reproductive suppression. All statistical tests were run on reproductive suppression scored in the strict sense.

4.2.2 Statistical Analysis

If more than one study provided information, results were averaged across study populations. Predator sizes were calculated as the average of male and female body mass. The dependence of life-history traits on female body mass was tested using $\log_{10}$ transformed data. All data were used in the analysis (i.e., potential outliers were not removed). To facilitate comparison with previously published studies, we report two sets of slopes: (1) linear (ordinary least squares) regression and (2) reduced major axis (RMA) regression using the regression equation $\log_{10}$ (species means of life-history variable) = intercept + slope $\times$ $\log_{10}$ (species means of female body mass). In this chapter we refer to the RMA values. Partial correlation coefficients were calculated to identify significant relationships between life-history traits while simultaneously correcting for the influence of other factors known or presumed to have an effect. Logistic regressions were used to identify factors that significantly influenced the likelihood of reproductive suppression being present in a species. Analyses were conducted with SYSTAT 5.0 and LOGIT 2.0 (Steinberg & Colla 1990) on a personal computer following recommendations by Sokal and Rohlf (1981) and Conover (1980). Results are given as means $\pm$ standard deviations. All P-values are for two-tailed tests unless specified otherwise. Residuals of linear regressions were tested for normality by using the Lilliefors test as implemented (with corrected P-values: Wilkinson 1990) in SYSTAT. Residuals for all models were normally distributed.

In comparative analyses it is important to identify phylogenetic patterns and to deal with the problem of statistical independence of species means (Harvey & Pagel 1991). Briefly, this problem arises because we cannot assume that data from one species are independent of data from other species. The species may share common ancestors, and thus a trait may be similar in two species because of common ancestry rather than as a result of independent evolutionary processes. However, our understanding of canid phylogeny is incomplete, and the interpretation of genetic relationships and divergence times is problematic. Because of the tentative nature of the current phyloge-

Table 4.1. The allometry of life-history traits and reproductive output in Canidae[a]

L-history trait	Unit	Linear regression			n	F	P	RMA Slope
		i-cept	Slope	r^2				
Neonate mass	g	-0.81	0.78 ± 0.05	0.925	20	221.5	<0.0001	0.81
Litter size	—	-0.33	0.27 ± 0.06	0.401	32	20.1	0.0001	0.44
Litter mass	g	-1.09	1.04 ± 0.08	0.897	20	156.5	<0.0001	1.10[b]
Gestation	d	1.40	0.10 ± 0.01	0.707	25	55.5	<0.0001	-0.11
Age at eye opening	d	1.18	-0.04 ± 0.08	0.014	18	0.2	0.64	0.22
Age of teeth eruption	d	0.90	0.05 ± 0.11	0.018	13	0.2	0.67	0.38
Age when first eating solids	d	0.84	0.14 ± 0.07	0.091	12	4.4	0.062	0.31
Age at weaning	d	1.66	0.05 ± 0.07	0.027	17	0.4	0.53	0.28
Age when reaching adult mass	d	0.88	0.36 ± 0.06	0.706	16	33.6	<0.0001	0.43
Age at sexual maturity	d	2.29	0.05 ± 0.03	0.124	16	2.0	0.18	0.14
Longevity	mo	1.35	0.22 ± 0.06	0.398	21	12.6	0.002	0.34
Lifetime litter mass	g × mo	-0.90	1.28 ± 0.12	0.875	17	105.3	<0.0001	1.35[c]
Lifetime litter size	mo	-0.05	0.49 ± 0.10	0.551	21	23.3	0.0001	0.68

[a]Data points were the simple species means.
[b]Slope not significantly different from 1 ($F_{1,18} = 1.42$, P = 0.25).
[c]Slope significantly higher than 1 ($F_{1,15} = 4.91$, P = 0.043).

netic trees for the entire family, we confined discussion of phylogenetic dependence to Appendix 4. Table 4.A4 and Figure 4.A1 contain the available information on canid phylogeny for the whole family (Wayne et al. 1989) and the results of the rerun of the allometric analysis that takes into account phylogenetic dependence. As the results show, correcting for phylogenetic dependence changes some results. Figure 4.A2 contains the available phylogenetic information on two branches of the canid family known as the wolflike and foxlike canids, respectively (Geffen et al. 1992; Wayne 1993). The two branches are well separated by morphological and molecular criteria. Tables 4.A5 and 4.A6 contains an allometric analysis of life-history traits for these two branches, correcting for phylogenetic dependence. There is also considerable variation in the incidence of reproductive suppression between these two branches, suggesting that allometric analyses are useful.

4.3 Results

4.3.1 Allometry of Life-History Traits

All parameters of reproductive output scaled with body mass (Table 4.1). Per breeding attempt, larger canids have larger neonates and a larger number of offspring, and litter mass (the product of the two) scales with female body mass with an exponent of 1.10. Of time-related life-history traits, only gestation, age when reaching adult body mass (a measure of developmental speed), and longevity showed significant dependence on body mass. All other development-related traits did not scale allometrically. All species can become sexually mature and are able to breed in their first year. The considerable variation in the difference in age between reaching adult body mass and sexual maturity suggests that age at sexual maturity may not be related to growth.

Is lifetime reproductive output independent of body mass (Gordon 1989)? Lifetime litter mass (the product of litter mass and longevity) scaled with an exponent of 1.35 ± 0.12. This exponent was significantly higher than unity, a result not seen before in any other mammalian family (see Table 4.1).

4.3.1.1 Group size and feeding ecology

Larger canids live in larger groups (Table 4.2; partial correlation of group size and female body mass correcting for maximum prey size, $c_p = 0.55$, $P = 0.0024$). Species that live in larger groups have a larger maximum prey size (correcting for female body mass, $c_p = 0.40$, $P = 0.03$) and are more likely to eat vertebrates (mean group size 5.05 ± 2.9, $n = 11$) than are invertebrate

Table 4.2. Allometry of feeding, breeding, and alloparental behavior in canids[a]

Species	Group size	Diet	Coop. hunting?	Territorial?	Nonmonogamy observed?	Breed. M provisions preg F?	Communal denning, nursing?	Multiple litters?	Multiple litters survive?	Field study of rearing?	Breed. M provisions pups?	Nonreproductive adults (NRA) in group?	Sex that stays as NRA?	Helpers present?	Sex of helper
Vul can	2.6	I	No	Yes	PG uncert.	—	No	Poss.	—	Yes	—	Yes	F	uncert.	—
Vul zer	—	O	No	No	—	—	No	No	—	—	Yes	No	—	—	—
Uro lit	2.0	I	No	Yes	—	—	—	—	—	Yes	—	—	—	—	—
Vul rue	2.0	I	No	Yes	—	—	—	—	—	—	—	—	—	—	—
Vul pal	—	O	—	—	—	—	—	—	—	—	—	—	—	—	—
Vul mac	2.0	V	—	No	PG	—	—	Yes	Yes	Yes	Yes	Yes (n = 1)	F	uncert.	—
Vul cor	—	V	—	—	—	—	—	Yes	—	—	—	—	—	—	—
Vul vel	2.0	O	—	No	PG?	—	—	—	—	Yes	—	poss.	—	—	—
Vul ben	3.0	O	—	—	PG	—	CD, CN	poss.	—	Yes	Yes	poss.	(F)	—	—
Vul cha	2.0	O	—	—	—	—	—	Yes	—	Yes	Yes	poss.	—	No	—
Pse gri	3.0	O	—	Yes	PG	—	CD, CN	Yes	—	Yes	Yes	Yes	F	Yes	F
Vul fer	2.0	V	—	—	PG	—	—	—	—	—	—	—	—	uncert.	—
Pse vet	—	O	—	—	—	—	—	—	—	—	—	—	—	—	—
Uro cin	2.0	O	—	—	PG	—	CD	Yes	—	Yes	Yes	poss.	(F)	—	—
Alo lag	2.4	V	No	Yes	PG	Yes	Yes	Yes	Yes	Yes	Yes	Yes	F	Yes	F
Oto meg	2.8	I	No	No	PG	—	Yes	Yes	—	Yes	Yes	Yes	F	Yes	MG

Species		Diet											
Cer tho	2.0	O	No	poss.	—	No	No	—	Yes	Yes	No	—	No	—
Pse gym	—	O	—	Yes	—	—	—	Yes	Yes	Yes	No	F	No	—
Vul vul	3.1	O	No	Yes	PG	Yes	Yes	Yes	Yes	Yes	Yes	—	Yes	F
Pse sec	—	O	—	Yes	—	—	—	—	—	Yes	—	—	—	—
Nyc pro	2.0	O	No	Yes	—	No	No	—	Yes	Yes	No	—	No	—
Can aur	2.5	V	Yes	Yes	PG	No	No	Yes	Yes	Yes	Yes	F+M	Yes	F+M
Spe ven	7.0	V	poss.	—	(n = 1)	No	No	—	—	Yes	Yes	F+M	—	—
Can mes	2.7	V	Yes	Yes	PG (n = 1)	CD	Yes	Yes	Yes	Yes	Yes	F+M	Yes	F+M
Can adu	—	O	—	—	—	—	—	—	—	—	—	—	—	—
Pse cul	—	O	—	—	—	—	—	—	—	—	—	—	—	—
Ate mic	—	—	—	—	—	—	—	—	—	—	—	—	—	—
Can lat	4.4	V	Yes	Yes	MM	Yes	Yes	Yes	Yes	Yes	Yes	F+M	Yes	F+M
Can sim	7.0	V	No	Yes	—	No	No	Yes	Yes	Yes	Yes	F+M	Yes	F+M
Cuo alp	8.3	V	Yes	Yes	—	Yes	Yes	Yes	Yes	Yes	Yes	F+M	Yes	F+M
Chr bra	2.0	O	No	Yes	(n = 1)	No	No	—	Yes	Yes	No	—	No	—
Lyc pic	8.2	V	Yes	No	PA, MM	Yes	Yes	Yes	Yes	Yes	Yes	F+M	Yes	F+M
Can lup	9.0	V	Yes	Yes	PG (n = 1) PA, MM	Yes	Yes	Yes	Yes	Yes	Yes	F+M	Yes	F+M

[a] F female; M male. Diet: I insectivore; O omnivore; V vertebrate eater. Mating pattern: MM multiple pairs; PA polyandry; PG polygyny; MG male guarding.

eaters (mean 2.47 ± 0.42, $n = 3$) or omnivores (mean 2.31 ± 0.5, $n = 10$; ANOVA, $F_{2,21} = 5.41$, $P = 0.013$). There was no relationship between maximum prey size and female mass (correcting for group size, $c_p = 0.20$, $P = 0.19$). Data on benefit–cost ratios of foraging in relation to body mass were not available for analysis.

4.3.1.2 Cooperative Hunting

Seventeen of the small canids are omnivorous or insectivorous and given current knowledge, are solitary foragers (see Table 4.2). Even small canids that consume vertebrates and that may kill prey 10 times or more greater than their body mass are solitary foragers. Cooperative hunting is therefore the domain of larger canids (logistic regression of incidence of cooperative hunting on female mass, $G = 6.80$, $df = 1$, $P = 0.009$). Maximum prey size to predator size ratios (τ) were significantly higher for cooperative hunters than for solitary foragers (Mann-Whitney $U = 1$, $P = 0.0017$). Diet breadth was independent of body mass in solitary foragers (Spearman's $\rho = 0.018$, $n = 10$, NS) but increased significantly with body mass in cooperative hunters ($\rho = 0.89$, $n = 6$, $P < 0.05$), suggesting that diet breadth can be extended through cooperative hunting. The largest value recorded for τ was 25 (moose at 800 kg killed by gray wolves). Two of the largest canids, the simien wolf and the maned wolf, feed strictly on small food items. Simien wolves live in large groups but feed exclusively on high-density rodent populations ($3,000-4,000$ kg/km^2, Gottelli & Sillero-Zubiri 1990). Maned wolves are omnivorous, feed on fruits and rodents, and live in groups of two (Dietz 1984). Thus, although larger canids exist in larger groups and larger groups are more likely to feed on large vertebrates, increasing body mass does not automatically imply cooperative hunting or consumption of larger prey. Clear demonstrations that cooperative hunting and carcass defense (e.g., in larger group sizes) increase individual intake rate are still missing, although limited data suggest that cooperative hunters are more successful than single hunters (Wyman 1967; Lamprecht 1978).

4.3.1.3 Parental and alloparental input

Canids are the only carnivore family that provision group members by regurgitating. Males in six species have been observed to provision their pregnant mates (see Table 4.2). Canids produce their young in dens, and typically the pups do not emerge until they are three weeks old. Both parents feed and guard their offspring through the period of dependency (see Table 4.2). Males also provision their mates during lactation. Alloparental input (i.e., guarding and provisioning) typically consists of both provisioning pups and the lactating mother and guarding the pups. Only in bat-eared foxes has alloparenting behavior been limited to guarding. The incidence of alloparental behavior increases with female body mass (logistic regression, $G = 4.77$, $df = 1$, $P = 0.029$). Sex of helpers varies with body mass: Smaller canids have only female helpers, whereas larger canids have both male and female helpers ($U = 0$, $P = 0.006$).

Nonpregnant domestic dogs and gray wolves can exhibit the same hormonal profiles as their pregnant counterparts and go through a pseudopregnancy that can culminate in pseudoparturition and lactation (Seal et al. 1979; Asa this volume). Helping through lactating by presumed nonpregnant subordinate females has been recorded in red foxes (von Schantz 1981) and Simien wolves (Gottelli & Sillero-Zubiri pers. comm.). Communal nursing has been observed in Bengal foxes, bat-eared foxes, arctic foxes, gray zorros, red foxes, coyotes, African wild dogs, and gray wolves (see Table 4.2).

4.3.1.4 Pup survival

What is the pattern of pup survival? Species may be considered obligatory pair breeding if only two adults are involved and obligatory cooperative breeding if more than two adults are required to raise pups successfully. Moehlman (1986) suggested that in small canids single parents might be able to raise pups successfully. In kit foxes ($n = 2$: O'Neal 1987) and red foxes ($n = 1$: Zabel & Taggart 1989), females succeeded in raising pups on their own. When medium-sized canid females lost their mates, both the females and their pups died within a few days ($n = 2$, silverbacked jackals: Moehlman 1986). Among large canids, African wild dogs are usually considered to be obligatory cooperative breeders, but a pair of wild dogs without helpers did successfully raise four pups ($n = 1$, R: Burrows pers. comm. 1992). Recently, Boyd and Jimenez (1994) observed litters being successfully raised by a single parent ($n = 3$) in a low-density gray wolf population that was colonizing an area of relatively high-density prey populations.

Do helpers significantly increase reproductive success by increasing the number of pups raised to independence? Significant positive effects of helpers have been demonstrated in silverbacked (Moehlman 1979) and golden jackals (Moehlman 1986), nonsignificant positive effects in coyotes (Bekoff & Wells 1986) and African wild dogs (Malcolm & Marten 1982), and variation in fitness effects from negative to positive with resource levels in gray wolves (Harrington, Mech, & Fritts 1983).

4.3.2 Modeling Parental and Alloparental Input

In this section we use the framework of cost–benefit analysis (Trivers 1974; Lazarus & Inglis 1978; Clutton-Brock 1991) in parental investment theory to (1) explore how we can use allometric relationships to make predictions about the factors that influence cooperative breeding and (2) consider how fitness consequences of alloparental contributions may vary and under what conditions alloparental care is likely to evolve. The lack of empirical results forces us to use simple models, primarily as a heuristic tool, to emphasize that the impressive variety and intraspecific flexibility in canid social organization may be derived from a few general principles. Benefits of parental care are usually measured in terms of the survival and future reproductive success of the offspring. Costs are the reduced fitness of the parent's future offspring due to its care of the current offspring (Trivers 1972) and may be approximated by measures of energetic expenditure or mortality between breeding seasons (Clutton-Brock 1991). The sum of parental and alloparental input is the total care and resources actually received by the offspring (Evans 1990). The distinction between parental input and reproductive effort (the resources spent by parents during a breeding attempt) is often subtle but crucial because offspring fitness depends on parental input rather than on parental effort per se. Parental fitness, however, depends in part on offspring fitness (i.e., input) and in part on the energy required to successfully raise the current batch of young (i.e., effort). To demonstrate this, imagine that the survival of jackal pups is a function of the number of regurgitations received per day (= input). In a good year, a given provisioning rate requires only a small hunting effort by the parents, whereas in a bad year the same number of regurgitations may require a considerable hunting effort with high risks involved. Reproductive effort would therefore vary tremendously between years, but parental input and offspring fitness would remain constant. Thus, benefits and costs of parental and alloparental care have to be considered per unit of parental input rather than in relation to parental effort (e.g., Parker, Mock, & Lamey 1989; Evans 1990).

In a field study, benefits and costs of parental and alloparental care could be evaluated in the following way. Benefits could be measured as survival of pups, and costs as a first approximation, as energetic expenditure (e.g., kilometers traveled to catch a prey item). A benefit curve in units of parental input could then be constructed by using data from several litters on pup survival in relation to the recorded average number of prey items consumed by the pup. The appropriate cost curve would be constructed by calculating the average number of kilometers that a parent or alloparent traveled for a given average number of prey items (Trivers 1974; Lazarus & Inglis 1978). Empirical benefit and cost curves are not yet available for any mammal

(Clutton-Brock 1991). In carnivores, an empirical benefit curve has been produced for the spotted hyena, *Crocuta crocuta* (Hofer & East 1993).

4.3.2.1 The allometry of parental input

Allometric analyses are useful if they lead to interesting hypotheses about characteristics whose effects are subsumed under body mass (Gittleman 1993). In our case, allometric patterns essentially ought to lead to hypotheses about changes in the cost and benefit curves of components of parental care in relation to parental input. The use of parental input as the key parameter permits us to focus on the ecological conditions of food acquisition and how they relate to body mass and to relate the variation in these conditions to litter survival. Figure 4.1 illustrates the principle. A parent maximizes its fitness when it optimizes parental care, that is, when it maximizes the difference between benefit and cost curves. For the cost curve *C1* the corresponding optimal parental input is $xc1$. In a "bad" year, however, it may cost the parent more to deliver the same amount of care per day, that is, the parent may now operate along the cost curve *C2*. The optimal parental input $xc2$ in this case is at a lower level of parental input that the previous level of $xc1$. Thus, when costs increase, optimal parental input will be shifted downward. Therefore, regurgitation rates may decline, not because the animal is unable to provide more food, but because it does not pay it to do so.

We suspect that the several hypotheses advanced to explain the role of diet choice, resource dispersion, resource abundance, resource depletion, and other ecological parameters in canid social organization and their allometric trends can be effectively reconsidered as sources of variation that determine the shape of the cost curves in relation to the delivered parental input. We offer some speculations as examples. We currently do not know how the addition of group members affects resource depletion and the cost of food acquisition. It may depend on the variance of resource availability (Carr & Macdonald 1986), the pattern of resource renewal (Waser 1981), or the mode of foraging (solitary vs. cooperative). To test the hypothesis that the allometric trend in the incidence of cooperative hunting is related to the evolution of reproductive output, we would need to show that cooperative hunting confers fitness benefits and explain how these vary with body mass. Benefits could mean, for instance, that cooperative hunters operate along cost curve *C1* in Figure 4.1 and that solitary hunters operate along *C2*. In this respect, a demonstration that individual intake rate increases with group size in cooperative hunters is neither necessary nor is its absence proof that cooperative hunting does not confer such fitness advantages. It would be perfectly feasi-

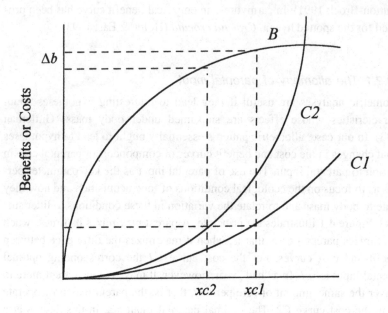

Care and Resources Received by Young

Figure 4.1. Costs and benefits of parental care in relation to parental input (i.e., the care and resources received by the offspring). Costs and benefits are ideally measured as Darwinian fitness. Approximations may include mortality to the next breeding season for costs and reproductive success per breeding season as benefits. Optimal parental care is the point at which the difference between benefit and cost curves reaches a maximum. For cost curve $C1$, the optimal parental input should be at level $xc1$. If costs vary between years, individuals, populations or species, then the optimal parental input will shift accordingly. For instance, if costs increase from $C1$ to $C2$, then there is a downward shift of the optimal parental input from $xc1$ to $xc2$, leading to a loss in benefits of Δb.

ble that cooperative hunting not increase individual intake rate but decrease foraging costs instead (e.g., by shared carcass defense against competitors). This would shift the optimal parental input to the right (in Figure 4.1) and lead to an increase in reproductive output. Cooperative hunting may also permit expansion of diet breadth in critical situations because the group can hunt prey of a size outside the range of a solitary forager. Thus, even if there is a lack of correlation between individual foraging intake and hunting group size, cooperative hunting may still confer fitness advantages.

Why should cooperative hunting scale allometrically? We do not really know, although the following aspects may be relevant. Because metabolic rates scale allometrically with an exponent of 0.71 (McNab 1989), hunting may be energetically more efficient in larger canids. Hence, larger cooperative hunters would operate on the lower cost curve, smaller ones on the higher cost curve. The number of potential competitors is a function of body mass, because the number of species declines sharply with body mass (Van Valen 1973); larger species would have fewer competitors.

4.3.2.2 The consequences of alloparental input

Our second point is simple but important. The effect of alloparental input may depend on the shape of the cost and benefit curves of care behavior and the decisions made by breeders (Table 4.3). The curves also indicate the circumstances under which alloparental behavior is likely to occur. Consider first the model in Figure 4.2(a), which uses conventional cost–benefit curves. The breeder's optimal parental input without alloparental contributions is at xb less than the maximum parental input possible. We call the helper's contributions (e.g., regurgitations) xh. Two situations may arise: (1) The breeder's input remains unchanged and the helpers' input is added to the breeders' input. Then the total input to the litter would equal $xb + xh$, and reproductive success relative to the situation without alloparental input would increase by Δb. A positive correlation between the number of helpers and the number of surviving pups would be an example of this (e.g., silver-backed jackals: Moehlman 1979); (2) alternatively, the breeder reduces its input by an amount equivalent to that of the helper's contribution. Hence, the total input would be $xb - xh + xh = xb$, and breeding success would be equivalent to that of a breeder without alloparental care. In this case, the breeder has a benefit in terms of reduced costs of parental care; the cost of parental care to the breeder is decreased by the amount $c2$ (load lightening). If helper and breeder operate along the same cost curve, then the increase in mortality incurred by the helper due to alloparental input $c1$ is considerably smaller than the decrease of the breeder's mortality (Crick 1992). This asymmetry would imply a positive contribution to the future indirect component of the helper's inclusive fitness in proportion to the degree of relatedness between helper and breeder (Mumme, Koenig, & Ratnieks 1989; Creel 1990). What do these alternatives imply for the design and interpretation of experimental studies? In a system where alloparental care increases the survival of young, removing helpers should result in a significant decrease in pup survival. In a system, however, where alloparental care serves as load lightening, removal of an alloparent might appear to make no significant difference in the input to pups. In this

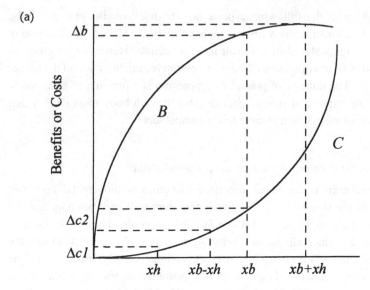

Care and Resources Received by Young

Figure 4.2. The effect of alloparental contributions on fitness depends on the shape of the cost and benefit curve. Benefits B and costs C, ideally measured in terms of Darwinian fitness, may be approximated by breeding success in the current season and mortality to the next breeding season, respectively. Benefits and costs are plotted in relation to parental and alloparental input (care and resources received by young). The alloparental input (e.g., provisioning rate) is xh, the parental contribution xb. When helpers contribute, breeders may (1) continue their input at the same level or (2) reduce their input to the level of $xb - xh$. In the first case, breeding success will increase by Δb; in the second case load-lightening occurs and the breeder gains a reduction in mortality of the amount of $\Delta c2$. Whether helping is favored depends on the amount of increase in helper mortality $\Delta c1$ due to helping compared to inclusive fitness benefits proportional to Δb or $\Delta c2$. In (a) the benefit curve is convex, and the cost curve is concave, the optimal parental input xb is smaller than the maximum possible parental input, and the rise in helper mortality $\Delta c1$ is small in relation to $\Delta c2$ or Δb. Helping is favored independent of the action of the breeder. In (b) the benefit curve is concave and the cost curve is convex, the optimal parental input xb is equivalent to the maximum possible parental input, and $\Delta c1$ is similar to Δb and is large compared to $\Delta c2$. Helping is more likely to be favored if the breeder chooses to keep its input constant at xb. In (c) both benefit and cost curves are convex, the optimal parental input xb is equivalent to the maximum possible parental input, $\Delta c1$ is large compared to both $\Delta c2$ and Δb. Helping is not favored, and whether it occurs is independent of the action of the breeder.

case, it would be premature to conclude that alloparental care has no fitness effects, because detecting this effect would require information on the breeder's subsequent mortality and reproductive success. So, the lack of a positive correlation between the number of surviving pups and number of

(b)

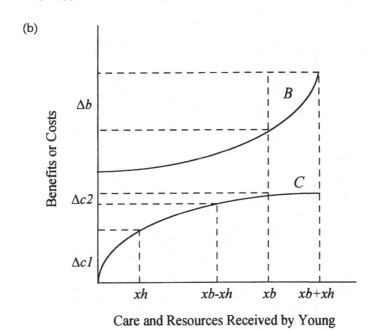

Care and Resources Received by Young

(c)

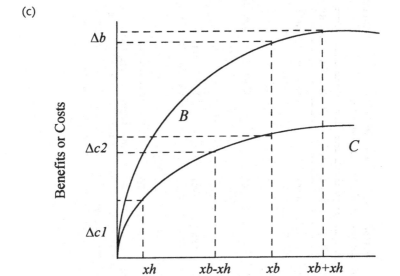

Care and Resources Received by Young

Table 4.3. *Fitness benefits, breeder decisions, and the incidence of alloparental care depend on the precise shape of cost and benefit curves for parental and alloparental care*

Figure	Shape of cost/benefit curves	Breeder's decision	Fitness consequences	Is alloparental care favored?
Figure 4.2(A)	Benefit: convex Cost: concave	Load lightening[a] Constant breeder input[b] $\Delta b \approx \Delta c2$	(a) Increased survival or future reproductive output of breeder (b) Higher pup survival	Positive: helping favored $\Delta c1$ (cost) $< \Delta b$ or $\Delta c2$ (benefits)
Figure 4.2(B)	Benefit: concave Cost: convex	Constant breeder input $\Delta b > \Delta c2$	Higher pup survival	Neutral: helping sometimes favored $\Delta c1$ (cost) $\approx \Delta b$ or $\Delta c2$ (benefits)
Figure 4.2(C)	Benefit: convex Cost: convex	Load lightening[a] Constant breeder input[b] $\Delta b \approx \Delta c2$	(a) Increased survival or future reproductive output of breeder (b) Higher pup survival	Negative: helping not favored $\Delta c1$ (cost) $> \Delta b$ or $\Delta c2$ (benefits)

[a]Load lightening: breeder reduces its input when alloparental care occurs compared to the case without alloparental care; hence, total input is the same for both cases of alloparental and no alloparental care.

[b]Constant breeder input: breeder's input remains the same whether or not alloparental care occurs; hence, total input in the case with alloparental care exceeds that of the case without alloparental care.

helpers does not necessarily indicate that helpers have no effect, that is, they simply bide their time until a suitable opening for dispersal appears (see Brown 1987).

If the cost and benefit curves are exchanged in shape (see Figure 4.2(B)), the optimal parental input for the breeder xb always corresponds to the maximum input possible (if we exclude the trivial case of no parental care). It is difficult to present it in the figure, and so for convenience xb is simply set as the maximum possible for a breeder without alloparental input. Load lightening may be a less interesting option for the breeder than keeping its contribution constant ($\Delta b > \Delta c2$). Because of the large cost increase $c1$ associated with helping, helping is less likely to occur ($\Delta c1$ large compared to $\Delta c2$ or Δb) where xb would remain constant under natural conditions. Experimental manipulation of the breeder to adopt load lightening should then lead to termination of alloparental care because the helper is then in the unfavorable situation of gaining only a fraction of $c2$ compared to the previously expected payoff from b. If both benefit and cost curves are convex (see Figure 4.2(C)), helping may pay even less. Now, independent of the breeder's decision, the expected payoff to the helper Δb or $\Delta c2$ is small relative to the increase in its mortality of $\Delta c1$ and helping is unlikely.

How do canid breeders adjust to alloparental input? How important is the future component of indirect fitness? We do not (yet) have canid studies that provide data to determine cost and benefit curves, or relate the details of food acquisition to fitness consequences. The incidence of alloparental care and the reaction of breeders could vary between years, individuals, populations or species, if the shape of the cost and benefit function of parental care changes.

4.3.2.3 Reproductive suppression

Reproductive suppression was recorded in 44 percent (strict sense) and 64 percent (broad sense) of the 25 species for which information was available (Table 4.4). The incidence of reproductive suppression per study population varied widely but stayed below 100 percent in all well-studied species in groups with more than two adults (see Table 4.4). The likelihood of a species showing reproductive suppression in the strict sense increased with female body mass (logistic regression, $t = 1.90$, $P = 0.057$) and was higher for vertebrate than nonvertebrate eaters ($t = 2.24$, $P = 0.025$; overall model $G = 17.16$, $df = 2$, $P = 0.0002$). Species with reproductive suppression had significantly higher litter sizes ($Cp = 0.394$, $P = 0.013$ one-tailed, $n = 25$) and higher litter masses ($Cp = 0.326$, $P = 0.047$ one-tailed, $n = 18$) after correcting for female body mass.

Table 4.4. *Incidence of reproductive suppression in canids*

Species	Evidence concerning reproduction suppression		Scored as reproductively suppressed?		Remarks
	In favor	Against	Extra adults[a]	Helpers[b]	
Vul can	Yes	Yes	Yes	No	Three of 5 families had extra F (1 lactating).
Vul zer	Missing	Missing	Missing	Missing	
Uro lit	Missing	(Yes)	No	No	Mating system monogamous only.
Vul rue	Missing	(Yes)	No	No	Group size and spacing indicates monogamy only.
Vul pal	Missing	Missing	Missing	Missing	
Vul mac	Yes	Yes	Yes	No	$n = 1$ extra F; $n = 3$ multiple litters.
Vul cor	Missing	Missing	No	No	
Vul vel	Uncertain	Uncertain	Yes	No	Possible extra F.
Vul ben	Uncertain	Yes	No	No	Observations of group sizes 3 to 5 of unaged individuals; $n = 1$ two extra lactating F.
Vul cha	Uncertain	Yes	Yes	No	$n = 1$ observation of extra adults that "steal food"; $n = 1$ multiple litters per group.
Pse gri	Yes	Yes	Yes	Yes	Extra F with ($n = 1$) and without ($n = 1$) pups that nursed pups of dominant F after her death.
Vul fer	Missing	Missing	Missing	Missing	
Pse vet	Missing	Missing	Missing	Missing	
Uro cin	Weak	Yes	No	No	6.4% of adult F do not reproduce; Multiple litters, polygyny.

Alo lag	Yes	Yes	Yes	Yes	F helpers; surviving multiple litters observed.
Oto meg	Yes	Yes	Yes	No	$n = 1$ extra M guarding? Multiple litters, communal denning and nursing, polygyny.
Cer tho	No	(Yes)	No	No	Group size and spacing indicates monogamy only.
Pse gym	No	(Yes)	No	No	No observations of extra adults.
Vul vul	Yes	Yes	Yes	Yes	F helpers; multiple litters, polygyny.
Pse sec	Missing	Missing	Missing	Missing	
Nyc pro	No	(Yes)	No	No	Group size indicates monogamy only.
Can aur	Yes	Yes	Yes	Yes	F & M helpers; $n = 1$ multiple litter.
Spe ven	Yes	No	Yes	Yes	Reproductive suppression of extra F in captivity.
Can mes	Yes	Yes	Yes	Yes	F & M helpers; $n = 1$ multiple litter.
Can adu	Missing	Missing	Missing	Missing	
Pse cul	Missing	Missing	Missing	Missing	
Ate mic	Missing	Missing	Missing	Missing	
Can lat	Yes	Yes	Yes	Yes	F & M helpers; multiple monogamy.
Can sim	Yes	No	Yes	Yes	F & M helpers.
Cuo alp	Yes	Yes	Yes	Yes	F & M helpers; multiple litters; extra F nursing.
Chr bra	No	No	No	No	Group size and spacing indicate monogamy only.
Lyc pic	Yes	Yes	Yes	Yes	F & M helpers; multiple litters, extra F nursing, polyandrous matings.
Can lup	Yes	Yes	Yes	Yes	F & M helpers; $n = 1$ multiple litter; $n = 3$ multiple monogamy.

[a] Score as reproductively suppressed if extra adults were observed.
[b] Score as reproductively suppressed only if extra adults were observed helping.

4.3.2.4 Resources, philopatry, and dispersal

Few detailed studies identify the cause(s) of emigration by young that could
have been retained as alloparentals. In contrast to the often pronounced suggestion that parents aggressively drive young out of the territory, a recent
study of red foxes by Harris and White (1993) demonstrated that emigration
is associated with a decline in affiliative behavior by the breeding pair. In
silver-backed jackals, fathers regurgitated food to helpers, and both parents
groomed helpers. Agonistic behavior that could be interpreted as one individual threatening and driving another family member away from the den site
occurred only between same-sex helpers (Moehlman 1983). In some cases, a
decline in resources in the territory led to increased mortality of pups rather
than starvation of helpers (wolves: Harrington et al. 1983; African wild dogs:
Malcolm & Marten 1982). Whether and how declines in resource levels are a
function of resource depletion by alloparental individuals is currently
unknown. Concurrent with the trend that smaller canids have a tendency
toward polygyny and/or female helpers, males are the sex that disperse sooner
and farther. Golden and silver-backed jackals and coyotes exhibit an equal
sex ratio among helpers and dispersers. The smaller- and medium-sized
canids tend to delay dispersal for no more than a year. In most of the larger
canids (e.g., simien wolves, dholes, African wild dogs, and gray wolves), dispersal is biased in favor of females, although both sexes may stay and help
and some individuals remain with their natal pack throughout their life (Table
4.5).

Table 4.5. *Philopatry and dispersal in canids[a]*

Species	NRA	EBA	Helpers	Dispersers
Vul can	F	F(PG)	—	—
Vul zer	—	—	—	—
Uro lit	None	—	—	—
Vul rue	—	—	—	—
Vul pal	—	—	—	—
Vul mac	F	F(PG)	?	Males dispersed farther than females (1)
Vul cor	—	?(ml)	—	—
Vul vel	?F	F(PG)?	—	—
Vul ben	F	F(PG)	—	—
Vul cha	?–?	?(ml)	No	—
Pse gri	F	F(PG)	F	—
Vul fer	—	—	—	—
Pse vet	—	—	—	—

Table 4.5. (*cont.*)

Species	NRA	EBA	Helpers	Dispersers
Uro cin	?F?	F(PG)	—	Only juvenile males disperse (2)
Alo lag	F	F(PG)	F	Males dispersed farther than females (3)
Oto meg	F	F(PG)	M	—
Cer tho	No	—	No	—
Pse gym	No	—	—	—
Vul vul	F	F(PG)	F	Significantly more males disperse and they disperse farther (4, 5, 6, 7); in "bad" years males disperse more and farther, otherwise = (8)
Pse sec	—	—	—	—
Nyc pro	No	—	No	—
Can aur	F&M	F(PG)	F&M	—
Spe ven	Yes?	—	—	—
Can mes	F&M	F(PG)	F&M	Females & males disperse equally (9)
Can adu	—	—	—	—
Pse cul	—	—	—	—
Ate mic	—	—	—	—
Can lat	F&M	F&M (MM)	F&M	Females & males disperse equally (10,11) Males disperse farther (12,13) Females disperse farther (14, 15, 16,)
Can sim	F&M	—	F&M	Females disperse at 2 yrs, males don't disperse (17)
Cuo alp	F&M	?(ml)	F&M	—
Chr bra	No	—	No	—
Lyc pic	F&M	F&M (PA, MM)	F&M	More females disperse (18) Females and males disperse equally (19)
Can lup	F&M	F&M	F&M	Females disperse farther (20)

[a]NRA nonreproductive adult; EBA extra breeding adults; F female; M male; MM multiple pairs; PG polygyny; PA polyandry; ml multiple litters; dash (—) no data.
[b](1) O'Neal et al. 1987, (2) Nicholson et al. 1985, (3) Hersteinsson & Macdonald 1982, (4) Ables 1975, (5) Storm et al. 1976, (6) Macdonald 1980, (7) Mulder 1985, (8) Linstrom 1988, (9) Ferguson et al. 1983, (10) Hawthorne 1971, (11) Bekoff & Wells 1982, (12) Gipson & Selander 1972, (13) Berg & Chesness 1978, (14) Knowlton 1972, (15) Nellis & Keith 1976, (16) Bowen 1978, (17) Gottelli & Sillero-Zubiri 1992, (18) Frame & Frame 1977, (19) Fuller et al. 1992, (20) Keith 1983.

4.4 Discussion

The positive allometry of lifetime litter mass indicates that larger canids invest more than smaller canids in prepartum reproduction (see Table 4.1). An impressive number of traits scale allometrically, not just life-history traits, but also ecological and behavioral variables (diet, maximum prey size, group size, cooperative hunting). How do these trends correlate with evidence of cooperative rearing of young, reproductive suppression, and feeding ecology?

In hypothesis 1, we asked: Do larger canids produce larger litters because they can rely on recruiting a sufficient number of alloparents by suppressing the alloparents' reproduction? Consistent with the predictions, (a) group size scaled with body mass, (b) the incidence of alloparental care scaled with body mass, and (c) the incidence of reproductive suppression scaled with body mass.

In hypothesis 2, we asked: Do larger canids produce larger litters because the size and type of food they consume improves the economy of provisioning in that it increases the benefit–cost ratio of foraging for provisioning? Consistent with the predictions, (d) larger canids are more likely to eat large (i.e., vertebrate) items, and (e) the incidence of cooperative hunting scales with body mass. There was insufficient information to evaluate whether (f) the benefit–cost ratio of foraging increases with body mass.

In hypothesis 3, we asked: Do larger canids produce larger litters because the resources on a breeding pair's home range do not constrain survival and recruitment of alloparents? Prediction (g) that cooperative hunting scaled with maximum prey size was confirmed, but there were insufficient data in the literature to test whether in larger canids (h) breeding individuals are less likely to reject additional group members, (i) the occurrence of starvation in alloparents declines with body mass, (j) amount of resources per home range increases, and (k) rates of resource depletion decrease.

We conclude that increased reproductive output in canids may be an evolutionary consequence of selection that favored reproductive suppression as a cause of helper recruitment. The positive allometry of reproductive output is likely to be a consequence of changes in feeding ecology, diet breadth, maximum prey size, and the incidence of cooperative hunting that all scale allometrically.

Analyses of allometric trends with behavior and ecology showed that larger canids live in larger groups and that species that live in larger groups are more likely to eat vertebrates and to have a larger maximum prey size. Larger canids also are more likely to hunt cooperatively. Concurrently, larger canids have larger lifetime litter masses and a higher incidence of alloparental behavior. If reproductive suppression is scored in a strict sense (when the social and reproductive status of nonreproductive adults is known), then larger canids exhibit more reproductive suppression, and species with reproductive suppression have higher litter masses. Thus, there are clear indications that as energetic costs of reproduction are going up and the reproductive tactic is to produce more young per breeding attempt, there is a higher incidence of alloparental behavior and reproductive suppression. Creel tested the hypothesis that in carnivores cooperative breeding and reproductive suppression were related to high energetic costs of reproduction (e.g., gestation and lactation in 16 species) and found that litter mass, litter growth rate, and total investment were higher in communally breeding carnivores that were scored as reproductively suppressed (Creel & Creel 1991).

Our scoring of occurrence of reproductive suppression is less than satisfactory. Given the across-taxa data that all species can obtain sexual maturity before 12 months of age, the assumption has been made that this is always true in nature. Field studies that recorded information on provisioning, growth rate, and age of sexual maturity might reveal that maturation rates varied with nutrition and that some alloparents were not sexually mature and hence not reproductively suppressed. However, this type of assessment is complicated because young canid females can be hormonally suppressed. Furthermore, it is important to distinguish between reproductive suppression due to (1) suppression of endocrine cycles and (2) suppression of reproductive behavior (Packard et al. 1985). In gray wolves, female age at first ovulation can vary between 10 and 22 months, and both social and environmental factors may delay puberty. However, once a female has cycled, anestrus is rare, and most reproductive failure in adult females is attributed to lack of copulation. Subordinate females exhibit normal estrus cycles and ovulation, and in captive packs four nonpregnant females had serum hormone concentrations through the luteal stage similar to those in females that produced litters (Seal et al. 1979, 1987). These females were hormonally primed to lactate and produce milk, possibly for another female's offspring. In the field there have been observations of subordinate canid females showing no evidence of having produced their own pups but nursing a dominant's pups (red foxes: von Schantz 1981; Simien wolves: Sillero-Zubiri 1994). Unless we know the reproductive history of a female, it is possible, although misleading, (1) to attribute pups to a female that was pseudopregnant, or (2) to score alloparental care by nonbreeders as communal nursing. Evidence of parturition is critical to ascribing maternity.

Male canids also pose problems in terms of assessing reproductive activity. Information on their sexual maturation in the field is limited and not linked to behavioral and nutritional status. Subordinate males may get copulations, and the assumed paternity of a litter of pups may be incorrectly attributed to the dominant male. It is also possible that there could be multiple paternity in a litter.

The recruitment of helpers, reproductive conflict, and reproductive suppression are based not only on the energetic costs of reproduction that allometric trends have illuminated but also on ecological and demographic factors that affect the costs and benefits of different reproductive tactics and limit the breeding options of subordinate individuals (Brown 1982; Emlen 1982a, 1982b). If helpers do improve pup survival, then breeding individuals will benefit most if subordinates stay and help rather than disperse and produce grandpups. However, subordinate tactics will vary, depending on the probability of dispersing successfully and/or resource limitations on the natal territory. Hence, one would expect intraspecific variation in reproductive tactics and mating systems that would depend on demography, available territories, food abundance, the energetics of food acquisition, and the cost–benefits of provisioning pups. It is not surprising that gray wolves as a species have been observed to be monogamous, polygynous, and polyandrous, but our understanding is limited as to how ecological factors affect the costs and benefits of these different reproductive tactics.

Allometric analyses are useful for determining life-history traits that are related to body mass, but they are just one part of the equation for understanding the variation in canid social systems. Ecological constraints and their effect on feeding, spacing, and social and reproductive behavior are poorly understood. Interspecific variation in canids indicates that food type (fruit, invertebrate, vertebrate) and body mass may influence group size and the potential for cooperative breeding. The maned wolf is a large canid. On the basis of its body mass alone one might predict that it would tend to live in large groups, have large litters, and have alloparents. But it is a solitary forager, an omnivore that feeds on fruit and rodents, has no alloparents, and produces on average two pups. By contrast, the similar-sized Simien wolf, a solitary forager that feeds almost exclusively on rodents available in very high biomass, lives in large groups, has a mean litter size (pups out of den) of 4.0, and has both male and female helpers. What is the difference? It probably involves both the nutritional value of the food ingested and the energetics of acquiring food. The fitness benefits of helping and the probability of successful dispersal are also important (see also Lucas, Waser, & Creel this volume).

Dispersal is difficult to measure, but detailed data on the energetics of forag-ing, input to pups, and pup survival can be recorded. Are some species totally limited by ecology/resources such that they cannot use fitness-related traits?

Such questions also need to be investigated intraspecifically. Relative prey size may be an important determinant of coyote group size. In habitats where mule deer (*Odocoileus hemionus*) and elk (*Cervus elaphus*) were important food items, coyotes had delayed pup dispersal and had larger groups (Bowen 1978, 1981; Berkoff & Wells 1982, 1986). Bowen found a strong correlation between pack size, territory size, and the percentage of mule deer in the winter diet. He attributed this relationship to an increase in individual fitness with an increase in food acquisition efficiency that resulted from a combination of searching, capture, and defense of food. Messier and Barrette (1982) concluded from their study at a northern latitude (46°) that large prey facilitated group liv-ing but that the major selective force for larger social groups was delayed dis-persal due to a delay in maturation and the saturation of habitat. In localities where small rodents were the main prey, group size tended to be smaller, and dispersal occurred earlier (Bekoff & Wells 1982). However, a scenario in which prey size determines group size is confounded by populations that feed primar-ily on high-density populations of small rodents, in habitats in which dispersal may be difficult and groups are relatively large (Andelt 1982). Quantitative information on the costs and benefits of foraging is needed to understand how resources and canid density affect their behavior and breeding.

Information on the costs and benefits of postpartum input from parents and alloparents is limited. More field studies are needed that go beyond recording the abundance, type, and availability of food resources to determining the energetics of food acquisition and individual effort in terms of costs and bene-fits. Input to pups needs to be evaluated in terms of feeding rate per pup and how this varies with and without alloparents. It is critical, then, to link these data to a measure of fitness: pup survival. At the same time, it is critical that such data be collected with sensitive and benign methods – first and foremost, because we ethically owe such respect to study species, and second because otherwise we will be recording artifacts of interference.

It would be useful if we had better information on how pairs form and how successful they are in their first breeding attempt. Data for silverbacked and golden jackals suggest that pairs without helpers successfully raise only one pup. Thus, a possible scenario is that in their second breeding year the parental pair can recruit one helper. There could then be a yearly increase in helpers and surviving pups, but there would presumably be a point of dimin-ishing benefits, given that jackals rarely produce more than six pups. If a pair had poor success in raising pups, then opportunities for enhancement by allo-parents diminish. Of course, surviving pups will not always stay, and whether or not they do remain philopatric depends on demography and the probability of dispersing successfully.

The smaller canids are observed to be monogamous and/or polygynous, and when there is a nonbreeding adult and/or alloparent, the sex is female. Smaller canids tend to have fewer pups and may require less paternal input. In two species, females on their own were observed to raise pups successfully. By contrast, the larger canids exhibit monogamy, multiple monogamy, rarely polygyny, and with the two largest canids, polyandry. Both sexes display allo-parental care, but the largest canids show a bias toward males. Do these trends reflect differences in competition by females for males, and vice versa? Data on parental investment, its costs and benefits, are needed if we are to deter-mine if sexual selection is operating (Fisher 1930; Trivers 1972; Moehlman 1986) and how mating systems are determined in canids.

The model in Figure 4.3 summarizes the significant relationships between canid body mass, cooperative breeding, and reproductive suppression. Female body mass has a positive correlation with (1) group size, cooperative hunting, and the importance of vertebrate prey in the diet; (2) alloparental care and reproductive suppression; and (3) litter size and litter mass. In turn, vertebrate prey has a positive correlation with reproductive suppression, and larger groups tend to take larger prey. Finally, reproductive suppression has a posi-tive correlation with litter size and litter mass. What does this tell us? Body mass and feeding ecology both play an important role in determining canid social behavior and reproduction. But it is also clear that we do not have suffi-cient information to test how the costs and benefits of food acquisition can affect the relationship between reproductive suppression and litter size. The arctic fox is an intriguing example in that it has larger than expected litter sizes, having a mean litter size that is comparable to that of wild dogs. The key to this flexibility is resources. When lemmings are abundant, arctic foxes produce very large litters. When prey are not abundant, arctic foxes produce small litters and may abandon them (Ovyaniskov pers. comm. 1992). But we do not have quantitative information on the cost of acquiring lemmings when they are superabundant. These are the data that would allow us to determine the benefits of different provisioning rates. Selection for large litter sizes in an environment that has period bursts of high productivity of resources that are energetically cheap to acquire could allow arctic foxes to evolve a breeding system that does not require many alloparents. By contrast, the maned wolf is a large canid with a relatively small litter size. It lives in an environment where large prey are not available. The prey biomass of their habitat may

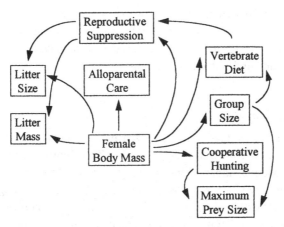

Figure 4.3. Significant relationships among canid body mass, cooperative breeding, and reproductive suppression.

limit their group size, and the cost of acquiring rodents and fruit may be relatively high, such that they do not have the option of feeding large litters. Simien wolves are similar in body mass to maned wolves, and they also feed primarily on small prey: rodents. But their prey is available at a significantly high biomass, and these canids live in large groups and cooperatively breed. It would be useful have quantitative data on what it costs a Simien wolf to capture this very abundant prey.

As better information becomes available on individual reproductive status, parentage of litters, the costs of pup provisioning, the relative input of parents with and without helpers, and fitness benefits in terms of pup survival, we predict that even greater intraspecific variability in canid reproductive tactics will become apparent. With more intensive field studies, we may discover that most canids can exhibit cooperative breeding and reproductive suppression, and it will be possible to correlate the occurrence of these traits with variation in food type and food acquisition rates, demography, and pup survival. We may discover how some species are limited by their ecology and food resources and are constrained in their reproductive options. Such information will assist the conservation of endangered species. Are larger species like the African wild dog, the dhole, and the Simien wolf more dependent on large social groups for the successful raising of offspring? If so, is this due to energetic constraints on an individual's ability to gather food and provide sustenance to dependent pups? Given that canids have an amazing ability to produce large litters, are they better equipped reproductively than most carnivores to recover from environmental disasters, disease, and human impact? More detailed information on the variability that a canid species can exhibit will greatly assist the formulation of models that accurately predict the viability of a population under changing circumstances.

4.5 Conclusions

Canids form a family that is characterized by obligatory monogamy and a high degree of cooperative breeding behavior. This family is unusual in that lifetime reproductive output shows positive allometry. As detailed field studies of breeding behavior accumulate, it becomes clear that breeding behavior is much more flexible than previously considered or recorded. Paternity analyses have yet to be done and will probably reveal a more flexible mating structure than previously envisaged, similar to that of many socially monogamous birds. Reproductive suppression may be usefully viewed as one possible outcome of the interaction of reproductive tactics among various group members. Variation in reproductive tactics between individuals arises from a variety of sources including age, demography, nutrition, and resources. This suggests that limiting the discussion of reproductive suppression in a species to a simple qualitative decision ignores the variation in reproductive tactics that different individuals might employ. However, there are few studies that consider in a rigorous fashion the costs and benefits of parental and alloparental care. If we are to achieve a better understanding of the evolution of cooperative breeding, field studies will need to go beyond pup survival as the only measure of the impact of alloparental care.

Appendix

The importance of phylogeny in comparative studies has been stressed by Felsenstein (1985), Harvey and Pagel (1991), and Riska (1991), and they have provided us with statistical models that allow the incorporation of phylogenetic information into allometric analyses. We used a general phylogeny for canids proposed by Wayne et al. (1989; Figure 4A.1) and recent molecular studies of Wayne (1993; Figure 4.A2a) on *Canis*-like canids and Geffen et al. (1992; Figure 4.A2b) on *Vulpes*-like canids. We analysed the *Canis*-like and *Vulpes*-like phylogenies separately, because taxonomic studies have repeatedly proposed to separate the wolflike canids in the subfamily *Caninae* from the foxlike canids as the subfamily *Vulpinae* (Clutton-Brock, Corbett, & Hills

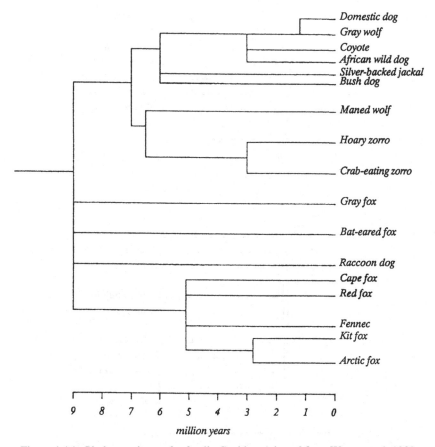

Figure 4.A1. Phylogenetic tree for family Canidae. Adapted from Wayne et al. 1989.

1976). We then calculated weighted difference scores, also known as independent comparisons, or "linear contrasts," (Harvey & Purvis 1991) for each bifurcation of the phylogenetic trees, using Felsenstein's (1985) procedure as modified by Pagel (1992) and implemented by Purvis (1991). As Harvey and Pagel (1991) show, a regression of a set of independent comparisons of a life-history variable on a set of independent comparisons of body mass forced through the origin yields an unbiased estimate of the true allometric slope across taxa. We chose as the appropriate regression model reduced major axis (RMA) regression, because both dependent and independent variables were estimated with error (Harvey & Mace 1982).

(a)

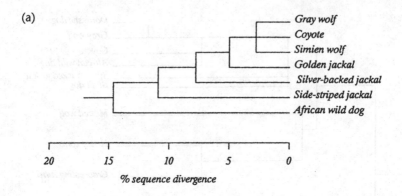

(b)

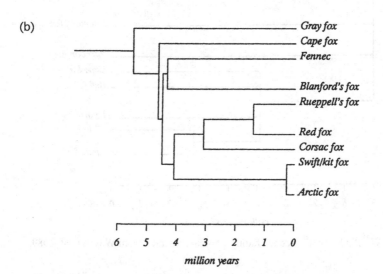

Figure 4.A2. (a) Phylogenetic tree for wolflike canids based on 736 pairs of mito-chondrial cytochrome b gene (adapted from Wayne 1993); (b) phylogenetic tree for foxlike canids based on presence–absence matrix of shared restriction fragments (adapted from Geffen et al. 1992).

These statistical models do not permit the analysis of the effect of several variables on a dependent variable, or the effect of a continuous variable on a dependent categorical variable. Our results of such analyses using standard procedures (e.g., partial correlation coefficients) must therefore be considered preliminary, and we hope that they will be replaced in the future by results gained with new techniques that permit the incorporation of phylogenetic dependence.

The reruns incorporating phylogenetic dependence indicate significant allometric relationships for most traits, including neonate mass, litter mass, gestation, age at teeth eruption, age when first eating solids, age at sexual maturity, and lifetime litter mass. Four traits (age at eye opening, at teeth eruption, when first eating solids, at sexual maturity) substantially improved in significance; four traits lost the significant allometric relationship. These included the important traits of litter size, lifetime litter size, longevity, and age when reaching adult mass. Furthermore, the positive allometry reported for lifetime litter mass is reduced to an exponent of 0.929 (Table 4.A4). These changes and losses of significance are most likely due to two effects. The first concerns the sampling problems created by the currently available phylogeny that uses information from less than 50 percent of the species. The second is an interesting divergence in trends between the two subfamilies *Caninae* and *Vulpinae* that tends to reduce any effect that might be real at lower levels of the taxonomic hierarchy.

We explored the sampling effect by creating an alternative "phylogeny" that uses all species but suffers from a reduced resolution below the genus level. This "phylogeny" is the taxonomic classification of species into genus (following Wozencraft 1993) and subfamily/tribe categories (following Wayne 1993) and can be used in an alternative implementation of Felsenstein's (1985) procedure, the phylogenetic regression method of Grafen (1989, version 1.03 R8). This produced an exponent of 1.109 for lifetime litter mass ($F_{1,4} = 21.62$, $p = 0.01$, $n = 17$ species) and significant allometric relationships for litter size (slope 0.280, $F_{1,6} = 7.02$, $p = 0.038$, $n = 31$), lifetime litter size (slope 0.466, $F_{1,4} = 9.63$, $p = 0.036$, $n = 20$) and age when reaching adult mass (slope 0.296, $F_{1,4} = 7.93$, $p = 0.048$, $n = 15$).

At the subfamily level, where we can use information of many more species, litter size is nonsignificant for *Caninae* (Table 4.A5), although it is a major factor in reproductive output in *Vulpinae* (Table 4.A6). In general, allometric relationships that are insignificant in *Caninae* become significant in *Vulpinae* (litter size, gestation, age at eye opening, lifetime litter size). Litter mass and lifetime litter mass have significant allometric relationships in both subfamilies. In *Vulpinae* the exponent of the reduced major axis regression for lifetime liter mass is an incredible 1.931 ± 0.506. This suggests that there may have been strong selection pressure for increasing the number of offspring per reproductive attempt in the smaller canids that dominate the *Vulpinae*. By contrast, wolflike canids appear to be more conservative in output. They are a far more homogeneous group than the *Vulpinae* in that they all have reproductive suppression and alloparents. An expanded data set and additional analytical techniques may help us to resolve and interpret the significance of body mass and phylogeny in the evolution of cooperative breeding in canids.

Table 4A.1. *Taxonomies of extant canids*

Common name	Species name	Label	Wozencraft 1993	Wozencraft 1989	Corbet & Hill 1980
Arctic fox	*lagopus*	Al	*Alopex*	*Alopex*	*Alopex*
Small-eared zorro	*microtis*	Am	*Atelocynus*	*Dusicyon*	*Dusicyon*
Side-striped jackal	*adustus*	Cd	*Canis*	*Canis*	*Canis*
Golden jackal	*aureus*	Ca	*Canis*	*Canis*	*Canis*
Coyote	*latrans*	Cl	*Canis*	*Canis*	*Canis*
Gray wolf	*lupus*	Cw	*Canis*	*Canis*	*Canis*
Silver-backed jackal	*mesomelas*	Cm	*Canis*	*Canis*	*Canis*
Red wolf	*rufus*		*Canis*	*Canis*	*Canis*
Simien wolf	*simensis*	Cs	*Canis*	*Canis*	*Canis*
Crab-eating zorro	*thous*	Ct	*Cerdocyon*	*Dusicyon*	*Dusicyon*
Maned wolf	*brachyurus*	Cb	*Chrysocyon*	*Chrysocyon*	*Chrysocyon*
Dhole	*alpinus*	Cu	*Cuon*	*Cuon*	*Cuon*
African wild dog	*pictus*	Lp	*Lycaon*	*Lycaon*	*Lycaon*
Raccoon dog	*procyonoides*	Np	*Nyctereutes*	*Nyctereutes*	*Nyctereutes*
Bat-eared fox	*megalotis*	Om	*Otocyon*	*Otocyon*	*Otocyon*
Culpeo	*culpaeus*	Pc	*Pseudalopex*	*Dusicyon*	*Dusicyon*

Common name	Species	Code			
Darwin's zorro	*fulvipes*		*[included in griseus]*	*[included in griseus]*	*[included in griseus]*
Gray zorro	*griseus*	Pr	*Pseudalopex*	*Dusicyon*	*Dusicyon*
Azara's zorro	*gymnocercus*	Py	*Pseudalopex*	*Dusicyon*	*Dusicyon*
Sechuran zorro	*sechurae*	Ps	*Pseudalopex*	*Dusicyon*	*Dusicyon*
Hoary zorro	*vetulus*	Pv	*Pseudalopex*	*Dusicyon*	*Dusicyon*
Bush dog	*venaticus*	Sv	*Speothos*	*Speothos*	*Speothos*
Gray fox	*cinereoargenteus*	Uc	*Urocyon*	*Urocyon*	*Vulpes*
Island gray fox	*littoralis*	Ul	*Urocyon*	*Urocyon*	*Vulpes*
Bengal fox	*bengalensis*	Vb	*Vulpes*	*Vulpes*	*Vulpes*
Blanford's fox	*cana*	Va	*Vulpes*	*Vulpes*	*Vulpes*
Cape fox	*chama*	Vh	*Vulpes*	*Vulpes*	*Vulpes*
Corsac fox	*corsac*	Vo	*Vulpes*	*Vulpes*	*Vulpes*
Tibetan fox	*ferrilata*	Vf	*Vulpes*	*Vulpes*	*Vulpes*
Kit fox	*macrotis*	Vm	*[included in velox]*	*Vulpes*	*Vulpes*
Pale fox	*pallida*	Vp	*Vulpes*	*Vulpes*	*Vulpes*
Rüppell's fox	*rueppelli*	Vr	*Vulpes*	*Vulpes*	*Vulpes*
Swift fox	*velox*	Vs	*Vulpes*	*Vulpes*	*Vulpes*
Red fox	*vulpes*	Vv	*Vulpes*	*Vulpes*	*Vulpes*
Fennec	*zerda*	Vz	*Vulpes*	*Vulpes*	*Vulpes*

Table 4A.2. *Data used for allometric analyses of life-history traits of canids*

Species	Label	Female body mass (grams)	Male body mass (grams)	Sexual dimorphism (M/F)	Gestation period (days)	Neonate mass (grams)	Litter size	Age when eyes open (days)	Age when teeth erupt (days)	Age when eating first solids (days)	Age at weaning (days)	Age when reaching adult body mass (days)	Age at sexual maturity (days)	Lifespan (months)
Alo lag	Al	3,914	4,834	1.235	52	58.5	7.8	9	–	21	63	91	270	108
Ate mic	Am	9,060	8,607	0.950	–	–	–	–	–	–	–	–	–	–
Can adu	Cd	8,290	9,390	1.133	63	–	5.4	8	–	–	–	–	–	127
Can aur	Ca	5,780	6,640	1.149	–	189	5.7	9	11	21	63	210	330	168
Can lat	Cl	12,275	14,425	1.175	63	276	6.5	13.5	10	28	57	270	300	216
Can lup	Cw	31,850	40,650	1.276	63	425	6.0	11	15	28	56	–	300	192
Can mes	Cm	7,700	8,430	1.095	60	198.6	5.5	8	7	21	56	224	360	168
Can sim	Cs	13,200	16,300	1.235	60	–	4.0	–	–	–	70	–	–	–
Cer tho	Ct	4,350	6,870	1.579	56	140	4.5	14	7	26.7	90	150	270	121
Chr bra	Cb	21,900	23,980	1.095	64.3	413	2.9	9.75	11	32	105	210	300	180
Cuo alp	Cu	16,000	19,000	1.188	62	275	8.5	13	15	–	58	210	360	192
Pse cul	Pc	8,800	12,000	1.364	57.5	168	5.2	–	–	–	–	–	–	–
Pse ful	Pf	–	–	–	–	–	–	–	–	–	–	–	–	–
Pse gri	Pr	3,540	4,140	1.169	55.5	–	4.5	–	–	–	–	–	365	–
Pse gvm	Py	4,400	4,400	–	58	–	4.0	–	–	–	–	–	240	164

Pse sec	Ps	4,500	–	–	–	–	3.0	–	–	–	–	–	–	–
Pse vet	Pv	3,800	–	–	–	–	4.0	–	–	–	–	–	–	–
Vul zer	Vz	1,106	1,560	1.410	51.9	33.9	2.2	14.75	14	17	61	92	270	140
Lyc pic	Lp	22,230	21,770	0.979	72.4	303.7	10.9	13	21	–	70.5	360	330	132
Nyc pro	Np	4,955	4,510	0.910	61	124.1	6.4	10	14	17	70	150	300	168
Oto meg	Om	4,110	4,030	0.981	60	127.1	5.5	5	13	–	105	150	–	168
Spe ven	Sv	7,383	8,417	1.140	67	161.3	5.0	12	12	35	120	203	270	120
Uro cin	Uc	3,860	4,130	1.070	53	85.1	4.3	12.5	–	–	–	150	300	149
Uro lit	Ul	1,600	1,600	–	52	–	2.2	–	–	–	–	–	–	–
Vul ben	Vb	2,400	2,600	1.083	52	58.5	3.67	–	–	–	–	–	–	–
Vul can	Va	833	956	1.148	–	29.5	2.0	–	–	–	–	–	–	48
Vul cha	Vh	2,500	2,800	1.120	51.5	–	3.5	–	–	–	–	–	–	–
Vul cor	Vo	2,073	2,105	1.015	50	80	4.6	15	–	28	–	–	–	–
Vul fer	Vf	3,670	4,100	1.117	55	–	3.5	–	–	–	–	–	–	–
Vul mac	Vm	1,910	2,060	1.079	–	39.9	4.55	–	–	–	56	120	–	144
Vul pal	Vp	1,750	1,750	–	–	–	4.0	–	–	–	–	–	–	–
Vul rue	Vr	1,530	1,700	1.111	–	–	3.3	–	–	–	–	–	–	78
Vul vel	Vs	2,250	2,440	1.084	50	–	4.25	12.5	–	–	49	150	–	153
Vul vul	Vv	4,440	5,110	1.149	53	104	5.3	12.5	21	20	56	175	285	144

Table 4A.3. *Sources for data on life-history traits and breeding biology of canids*

Alo lag	Angerbjörn et al. 1991; Banfield 1974; Barabash-Nikiforov 1938; Garrott & Eberhardt 1987; Hersteinsson 1984; MacPherson 1969; Novikov 1962; Ognev 1962; Ovsyanikov in Ginsberg & Macdonald 1990; Stroganov 1969; Wakely & Mallory 1988; Walker 1975.
Ate mic	Ginsberg & Macdonald 1990; Hershkovitz 1961.
Can adu	Steven-Hamilton in van der Merwe 1953a.
Can aur	T. Fuller pers. comm.; Fuller et al. 1989; Heptner et al. 1974; Moehlman 1979, 1983, 1986, 1989; Seitz 1959; van Lawick & van Lawick-Goodall 1970.
Can lat	Bekoff & Jamieson 1975; Bowen 1981, 1982; Heptner et al. 1974; Hildebrand 1952; Knudsen 1976; Young & Jackson 1951.
Can lup	Fritts & Mech 1981; Harrington & Paquet 1982; Harrington et al. 1983; Mech 1970; Mech & Nelson 1989; Paquet et al. 1982; van Ballenberghe & Mech 1975.
Can mes	J. Ferguson pers. comm.; T. Fuller pers. comm.; Fuller et al. 1989; Kingdon 1977; Moehlman pers. com.; Moehlman 1979, 1983, 1986, 1989; van der Merwe 1953a, 1953b.
Cer tho	Brady 1978, 1979; Crandall 1964; Eisenberg 1981; Hennemann et al. 1983; Moehlman 1986, 1989; Montgomery & Lubin 1978.
Chr bra	Acosta 1972; Bartmann & Bartmann 1986; Bartmann & Nordhoff 1984; Beccaceci in Ginsberg & Macdonald 1990; Biben 1983; Brady & Ditton 1979; Crandall 1964; Dietz 1984; Hämmerling & Lippert 1975; G. von Hegel pers. comm.; Moehlman 1986; Schneider et al. 1979.
Cuo alp	Cohen 1978; Davidar 1975; Johnsingh 1982; Sosnovskii 1967.
Lyc pic	R. Burrows pers. comm.; Frame et al. 1979; Kingdon 1977; Malcolm & Marten 1982; Malcolm & van Lawick 1975; van Heerden & Kuhn 1985; van Lawick 1973.
Nyc pro	Heptner & et al. 1967; Ikeda 1983; Ikeda et al. 1979; Kauhala 1992; Schneider 1950; Stroganov 1962.
Oto meg	Masonust 1986: Nel 1978: 1984: Nel et al. 1984: Stuart 1981.

Pse cul	Crespo 1975; Crespo & de Carlo 1963; Fuentes & Jaksic 1979; Gittleman 1985a; Jaksic et al. 1980, 1981, 1983; Medel & Jaksic 1988; Moehlman 1986.
Pse ful	Medel et al. 1990.
Pse gri	Ginsberg & Macdonald 1990; Jaksic et al. 1980, 1983.
Pse gym	Crespo 1971, 1975; Ginsberg & Macdonald 1990.
Pse sec	Asa & Wallace 1990; Birdseye 1956; Ginsberg & Macdonald 1990.
Pse vet	Coimbra-Filho 1966.
Spe ven	Biben 1983; Collier & Emerson 1973; Crandall 1964; Defler 1986; Drüwa 1977; Ginsberg & Macdonald 1990; Hershkovitz 1957; Jantschke 1973; Porton et al. 1987.
Uro cin	Banfield 1974; Crandall 1964; Hildebrand 1952; Layne & McKeon 1956; Trapp & Hallberg 1975.
Uro lit	Laughrin 1977.
Vul ben	Acharjyo & Misra 1976; Gittleman 1984; Johnsingh 1978; Roberts 1977.
Vul can	Geffen & Macdonald 1992; Harrison & Bates 1989; Mendelssohn et al. 1987.
Vul cha	Bester 1982; Bothma 1966; Nel 1984.
Vul cor	Altmann & Recker 1971; Dathe 1966; D. Langwald pers. comm.; Mitchell 1977; Sidorov & Botvinkin 1987; Stroganov 1962.
Vul fer	Mitchell 1977; G.B. Schaller pers. comm.
Vul mac	Hildebrand 1952.
Vul pal	Bueler 1973; Rosevear 1974.
Vul rue	Ginsberg & Macdonald 1990; Kowalski 1988; Lindsay & Macdonald 1986; Mendelssohn et al. 1987; E. Olfermann pers. comm.; Roberts 1977; Rosevear 1974.
Vul vel	Crandall 1964; Scott-Brown et al. 1987.
Vul vul	Dekker 1983; 1989, 1990; Henry 1986; Hofer 1986; Hoffmann & Kirkpatrick 1954; Layne & McKeon 1956; Lindström 1988, 1989; Lloyd 1980; Macdonald 1979, 1980, 1981; Mulder 1985; Tembrock 1958; Voigt 1987; von Schantz 1984a; Zabel & Taggart 1989.
Vul zer	Bronx Zoo pers. comm.; Crandall 1964; Gauthier-Pilters 1962, 1967; Gangloff 1972; Keller-Wildi 1964; Koenig 1970; Noll-Banholzer 1979; Petter 1957; Rosenthal 1974; Rosevear 1974; Saint-Girons 1962; Volf 1957; Weiher 1976.

Table 4A.4. *The allometry of life-history traits and reproductive output in Canidae, corrected for phylogenetic dependence*

Life-history trait	Unit	Linear slope ± SE	Regression				RMA slope ± SE
			r^2	n	F	P	
Neonate size	g	0.623 ± 0.051	0.962	7	151.1	<0.0001	0.635 ± 0.051
Litter size		0.179 ± 0.173	0.133	8	1.08	0.33	0.491 ± 0.209
Litter mass	g	0.833 ± 0.192	0.758	7	18.77	0.005	0.957 ± 0.199
Gestation	d	0.069 ± 0.017	0.725	7	15.86	0.007	0.081 ± 0.018
Age at eye opening	d	−0.192 ± 0.089	0.438	7	4.68	0.074	0.290 ± 0.215
Age of tooth eruption	d	0.330 ± 0.098	0.694	6	11.37	0.02	0.396 ± 0.102
Age when first eating solids	d	0.106 ± 0.038	0.608	6	7.75	0.039	0.136 ± 0.041
Age at weaning	d	0.049 ± 0.075	0.065	7	0.42	0.54	0.191 ± 0.095
Age when reading adult mass	d	0.156 ± 0.144	0.164	7	1.18	0.32	0.386 ± 0.172
Age at sexual maturity	d	0.049 ± 0.016	0.643	6	9.00	0.03	0.061 ± 0.017
Longevity	mo	0.026 ± 0.122	0.007	7	0.05	0.84	0.301 ± 0.166
Lifetime litter mass	g × mo	0.847 ± 0.156	0.832	7	29.65	0.002	0.929 ± 0.159
Lifetime litter size	mo	0.228 ± 0.147	0.288	7	2.42	0.17	0.426 ± 0.167

Data points were standardized linear contrasts (independent comparisons) created by the CAIC program (version 1.2) of Purvis (1991), following Felsenstein (1985) and Pagel (1992). This analysis complements the results listed in Table A4.1. Phylogeny from Figure 4A.1.

Table 4A.5. *The allometry of life-history traits and reproductive output in the wolflike (mostly Canis) cansid, a subgroup of the Canidae, corrected for phylogenetic dependence*

Life-history trait	Unit	Linear slope ± SE	Regression r^2	n	F	P	RMA slope ± SE
Neonate size	g	0.474 ± 0.047	0.962	5	101.6	0.001	0.483 ± 0.047
Litter size		0.093 ± 0.154	0.058	7	0.37	0.57	0.388 ± 0.195
Litter mass	g	0.559 ± 0.134	0.812	5	17.33	0.014	0.620 ± 0.138
Gestation	d	0.026 ± 0.031	0.119	6	0.68	0.45	0.075 ± 0.038
Age at eye opening	d	0.023 ± 0.131	0.006	6	0.03	0.87	0.295 ± 0.179
Age of tooth eruption	d	0.379 ± 0.142	0.640	5	7.10	0.056	0.473 ± 0.150
Age when first eating solids	d	0.089 ± 0.113	0.171	4	0.62	0.49	0.215 ± 0.134
Age at weaning	d	−0.063 ± 0.113	0.059	6	0.31	0.60	0.261 ± 0.184
Age when reaching adult mass	d	0.415 ± 0.051	0.956	4	65.89	0.004	0.425 ± 0.051
Age at sexual maturity	d	−0.020 ± 0.050	0.039	5	0.16	0.71	0.103 ± 0.080
Longevity	mo	−0.012 ± 0.096	0.003	6	0.02	0.90	0.216 ± 0.140
Lifetime litter mass	g × mo	0.527 ± 0.193	0.651	5	7.48	0.052	0.653 ± 0.203
Lifetime litter size	mo	0.071 ± 0.164	0.036	6	0.19	0.68	0.374 ± 0.213

Data points were standardized linear contrasts (independent comparisons) created by the CAIC program (version 1.2) of Purvis (1991), following Felsenstein (1985) and Pagel (1992). Phylogeny from Figure 4.A2(a).

Table 4A.6. *The allometry of life history traits and reproductive output in the foxlike (mostly Vulpes) canids, a subgroup of the Canidae, corrected for phylogenetic dependence*

Life-history trait	Unit	Linear slope ± SE	Regression				RMA slope ± SE
			r^2	n	F	P	
Neonate size	g	0.583 ± 0.151	0.749	6	14.89	0.012	0.674 ± 0.156
Litter size		0.549 ± 0.104	0.800	8	27.93	0.001	0.614 ± 0.107
Litter mass	g	1.185 ± 0.221	0.852	6	28.76	0.003	1.284 ± 0.225
Gestation	d	0.045 ± 0.017	0.574	6	6.74	0.048	0.059 ± 0.018
Age at eye opening	d	−0.340 ± 0.121	0.663	5	7.86	0.049	0.418 ± 0.398
Age when first eating solids	d	−0.132 ± 0.249	0.123	3	0.28	0.65	0.376 ± 0.437
Age at weaning	d	0.130 ± 0.128	0.341	3	1.04	0.42	0.222 ± 0.143
Age when reaching adult mass	d	−0.092 ± 0.326	0.026	4	0.08	0.8	0.573 ± 0.504
Age at sexual maturity	d	0.039 ± 0.046	0.259	3	0.7	0.49	0.076 ± 0.053
Longevity	mo	0.457 ± 0.364	0.240	6	1.58	0.27	0.933 ± 0.422
Lifetime litter mass	g × mo	1.666 ± 0.488	0.745	5	11.66	0.027	1.931 ± 0.506
Lifetime litter size	mo	1.023 ± 0.319	0.672	6	10.25	0.024	1.247 ± 0.335

Data points were standardized linear contrasts (independent comparisons) created by the CAIC program (version 1.2) of Purvis (1991), following Felsenstein (1985) and Pagel (1992). Phylogeny from Figure 4A.2 (b).

Acknowledgments

We are grateful to the Wildlife Conservation Society and the Max Planck Gesellschaft for financial support; M. L. East, B. H. Figenschou, R. Klein, M. Rowan, and A. Türk for assistance; and A. Angerbjörn, D. Langwald, J. Ferguson, T. K. Fuller, G. von Hegel, E. Olfermann, N. Ovsyanikov, G. B. Schaller, B. Van Valkenburgh, and R. K. Wayne for access to unpublished data. We thank J. Gittleman for a helpful review of this chapter.

References

Ables, E. C. (1975). Ecology of the red fox in North America. In *The wild canids*, ed. M. W. Fox, pp. 216–236. New York: Van Nostrand Reinhold.

Acharjyo, L. N., Misra, R. (1976). A note on the breeding of the Indian fox (*Vulpes bengalensis*) in captivity. *J. Bombay Natural History Soc.* 73:208–208.

Acosta, A. (1972). Hand-rearing a litter of maned wolves *Chrysocyon brachyurus* at Los Angeles Zoo. *Int. Zoo Yearbook* 12:170–174.

Altmann, D., & Recker, W. (1971). Verhaltensanalyse der Ontogenese von Steppenfchsen, *Vulpes corsac*. *Zoologischer Garten Leipzig N.F.* 41:1–6.

Andelt, W. F. (1982). *Behavioral ecology of coyotes on Welder Wildlife Refuge, south Texas*. PhD dissertation, Colorado State University, Ft. Collins.

Angerbjörn, A., Arvidson, B., Noren, E., & Stromgren, L. (1991). The effect of winter food on reproduction in the arctic fox, *Alopex lagopus* – a field experiment. *J. Anim. Ecol.* 60:705–714.

Asa, C. S., & Wallace, M. P. (1990). Diet and activity pattern of the sechuran desert fox (*Dusicyon sechurae*). *J. Mamm.* 71:69–72.

Banfield, A. W. F. (1974). *The mammals of Canada*. Toronto: University of Toronto Press.

Barabash-Nikiforov, I. (1938). Mammals of the Commander Islands and the surrounding sea. *J. Mamm.* 19:423–429.

Bartmann, W., & Bartmann, C. (1986). Mähnenwölfe (*Chrysocyon brachyurus*) in Brasilien – ein Freilandbericht. *Zeitschrift des Kölner Zoo* 29:165–176.

Bartmann, W., & Nordhoff, L. (1984). Paarbindung und Elternfamilie beim Mähnen-wolf (*Chrysocyon brachyurus*, Illiger, 1811). *Zeitschrift des Kölner Zoo* 27:63–71.

Bekoff, M., & Jamieson, R. (1975). Physical development in coyotes (*Canis latrans*), with a comparison to other canids. *J. Mamm.* 56:685–692.

Bekoff, M., & Wells, M. (1980). The social ecology of coyotes. *Sci. Amer.* 242:130–148.

Bekoff, M., & Wells, M. (1982). Behavioral ecology of coyotes: Social organization, rearing patterns, space use, and resource defense. *Z. Tierpsychol.* 60:281–305.

Bekoff, M., & Wells, M. (1986). Social ecology and behavior of coyotes. *Adv. Study Behav.* 16:251–338.

Bekoff, M., Diamond, J., & Mitton, J. B. (1981). Life-history patterns and sociality in canids: Body size, reproduction, and behavior. *Oecologica* 50:386–390.

Berg, W. E., & Chesness, R. A. (1978). Ecology of coyotes in Northern Minnesota. In *Coyotes: Biology, behavior, and management*, ed. M. Bekoff, pp. 229–246. New York: Academic Press.

Bester, J. L. (1982). *Die gedragsekologie en bestuur van die silwerfos* Vulpes chama *(A. Smith) met spesiale verwysing na die Orange Vrystaat*. MSc thesis, University of Pretoria.

Biben, M. (1983). Comparative ontogeny of social behavior in three South American canids, the maned wolf, crab-eating fox and bush dog: Implications for sociality. *Anim. Behav.* 31:814–826.

Birdseye, C. (1956). Observations on a domesticated Peruvian desert fox, *Dusicyon. J. Mamm.* 34:284–287.

Bothma, J. P. (1966). Food of the silver fox *Vulpes chama. Zool. africana* 2:205–210.

Bowen, W. D. (1978). *Social organization of the coyote in relation to prey size.* PhD dissertation, University of British Columbia, Vancouver, B.C.

Bowen, W. D. (1981). Variation in coyote social organization: the influence of prey size. *Can. J. Zool.* 59:639–652.

Bowen, W. D. (1982). Home range and spatial organization of coyotes in Jasper National Park, Alberta. *J. Wildlife Mgt.* 46:201–216.

Boyd, D. K., & Jimenez, M. D. (1994). Successful rearing of young by wild wolves without mates. *J. Mamm.* 75:14–17.

Brady, C. A. (1978). Reproduction, growth and parental care in crab-eating foxes *Cerdocyon thous* at the National Zoological Park, Washington. *Int. Zoo Yearbook* 18:130–134.

Brady, C. A. (1979). Observations on the behaviour and ecology of the crab-eating fox (*Cerdocyon thous*). In *Studies of vertebrate ecology in the Northern Neotropics,* ed. J. F. Eisenberg, pp. 161–171. Smithsonian Institution Press: Washington, D.C.

Brady, C. A. & Ditton, M. K. (1979). Management and breeding of maned wolves *Chrysocyon brachyurus* at the National Zoological Park, Washington. *Int. Zoo Yearbook* 19:171–176.

Brown, J. L. (1982). Optimal group size in territorial animals. *J. Theo. Biol.* 95:793–810.

Brown, J. L. (1987). *Helping and communal breeding in birds.* Princeton: Princeton University Press.

Bueler, L. (1973). *Wild dogs of the world.* New York: Stein and Day.

Calder, W. A. III. (1984). *Size, function, and life history.* Cambridge, Mass.: Harvard University Press.

Carr, G., & Macdonald, D. W. (1986). The sociality of solitary foragers: a model based on resource dispersion. *Anim. Behav.* 34:1540–1549.

Clutton-Brock, J., Corbett, G. B., & Hills, M. (1976). A review of the family Canidae with a classification by numerical methods. *Bull. Brit. Mus. Zool.* 29:119–199.

Clutton-Brock, T. H. (1991). *The evolution of parental care.* Princeton: Princeton University Press.

Coimbra-Filho, A. F. (1966). Notes on the reproduction and diet of Azara's fox *Cerdocyon thous azarae* and the hoary fox *Dusicyon vetulus* at Rio de Janeiro Zoo. *Int. Zoo Yearbook* 6:168–169.

Collier, C., & Emerson, S. (1973). Hand-raising bush dogs *Speothos venaticus* at the Los Angeles Zoo. *Int. Zoo Yearbook* 13:139–140.

Conover, W. J. (1980). *Practical nonparametric statistics.* 2nd ed. New York: Wiley.

Crandall, L. S. (1964). *The management of wild mammals in captivity.* Chicago: University of Chicago Press.

Creel, S. R. (1990). The future components of inclusive fitness: Accounting for interactions between members of overlapping generations. *Anim. Behav.* 40: 127–134.

Creel, S. R., & Creel, N. M. (1991). Energetics, reproductive suppression and obligate communal breeding in carnivores. *Behav. Ecol. Sociobiol.* 28:263–270.

Crespo, J. A. (1971). Ecologia del zorro gris *Dusicyon gymnocercus* en la provincia de la Pampa. *Rev. Mus. Arg. Cient. Natural (Ecology)* 1:147–205.

Crespo, J. A. (1975). Ecology of the pampas gray fox and the large fox (Culpeo). In *The wild canids. Their systematics, behavioural ecology and evolution,* ed. M. W. Fox, pp. 179–191. New York: Van Nostrand Reinhold.

Crespo, J. A., & de Carlo, J. M. (1963). Estudio ecologico de una poblacio de zorros colorados (*Dusicyon culpaeus*) en el oeste de la provincia de Neuguen. *Rev. Mus. Arg. Cient. Natural Bs. As.* (*Ecology*) 1:1–55.

Crick, H. P. Q. (1992). Load-lightening in cooperatively breeding birds and the cost of reproduction. *Ibis* 134:56–61.

Dathe, H. (1966). Breeding the corsac fox *Vulpes corsac* at East Berlin Zoo. *Int. Zoo Yearbook,* 6:166–167.

Defler, T. R. (1986). A bush dog (*Speothos venaticus*) pack in the Eastern Llanos of Colombia. *J. Mamm.* 67:421–422.

Dekker, D. (1983). Denning and foraging habits of red foxes, *Vulpes vulpes,* and their interaction with coyotes, *Canis latrans,* in central Alberta. *Can. Field Nat.* 103:261–264.

Dekker, D. (1989). Population fluctuations and spatial relationships among wolves, *Canis lupus,* coyotes, *Canis latrans,* and red foxes, *Vulpes vulpes,* in Jasper National Park, Alberta. *Can. Field Nat.* 103:261–264.

Dekker, D. (1990). Population fluctuations and spatial relationships among wolves, coyotes, and red foxes in Jasper National Park. *Alberta Nat.* 20:15–20.

Dietz, J. M. (1984). Ecology and social organization of the maned wolf (*Chrysocyon brachyurus*). *Smithsonian Contributions to Zoology* 392:1–51.

Drüwa, P. (1977). Beobachtungen zur Geburt und natürlichen Aufzucht von Waldhunden (*Speothos venaticus*) in der Gefangenschaft. *Zoologischer Garten Neue Folge Jena* 47:109–137.

Eisenberg, J. (1981). *The mammalian radiations.* Chicago: University of Chicago Press.

Emlen, S. T. (1982a). The evolution of helping. I. An ecological constraints model. *Am. Nat.* 119:29–39.

Emlen, S. T. (1982b). The evolution of helping. II. The role of behavioral confict. *Am. Nat.* 119:40–53.

Evans, R. M. (1990). The relationship between parental input and investment. *Anim. Behav.* 39:797–798.

Ewer, R. F. (1973). *The carnivores.* Ithaca: Cornell University Press.

Felsenstein, J. (1985). Phylogenies and the comparative method. *Am. Nat.* 125:1–15.

Ferguson, J. W. H., Nei, A. J., & deWet, M. J. (1983). Social organization and movement patterns of black-backed jackals *Canis mesomelas* in South Africa. *J. Zool., Lond.* 199:487–502.

Fisher, R. A. (1930). *The genetical theory of natural selection.* Oxford: Oxford University Press.

Frame, L. H., & Frame, G. W. (1977). Female wild dogs emigrate. *Nature, Lond.* 263:227–229.

Frame, L. H., Malcolm J. R., Frame, G. W., & van Lawick, H. (1979). Social organization of African wild dogs (*Lycaon pictus*) on the Serengeti Plains, Tanzania. *Z. Tierpsychol.* 50:225–249.

Fritts, S. H., & Mech, L. D. (1981). Dynamics, movements, and feeding ecology of a newly protected wolf population in northwestern Minnesota. *Wildlife Monographs* 80:1–79.

Fuentes, E. R., & Jaksic, F. M. (1979). Latitudinal size variation of Chilean foxes: Tests of alternative hypotheses. *Ecology* 60:43–47.

Fuller, T. K., Biknevicius, A. R., Kat, P. W., Van Valkenburgh, B., & Wayne, R. K. (1989). The ecology of three sympatric jackal species in the Rift Valley of Kenya. *Af. J. Ecol.* 27:313–323.

Fuller, T. K., Kat, P. W., Bulger, J. B., Maddock, A. H., Ginsberg, J. R., Burrows, R., McNutt, J. W., & Mills, M. G. (1992). Population dynamics of African wild dogs. In *Wildlife 2001 populations,* ed. D. R. McCullough & R. H. Barrett, pp. 1125–1139. Amsterdam: Elsevier.

Gangloff, L. (1972). Breeding fennec foxes *Fennecus zerda* at Strasbourg Zoo. *Int. Zoo Yearbook* 12:115–116.

Garrott, R. A., & Eberhardt, L. E. (1987). Arctic fox. In *Wild furbearer management and conservation in North America,* ed. M. Novak, J. A. Baker, M. E., Obbard, & B. Malloch, pp. 394–406. Toronto: Ontario Ministry of National Resources.

Gauthier-Pilters, H. (1962). Beobachtungen an Feneks (*Fennecus zerda* Zimm.). *Z. Tierpsychol.* 19:440–464.

Gauthier-Pilters, H. (1967). The fennec. *African Wild Life* 21:117–125.

Geffen, E., & Macdonald, D. W. (1992). Small size and monogamy: spatial organization of Blanford's foxes, *Vulpes cana. Anim. Behav.* 44:1123–1130.

Geffen, E., Mercure, A., Girman, D. J., Macdonald, D. W., & Wayne, R. K. (1992). Phylogenetic relationships of the fox-like canids: mitochondrial DNA restriction fragment, site and cytochrome b sequence analyses. *J. Zool., Lond.* 228:27–39.

Ginsberg, J. R., & Macdonald, D. W. (1990). *Foxes, wolves, jackals, and dogs. An action plan for the conservation of canids.* Gland, Switzerland: IUCN/SSC Canid Specialist Group.

Gipson, P. S., & Selander, J. A. (1972). Home range and activity of the coyote (*Canis latrans frustor*) in Arkansas. *Proc. Southeast Assoc. Fish and Game Comm.* 26:82–95.

Girman, D. J., Kat, P. W., Mills, M. G. L., Ginsberg, J. R., Borner, M., Wilson, V., Fanshawe, J. H., Fitzgibbon, C., Lau, L. M., & Wayne, R. K. (1993). Molecular genetic and morphological analyses of the African wild dog (*Lycaon pictus*). *J. Hered.* 84:450–459.

Gittleman, J. L. (1984). *The behavioural ecology of carnivores.* PhD dissertation, University of Sussex. Brighton, England.

Gittleman, J. L. (1985a). Functions of communal care in mammals. In *Evolution: Essays in honour of John Maynard Smith,* ed. P. H. Greenwood, P. H. Harvey, & M. Slatkin, pp. 187–205. Cambridge: Cambridge University Press.

Gittleman, J. L. (1985b). Carnivore body size: ecological and taxonomic correlates. *Oecologia* 67:540–554.

Gittleman, J. L. (1986). Carnivore life history patterns: allometric, phylogenetic, and ecological associations. *Am. Nat.* 127:744–771.

Gittleman, J. L. (1993). Carnivore life histories: a re-analysis in the light of new models. *Symp. Zool. Soc. Lond.* 65:65–86.

Gordon, I. J. (1989). The interspecific allometry of reproduction: Do larger species invest relatively less in their offspring? *Funct. Ecol.* 3:285–288.

Gottelli, D., & Sillero-Zubiri, C. (1990). The simien jackal: ecology and conservation. Bale Mountains Research Project. *Wildlife Conservation International,* p. 106.

Gottelli, D., & Sillero-Zubiri, C. (1992). The Ethiopian wolf – an endangered endemic canid. *Oryx* 26:205–214.

Grafen, A. (1989). The phylogenetic regression. *Phil. Trans. Roy. Soc. Lond.* B326:119–156.

Harrington, F. H., Mech, L. D., & Fritts, S. H. (1983). Pack size and wolf survival: their relationship under varying ecological conditions. *Behav. Ecol. Sociobiol.* 13:19–26.

Harrington, F. H., & Paquet, P. C. (1982). *Wolves of the world: Perspectives of behavior, ecology, and conservation.* Noyes: Park Ridge, N.J.

Harris, S., & White, P. C. L. (1993). Is reduced affiliative rather than increased agonistic behaviour associated with dispersal in red foxes? *Anim. Behav.* 44:1085–1089.

Harrison, D. L., & Bates, P. J. J. (1989). Observations on two mammal species new to the Sultanate of Oman, *Vulpes cana* Blanford, 1877 (Carnivora: Canidae) and *Nycteris thebaica* Geoffroy, 1818 (Chiroptera: Nycteridae). *Bonner zoologische Beitrge* 40:73–77.

Harvey, P. H., & Mace, G. M. (1982). Comparisons between taxa and adaptive trends: problems of methodology, In *Current problems in sociobiology,* ed. King's College Sociobiology Group, pp. 343–361. Cambridge: Cambridge University Press.

Harvey, P. H., & Pagel, M. D. (1991). *The comparative method in evolutionary biology.* Oxford: Oxford University Press.

Harvey, P. H., & Purvis, A. (1991). Comparative methods for explaining adaptations. *Nature, Lond.* 351:619–624.

Hawthorne, V. M. (1971). Coyote movements in Sagehen Creek Basin, northeastern California. *Calif. Fish & Game* 57:154–161.

Hennemann, W. W., Thompson, S. D., & Konecny, M. J. (1983). Metabolism of crab-eating foxes, *Cerdocyon thous:* Ecological influences on the energetics of canids. *Physiol. Zool.* 56:319–324.

Henry, J. D. (1986). *Red fox: The catlike carnine.* Washington, D.C.: Smithsonian Institution Press.

Heptner, V. G., Naumov, N. P., Jürgenson, P. B., Sludski, A. A., Cirkova, A. F., & Bannikov, A. G. (1967). *Die Säugetiere der Sowjetunion.* Moskau: Seekühe und Raubtiere.

Heptner, V. G., Naumov, N. P., Jürgenson, P. B., Sludski, A. A., Cirkova, A. F., & Bannikov, A. G. (1974). *Die Sugetiere der Sowjetunion.* Bd II: Seekühe und Raubtiere. Jena: VEB Gustav Fischer Verlag.

Hershkovitz, P. (1957). A synopsis of the wild dogs of Colombia. Novedades Colomb. *Contrib. Cient.* 3:157–161.

Hershkovitz, P. (1961). On the South American small-eared zorro *Atelocynus microtis* Sclater (Canidae). *Fieldiana Zool.* 39:505–523.

Hersteinsson, P. (1984). *The behavioral ecology of the arctic fox* (Alopex lagopus) *in Iceland.* PhD dissertation, Oxford University.

Hersteinsson, P., & MacDonald, D. W. (1982). Some comparisons between red and arctic foxes, *Vulpes vulpes* and *Alopex lagopus,* as revealed by radiotracking. *Proc. Royal Soc.* 49:259–289.

Hildebrand, M. (1952). The integument in Canidae. *J. Mamm.* 33:419–428.

Hämmerling, F., & Lippert, W. (1975). Beobachtung des Geburts- und Mutter-Kind-Verhaltens beim Mähnenwolf über ein Nachtsichtgerät (Infrarot-Beobachtungsanlage). *Zoologischer Garten Neue Folge* 45:393–415.

Hofer, H. (1986). *Patterns of resource distribution and exploitation by the red fox* (Vulpes vulpes) *and the Eurasian badger* (Meles meles): *A comparative study.* PhD dissertation, Oxford University.

Hofer, H., & East, M. L. (1993). The commuting system of Serengeti spotted hyaenas: How a predator copes with migratory prey. III. Attendance and maternal care. *Anim. Behav.* 46:575–589.

Hoffmann, R. A., and Kirkpatrick, C. M. (1954). Red fox weights and reproduction in Tippecanoe County, Indiana. *J. Mamm.* 35:504–509.

Ikeda, H. (1983). Development of young and parental care of the raccoon dog *Nyctereutes procyonoides viverrinus* Temminck, in captivity. *J. Mamm. Soc. Japan.* 9:229–236.

Ikeda, H., Eguchi, K., & Ono, Y. (1979). Home range utilization of a raccoon dog

Nyctereutes procyonoides viverrinus, TEMMINCK, in a small islet in western Kyushu. *Japanese J. Ecol.* 29:35–48.

Jaksic, F. M., Greene, H. W., & Yañez, J. L. (1981). The guild structure of a community of predatory vertebrates in central Chile. *Oecologia* 49:21–28.

Jaksic, F. M., Schlatter, R. P., & Yañez, J. L. (1980). Feeding ecology of central chilean foxes, *Dusicyon culpaeus* and *Dusicyon griseus*. *J. Mamm.* 61:254–260.

Jaksic, F. M., Yañez, J. L., & Rau, J. R. (1983). Trophic relations of the southernmost populations of *Dusicyon* in Chile. *J. Mamm.* 64:693–697.

Jantschke, F. (1973). On the breeding and rearing of bush dogs *Speothos venaticus* at Frankfurt Zoo. *Int. Zoo Yearbook* 13:141–143.

Johnsingh, A. T. (1978). Some aspects of the ecology and behaviour of the Indian fox – *Vulpes bengalensis* (Shaw). *J. Bombay Nat. Hist. Soc.* 75:397–405.

Kauhala, K. (1992). *Ecological characteristics of the raccoon dog in Finland*. PhD dissertation, University of Helsinki, Finland.

Keith, L. B. (1983). Population dynamics of wolves. In *Wolves in Canada and Alaska; their status, biology and management*. Can. Wildl. Ser. Rep. Ser. 45:66–77.

Keller-Wildi, E. (1964). Todesmutige Wollknäuel – Zwergfüchse der Sahara im Hause. *Tier* 10:8–10.

Kingdon, J. (1977). *East African mammals. An atlas of evolution in East Africa*. Vol. IIIA: Carnivores. London: Academic Press.

Kleiman, D. G. (1977). Monogamy in mammals. *Q. Rev. Biol.* 52:39–69.

Kleiman, D. G., & Eisenberg, J. F. (1973). Comparisons of canid and felid social systems from an evolutionary perspective. *Anim. Behav.* 21:637–659.

Kleiman, D. G., & Malcolm, J. R. (1981). The evolution of male parental investment in mammals. In *Parental care in mammals,* ed. D. J. Gubernick & P. H. Klopfer, pp. 347–387. New York: Plenum Press.

Knowlton, F. F. (1972). Preliminary interpretations of coyote population mechanics with some management implications. *J. Wildl. Mgmt.* 36:367–382.

Knudsen, J. J. (1976). *Demographic analysis of a Utah–Idaho coyote population*. MS thesis, Utah State University, Logan.

Koenig, L. (1970). Zur Fortpflanzung und Jugendentwicklung des Wüstenfuchses (*Fennecus zerda* Zimm. 1780). *Z. Tierpsychol.* 27:205–246.

Kowalski, K. (1988). The food of the sand fox *Vulpes rueppelli* Schniz, 1825 in the Egyptian Sahara. *Folia Biologia* (*Krakow*) 36:89–94.

Kullberg, C., & Angerbjorn, A. (1992). Social behavior and cooperative breeding in Arctic foxes, *Alopex lagopus* (L.), in semi-natural environment. *Ethology* 90:321–335.

Lamprecht, J. (1978). On diet, foraging behavior and interspecific food competition of jackals in the Serengeti National Park, East Africa. *Z. Saugetierk.* 49:260–284.

Laughrin, L. (1977). *The Island fox: a field study of its behavior and ecology*. PhD dissertation, University of California, Santa Barbara.

Layne, J. N., & McKeon, W. H. (1956). Notes on the development of the red fox fetus. *New York Fish and Game Journal* 3:120–128.

Lazarus, J., and Inglis, I. R. (1978). The breeding behaviour of the pink-footed goose: parental care and vigilant behaviour during the fledging period. *Behaviour* 34:1791–1804.

Lindsay, I. M., and Macdonald, D. W. (1986). Behavior and ecology of the Rüppell fox, *Vulpes rueppelli*, in Oman. *Mammalia* 50:461–474.

Lindstedt, S. L., & Swain, S. D. (1988). Body size as a constraint of design and function. In *Evolution of life histories of mammals: Theory and pattern*, ed. M. S. Boyce, pp. 93–105. New Haven: Yale University Press.

Lindström, E. (1981). Reliability of placental scar counts in the red fox (*Vulpes vulpes* L.) with special reference to the fading of scars. *Mamm. Rev.* 11:137–149.

Lindström, E. (1988). Reproductive effort in the red fox, *Vulpes vulpes*, and future supply of a fluctuating prey. *Oikos* 52:115–119.

Lindström, E. (1989). Food limitation and social regulation in a red fox population. *Holarctic Ecol.* 12:70–79.

Lloyd, H. G. (1980). Habitat requirements of the Red fox. In *The red fox: Symposium on behaviour and ecology,* ed. E. Zimen, pp. 7–26. London: Junk.

Macdonald, D. W. (1979). Helpers in fox society. *Nature, Lond.* 282:69–71.

Macdonald, D. W. (1980). Social factors affecting reproduction among red foxes. *Vulpes vulpes.* In *The red fox: Symposium on behaviour and ecology,* ed. E. Zimen, pp. 123–175. London: Junk.

Macdonald, D. W. (1981). Resource dispersion and social organization of the red fox, *Vulpes vulpes. Proceedings of the Worldwide Furbearer Conference,* 918–949.

Macdonald, D. W. (1983). The ecology of carnivore social behavior. *Nature, Lond.* 282:69–71.

Macdonald, D. W., & Moehlman, P. D. (1983). Cooperation, altruism, and restraint in the reproduction of carnivores. In *Perspectives in ethology,* ed. P. Bateson & P. Kolpfer, pp. 433–467. New York: Plenum Press.

Macpherson, A. H. (1969). The dynamics of Canadian arctic fox populations. *Can. Wildl. Serv. Rept. Ser.* 56:1–49.

Malcolm, J., & Marten, K. (1982). Natural selection and the communal rearing of pups in African wild dogs (*Lycaon pictus*). *Behav. Ecol. Sociobiol.* 10:1–13.

Malcolm, J. R., & van Lawick, H. (1975). Notes on wild dogs hunting zebras. *Mammalia* 39:231–240.

Masopust, J. (1986). Bat-eared fox, *Otocyon megalotis* Desmarest, 1822 and its rearing at Zoological Garden Prague. *Gazella* 13:105–116.

McEwen, E. H., & Scott, A. (1956). Pigmented areas in the uterus of the arctic fox *Alopex lagopus innvitus. Proc. Zool. Soc., Lond.* 128:347–348.

McNab, B. K. (1989). Basal rate of metabolism, body size, and food habits in the order Carnivora. In *Carnivore behavior, ecology, and evolution,* ed. J. L. Gittleman, pp. 335–354. Ithaca: Cornell University Press.

Mech, L. D. (1970). *The wolf. The natural history of an endangered species.* Minneapolis: Minnesota University Press.

Mech, L. D., & Nelson, M. E. (1989). Polygyny in a wild wolf pack. *J. Mamm.* 70:675–676.

Medel, R. G., & Jaksic, F. M. (1988). Ecologia de los canidos sudamericanos: una revision. *Revista Chilena Hist. Nat.* 61:67–79.

Medel, R. G., Jimenez, J. E., Jaksic, F. M., Yañez, J. L., & Armesto, J. J. (1990). Discovery of a continental population of the rare Darwin fox, *Dusicyon fulvipes* (Martin, 1837) in Chile. *Biol. Conservat.* 51:71–77.

Mendelssohn, H., Yom-Tov, Y., Ilany, G., & Meninger, D. (1987). On the occurrence of Blanford's fox, *Vulpes cana,* Blanford, 1877, in Israel and Sinai. *Mammalia* 51:459–462.

Messier, F., & Barrette, C. (1982). The social system of the coyote (*Canis latrans*) in a forested habitat. *Can. J. Zool.* 60:1743–1753.

Mitchell, R. M. (1977). *Accounts of Nepalese mammals and analysis of the host-ectoparasite data by computer techniques.* PhD dissertation, Iowa State University, Ames.

Moehlman, P. D. (1979). Jackal helpers and pup survival. *Nature, Lond.* 277:382–383.

Moehlman, P. D. (1983). Socioecology of silverbacked and golden jackals (*Canis mesomelas* and *Canis aureus*). In *Advances in the study of mammalian behavior,* American Society of Mammalogists Special Publication No. 7, ed. J. F. Eisenberg and D. G. Kleiman, pp. 423–453. Lawrence, Kans.: American Society of Mammalogists.

Moehlman, P. D. (1986). Ecology of cooperation in canids. In *Ecological aspects of social evolution,* ed. D. I. Rubenstein & R. W. Wrangham, pp. 64–86. Princeton: Princeton University Press.

Moehlman, P. D. (1989). Intraspecific variation in canid social systems. In *Carnivore behavior, ecology, and evolution,* ed. J. L. Gittleman, pp. 143–163. London: Chapman & Hall.

Mulder, J. L. (1985). Spatial organization, movements and dispersal in a Dutch red fox (*Vulpes vulpes*) population: some preliminary results. *Terre Vie* 40:133–138.

Mumme, R. L., Koenig, W. D., and Ratnieks, F. L. W. (1989). Helping behaviour, reproductive value, and the future component of indirect fitness. *Anim. Behav.* 38:331–343.

Nel, J. A. J. (1978). Notes on the food and foraging behaviour of the bat-eared fox *Otocyon megalotis. Bull. Carnegie Mus. Nat. Hist.* 6:132–137.

Nel, J. A. J. (1984). Behavioural ecology of canids in the southwestern Kalahari. *Koedoe* Supplement 27:229–235.

Nel, J. A. J., Mills, M. G. L., van Aarde, R. J. (1984). Fluctuating group-size in bat-eared foxes (*Otocyon m. megalotis*) in the Kalahari. *J. Zool., Lond.* 202:327–340.

Nellis, C. H., & Keith, L. B. (1976). Population dynamics of coyotes in Central Alberta, 1964–1968. *J. Wildl. Mgmt.* 40:389–399.

Nicholson, W. S., Hill, E. P., & Briggs, D. (1985). Denning, pup-rearing, and dispersal in the gray fox in east-central Alabama. *J. Wildl. Mgmt.* 49:33–37.

Noll-Banholzer, U. (1979). Body temperature, oxygen consumption, evaporative water loss and heart rate in the fennec. *Comp. Biochem. Physiol.* A63:79–88.

Novikov, G. A. (1962). *Carnivorous mammals of the fauna of the USSR.* Jerusalem: Israel Program Sci. Transl.

Ognev, S. I. (1962). *Mammals of eastern Europe and northern Asia.* Vol. II. Carnivora (Fissipedia). Jerusalem: Israel Program Sci. Transl.

O'Neal, G. T., Flinders, J. T., & Clary, W. P. (1987). Behavioral ecology of the Nevada kit fox (*Vulpes macrotis nevadensis*) on a managed desert rangeland. In *Current mammalogy,* ed. H. H. Genoways, Vol. 1, pp. 443–481. New York: Plenum.

Packard, J. M., Seal, U. S., Mech, L. D., Plotka, E. D. (1985). Causes of reproductive failure in two families of wolves (*Canis lupus*). *Z. Tierpsychol.* 68:24–40.

Pagel, M. D. (1992). A method for the analysis of comparative data. *J. Theor. Biol.* 156:431–442.

Paquet, P. C., Bragdon, S., & McCusker, S. (1982). Cooperative rearing of simultaneous litters in captive wolves. In *Wolves of the world. Perspectives of behavior, ecology, and conservation,* ed. F. H. Harrington & P. C. Paquet, pp. 223–237. Park Ridge, N.J.: Noyes Publications.

Parker, G., Mock, D. W., and Lamey, T. C. (1989). How selfish should stronger sibs be? *Am. Nat.* 133:846–868.

Petter, F. (1957). La reproduction du fennec. *Mammalia* 21:307–309.

Porton, I. D., Kleiman, D. G., and Rodden, M. (1987). Aseasonality of bush dog reproduction and the influence of social factors on the estrous cycle. *J. Mamm.* 68:867–871.

Purvis, A. (1991). *Comparative analysis by independent contrasts, version 1.2: user's guide.* Oxford: Oxford University Press.

Riska, B. (1991). Regression models in evolutionary allometry. *Am. Nat.* 138:283–299.

Roberts, T. J. (1977). *The mammals of Pakistan.* London: Ernest Benn.

Rosenthal, M. (1974). Husbandry of the fennec fox. *American Association Zoological Parks & Aquariums Regional Conferences Proceedings,* 183–186.

Rosevear, D. R. (1974). *The carnivores of West Africa.* Trustees of the British Museum (Natural History). London.

Saint Girons, M. C. (1962). Notes sur les dates de reproduction en captivite du fennec, *Fennecus zerda* (Zimmermann 1780). *Mammalia* 21:307–309.

Schneider, H. E., Geidel, B., & Geidel, P. (1979). Bemerkungen zur künstlichen Aufzucht und zu Erkrankungen von Mähnenwölfen im Zoologischen Garten Dresden. In *Erkrankungen der Zootiere: Verhandlungsbericht des XXI Internationalen Symposiums über die Erkrankungen der Zootiere.* Dresden: Mulhouse, pp. 315–322.

Schneider, K. M. (1950). Zur gewichtsmäßigen Jugendentwicklung gefangengehaltener Wildcaniden nebst einigen zeitlichen Bestimmungen über ihre Fortpflanzung. *Verhandlungen der deutschen zoologischen Gesellschaft Marburg,* 271–285.

Scott-Brown, J. M., Herrero, S., & Reynolds, J. (1987). Swift fox. In *Wild furbearer management and conservation in North American,* ed. M. Novak, J. A. Baker, M. E. Obbard, & B. Malloch, pp. 432–441. Toronto: Ontario Ministry of Natural Resources.

Seal, U. S., Plotka, E. D., Mech, L. D., & Packard, J. M. (1987). Seasonal metabolic and reproductive cycles in wolves. In *Man and wolf,* ed. H. Frank, pp. 109–125. Dordrecht, The Netherlands: Junk.

Seal, U. S., Plotka, E. D., Packard, J. M., & Mech, L. D. (1979). Endocrine correlates of reproduction in the wolf. *Biol. Reprod.* 21:1057–1066.

Seitz, A. (1959). Beobachtungen an handaufgezogenen Goldschakalen (*Canis aureus algirensis* Wagner 1843). *Z. Tierpsychol.* 16:747–771.

Sidorov, G. N., & Botvinkin, A. D. (1987). The corsac fox (*Vulpes corsac*) in southern Siberia. *Zool. Zhumal.* 66:914–927.

Sillero Zubiri, C. (1994). *Behavioural ecology of the Ethiopian wolf,* Canis simensis. PhD dissertation, Oxford University, Oxford.

Skutch, A. F. (1935). Helpers at the nest. *Auk* 52:257–273.

Skutch, A. F. (1961). Helpers among birds. *Condor* 63:198–226.

Sokal, R. R., & Rohlf, F. J. (1981). *Biometry.* San Francisco: Freeman.

Steinberg, D., & Colla, P. (1991). *Logit: A supplementary module for Systat.* Systat: Evanston, Ill.

Storm, G. L., Andrews, R. D., Phillips, R. L., Bishop, R. A., Siniff, D.B., & Tester, J. R. (1976). Morphology, reproduction, dispersal, and mortality of a midwestern red fox population. *Wildl. Monographs* 49:1–82.

Stroganov, S. U. (1962). *Carnivorous mammals of Siberia.* Jerusalem: Israel Program Sci. Transl.

Stuart, C. T. (1981). Notes on the mammalian carnivores of the Cape Province. *South Africa Bontebok* 1:1–58.

Tembrock, G. (1958). Zur Ethologie des Rotfuchses (*Vulpes vulpes* [L.]), unter besonderer Bercksichtigung der Fortpflanzung. *Zoologischer Garten Leipzig N.F.* 24:289–532.

Trapp, G. R., & Hallberg, D. L. (1975). Ecology of the gray fox (*Urocyon cinereoargenteus*). In *The wild canids,* ed. M. W. Fox, pp. 164–178. New York: Van Nostrand.

Trivers, R. L. (1972). Parental investment and sexual selection. In *Sexual selection and the descent of man*. ed. B. Campbell, pp. 136–179. Chicago: Aldine.

Trivers, R. L. (1974). Parent–offspring conflict. *Am. Zool.* 14:249–264.

Van Ballenberghe, V., & Mech, L. D. 91975). Weights, growth, and survival of timber wolf pups in Minnesota. *J. Mamm.* 56:44–63.

van der Merwe, N. J. (1953a). The coyote and the black-backed jackal. *Fauna and Flora* 3:45–51.

van der Merwe, N. J. (1953b). The jackal. *Fauna and Flora* 4:2–83.

van Heerden, J., & Kuhn, F. (1985). Reproduction in captive hunting dogs *Lycaon pictus*. *S. Afr. J. Wildlife Res.* 15:80–84.

van Lawick, H. (1973). *Solo*. London: Collins.

van Lawick, H., & van Lawick-Goodall, J. (1970). *Innocent killers*. London: Collins.

Van Valen, L. (1973). Body size and numbers of plants and animals. *Evolution* 27:27–35.

Voigt, D. (1987). Red fox. In *Wild furbearer management and conservation in North America*, ed. M. Novak, J. A. Baker, M. E. Obbard, & B. Malloch, pp. 378–392. Toronto: Ontario Ministry of Natural Resources.

Volf, J. (1957). A propos de la reproduction du fennec. *Mammalia* 21:454–455.

von Schantz, T. (1981). Female cooperation, male competition, and dispersal in red fox *Vulpes vulpes*. *Oikos* 37:63–68.

von Schantz, T. (1984a). Carnivore social behavior – does it need patches? *Nature, Lond.* 307:388–390.

von Schantz, T. (1984b). "Non-breeders" in the red fox *Vulpes vulpes:* a case of resource surplus. *Oikos* 42:59–65.

Wakely, L. G., & Mallory, F. F. (1988). Hierarchial development, agonistic behaviours and growth rates in captive Arctic foxes. *Can. J. Zool.* 66:1672–1678.

Walker, E. P. (1975). *Mammals of the world*. 3rd ed. Baltimore: Johns Hopkins University Press.

Waser, P. M. (1981). Sociality or territorial defense? The influence of resource renewal. *Behav. Ecol. Sociobiol.* 8:231–237.

Wayne, R. K. (1993). Molecular evolution of the dog family. *Trends Gen.* 9:218–224.

Wayne, R. K., Benveniste, R. E., Janczewski, D. N., and O'Brien, S. J. (1989). Molecular and biochemical evolution of the carnivora. In *Carnivore behavior, ecology, and evolution*, ed. J. L. Gittleman, pp. 465–494. Ithaca: Cornell University Press.

Weiher, E. (1976). Hand-rearing fennec foxes *Fennecus zerda* at Melbourne zoo. *Int. Zoo Yearbook* 16:200–202.

Wilkinson, L. (1990). SYSTAT: *The system for statistics*. Evanston, Ill.: SYSTAT Inc.

Wozencraft, W. C. (1989). Appendix: classification of the recent carnivora. In *Carnivore behavior, ecology, and evolution*, ed. J. L. Gittleman, pp. 569–593. Ithaca: Cornell University Press.

Wozencraft, W. C. (1993). Order carnivora. In *Mammal species of the world*, ed. D. E. Wilson & D. M. Reeder, pp. 279–287. Washington, D. C.: Smithsonian Institution Press.

Wyman, J. (1967). The jackals of the Serengeti. *Animals* 10:79–83.

Young, S. P., & Jackson, H. H. T. (1951). *The clever coyote*. Lincoln: University of Nebraska Press.

Zabel, C. J., & Taggart, S. J. (1989). Shift in red fox, *Vulpes vulpes*, mating system associated with El Niño in the Bering Sea. *Anim. Behav.* 38:830–838.

5

Hormonal and Experiential Factors in the Expression of Social and Parental Behavior in Canids

CHERYL S. ASA

5.1 Introduction

Canids are frequently cited as being unusual among mammals for their tendency toward monogamy (Kleiman 1977). Yet, there are other features of their reproductive biology that are not typical of most mammalian species but that are important to understanding their unique reproductive systems. These features include monestrum, obligate pseudopregnancy in adult females that fail to conceive, incorporation of postpubertal offspring into the social unit, behavioral suppression of reproduction in subordinates so that only the dominant pair produces young, possible inbreeding avoidance, the production of altricial young, and parental care by other group members including adult males. This chapter will discuss the unique interplay of social organization and physiology that appears to have evolved in canids that may enhance both social accord and reproductive success.

5.2 The Basic Canid Social System

The primary social unit of canids is the male–female mated pair, with a strong tendency toward long-term allegiance, often for life (see Kleiman & Eisenberg 1973; Macdonald 1983; Moehlman 1989 for review). Even for species such as many of the foxes in which pairs separate during the nonbreeding season, the same animals are likely to be found together in successive years. Larger social groups typically are composed of the mated pair and young of the year or of previous years. Exceptions to the apparent rule of monogamy have been reported for the red, *Vulpes vulpes* (Macdonald 1983; Zabel & Taggart 1989) and kit, *V. macrotis* (Egoscue 1962) foxes. Observations of the red fox suggest that the occurrence of polygyny is probably related to food distribution.

There appears to be a continuum of degree of sociality that begins with the maned wolf (*Chrysocyon brachyurus*), which is solitary outside the breeding season (Biben 1983) and extends to the gray wolf (*Canis lupus*) and African hunting dog (*Lycaon pictus*) with their large, complex packs that remain together year-round (Mech 1970; Frame et al. 1979). Group size is generally considered to be related to foraging or hunting strategy (see Macdonald 1983 for discussion). Those species (including foxes and maned wolves) with a food base of small prey, such as rodents, invertebrates, or fruits, can effectively hunt alone and so tend to live in smaller social groups. Bat-eared foxes (*Otocyon megalotis*), however, are unusual in that they hunt for invertebrates in their small family groups (Lamprecht 1979).

The solitary foraging strategy of most foxes contrasts with species such as the gray wolf and the African hunting dog, which exhibit the coordinated hunting strategies required to bring down large ungulates and which live in large family groups incorporating postpubertal subordinates. However, although hunting dogs (Kruuk & Turner 1967) seem to be obligate cooperative hunters, gray wolves can be found in smaller groups or even hunting alone, particularly outside the breeding season (Murie 1944).

5.3 Social Aspects of Reproduction

5.3.1 Paternal Care

Several features that relate to reproductive strategies are distinctive components of canid social organization. One of these is the participation of the male in care of the young (Table 5.1). Some degree of paternal care has been observed in every canid species that has been studied. Not only is it common for the male to provision the female during her denning phase but to feed and help care for or guard pups as well.

Paternal care has been conceptually divided into direct and indirect care (Kleiman & Malcolm 1981), perhaps also reflecting the degree of commitment by the male. Direct care includes retrieval, carrying, grooming, providing food, babysitting, playing, and active defense. Indirect care refers to behaviors that contribute only indirectly to the well-being of the young, such as the acquisition and maintenance of territory or other resources.

The method that the larger canids use to bring food back to the den also is unusual. Stomach contents are regurgitated in response to the excited muzzle licking performed by the waiting mother or pups. Adults may also be stimulated to regurgitate to begging adolescents. Smaller species, however, are more likely to carry food items in their mouths (Malcolm 1985).

Table 5.1. *Categories of male parental investment in the family Canidae*

Species	Groom	Carry	Provide food	Defend	Baby-sit	Play	Guard	Care to female
Canis lupus			+	+	+	+	+	+
C. rufus[a]			+	+			+	
C. latrans			+	+	+	+	+	+
C. aureus			+	+	+	+	+	+
C. mesomelas			+	+	+	+	+	+
Lycaon pictus	+	+	+	+	+	+	+	+
Cuon alpinus			+	+				+
Alopex lagopus			+			+		+
Vulpes vulpes			+					+
V. chama[b]			+					
V. corsac				C				
V. macrotis[c]			+					
V. velox			+					
Fennecus zerda			+				C	C
Nyctereutes procyonoides			+					C
Otocyon megalotis			+		+	+	+	+
Dusicyon culpaeus			+					
D. griseus			+					
Cerdocyon thous			C	C	C			C
Chrysocyon brachyurus				C				
Speothos venaticus			C	C	C	C		

Recorded in field: +; recorded in captivity: C.
Source: Unless otherwise noted, adapted from Kleiman and Malcolm 1981.
[a]Riley and McBride 1975.
[b]Buehler 1973.
[c]Morrell 1972.

5.3.2 Incorporation of Postpubertal Young

Enlargement of the social unit beyond the male–female pair typically involves young that remain at least temporarily with their parents rather than disperse. Table 5.2 lists the species in which nonbreeding, subordinate young have been observed as part of the extended social group. Although the young of the maned wolf and some fox species disperse before the next breeding season (bat-eared fox: Lamprecht 1979; kit fox: Buehler 1973), in others such as red

Table 5.2. *Species in the family Canidae reported to have nonbreeding,*
subordinate alloparents

Common name	Scientific name	Reference
Red fox	*Vulpes vulpes*	Macdonald 1979
Corsac fox	*V. corsac*	Buehler 1973
Arctic fox	*Alopex lagopus*	Hersteinsson & Macdonald 1982
Gray wolf	*Canis lupus*	Murie 1944
Red wolf	*C. rufus*	Riley & McBride 1975
Coyote	*C. latrans*	Bekoff & Wells 1982
Dingo	*C. familiaris*	Corbett & Newsome 1975
Golden jackal	*C. aureus*	Moehlman 1983
Black-backed jackal	*C. mesomelas*	Moehlman 1983
Dhole	*Cuon alpinus*	Davidar 1975
African hunting dog	*Lycaon pictus*	Malcolm & Marten 1982

and corsac (*V. corsac*) foxes, young of the year may remain through winter
(Buehler 1973; Macdonald 1979).

Many species reports are incomplete in regard to this feature, but the mode
is clearly for the young, under certain circumstances and for varying lengths
of time, to remain with their parents. The adaptive value of the incorporation
of postpubertal young in the social group has been discussed at length
(Macdonald & Moehlman 1983; Moehlman 1983, 1989, and this volume). As
with the adult male, these other pack members participate in the care of their
younger siblings by guarding, bringing food and/or regurgitating, grooming,
and playing. However, even when postpubertal offspring remain in a pack, it
is very uncommon for any of them to mate. The parents usually remain the
sole breeding pair for the social unit.

5.3.3 Reproductive Suppression

Among canids, the mechanism for reproductive suppression has been demonstrated for only the gray wolf (Packard, Mech & Seal 1983; Packard et al.
1985), in which the parents use intimidation to prevent mating by the other
pack members. That is, subordinate pack members attain physiological
puberty and remain reproductively competent while part of the extended family but are prevented by their parents from mating. The adults may accom-
plish this behavioral suppression merely by direct stare if their dominance is
firmly established. However, in cases where subordinates, especially males,
challenge the position of the alpha male, fighting may ensue.

INTACT MALES

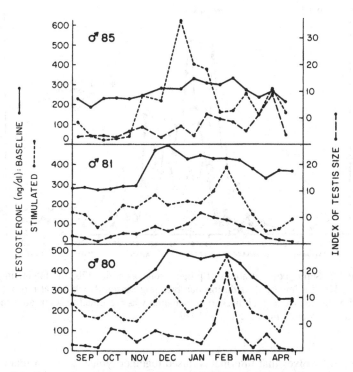

Figure 5.1. Seasonal patterns of baseline and LHRH-stimulated serum testosterone and index of testis size for three male wolves for one breeding season. From Asa et al. 1986.

Data are not available for all species, but results from gray wolves demonstrate that subordinate nonbreeding males and females do not differ from the dominant breeding pair on any parameter of reproductive physiology. Males past 22 months of age (Figure 5.1) display equivalent seasonal profiles of testosterone, testes size, and semen parameters, regardless of position in the social hierarchy (Asa et al. 1986, 1987, 1990; Seal et al. 1987). Likewise, all postpubertal females (Figure 5.2) have equivalent seasonal profiles of estrogen, progesterone, and luteinizing hormone (Seal et al. 1979; Asa et al. 1986 1987, 1990).

These results confirm that subordinate males and females are physically and physiologically able to reproduce. Although dominant wolves have been seen to prevent subordinates from mating by behavioral intervention, there may be other factors involved. For example, it has been noticed in captive

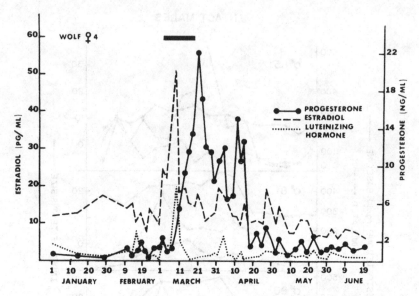

Figure 5.2. Serum concentrations of progesterone, estradiol-17b, and luteinizing hormone (LH) of a female wolf during one breeding season; black bar at top indicates period of sanguinos vaginal discharge; ovulation was presumed to occur after the LH peak in early March. From Seal et al. 1979.

colonies of wolves that littermates raised together typically do not mate with one another during the first season they are reproductively competent, that is, at about 22 months of age (Asa, unpublished). However, if they still do not have access to other wolves during subsequent seasons, they usually do mate with siblings.

Further evidence of a tendency to avoid breeding with close relatives comes from observations of a captive pack in which the dominant female died early in the breeding season (Asa unpublished). Although there were postpubertal daughters in the group, the alpha male did not choose to mate with either of them and no pups were produced that season. The alpha male was removed from the pack before the subsequent season, so it is not known if he would have ever chosen to mate with his daughters. These results suggest a mechanism in addition to behavioral intimidation that may serve to reduce the likelihood of reproduction by other than the dominant pair. This tendency to avoid mating with close relatives, sometimes referred to as an incest taboo, has been studied in a number of species (for example, see Bateson 1983) and is a component of optimal outbreeding.

5.4 Reproductive Physiology

5.4.1 Seasonal Reproduction

5.4.1.1 Monestrum and the female cycle

Reproduction is seasonal for canids in temperate latitudes, with mating occurring in late winter or early spring and pups being born about two months later. Photoperiod has been assumed to mediate the timing of reproduction in these species. However, removal of the pineal glands or of the suprachiasmatic nucleus, pathways regulating seasonal breeding in other mammalian species, failed to alter the timing of reproduction in male or female gray wolves (Asa et al. 1987), suggesting yet another reproductive anomaly for this family. Pilot studies indicated that melatonin, a hormone involved in the pineal pathway in other species, does affect the onset of breeding in captive gray wolves (Seal pers. comm. 1981), suggesting some similarity to mammalian models.

Seasonal reproduction is not unusual; many species breed only once a year. However, canids (Table 5.3) have only one ovulatory cycle per breeding season, a condition known as "monestrum," a characteristic that *is* very unusual. The only other mammalian species for which monestrum is reported are temperate-region bats in the families Rhinopomatidae, Megadermatidae, and Rhinolophidae and some in the Pteropodidae, Vespertilionidae, and Molossidae (Yalden & Morris 1975) and the European roe deer, *Capreolus capreolus* (Schams et al. 1980; Flint et al. 1994). Monestrum appears related to delays in fertilization (bats) or implantation (bats and roe deer) and not to aspects of the social system in these species.

Polyestrum, which is typical for other mammals, involves successive cycles of estrus and ovulation, either continuous or seasonal. That is, if a polyestrous female does not conceive on her first cycle, she has another chance. In fact, she probably has many more chances, thus increasing the probability of pregnancy. In contrast, for monestrous female canids, there is only one estrus and thus only one opportunity per season to conceive. The only exception noted to date is the bush dog, *Speothos venaticus*, which displays polyestrous cycles in captivity (Porton, Kleiman, & Rodden 1987).

Even for canid species at tropical latitudes, breeding typically occurs only once a year, probably cued by rainfall or food availability. The crab-eating fox (*Cerdocyon thous*) and domestic dog (*Canis lupus familiaris*) are exceptions. The crab-eating fox has two litters per year (Brady 1978), and the dog has a variable number with a mode of two that is apparently related to breed

Table 5.3. *Species in the family Canidae reported to be monestrous*

Common name	Scientific name	Reference
Red fox	*Vulpes vulpes*	Rowlands & Parkes 1935
Swift fox	*V. velox*	Kilgore 1969
Kit fox	*V. macrotis*	Egoscue 1956, 1962
Arctic fox	*Alopex lagopus*	Moller 1973
Andean fox	*Dusicyon culpaeus*	Crespo 1975
Pampas fox	*D. gymnocercus*	Crespo 1975
Gray wolf	*Canis lupus*	Seal et al. 1979
Coyote	*C. latrans*	Gier 1975
Domestic dog	*C. familiaris*	Rowlands & Hancock 1949
Maned wolf	*Chrysocyon brachyurus*	Rowlands & Hancock 1949

(Christie & Bell 1971b). In addition, the fennec fox (*Fennecus zerda*) and African hunting dog can produce a second litter but only if the first is lost (Gauthier-Pilters 1967; Frame et al. 1979). Yet, even in these cases, there are only one estrus and one ovulation during each breeding period.

Although systematic data have not been reported in the literature, breeders of domestic dogs have long been aware that females living in proximity to each other have relatively synchronous (within a few weeks) estrous periods, even though they are not seasonal breeders. These observations suggest that the estrous synchrony exhibited by wild canids may be socially facilitated and not be solely entrained by environmental factors such as photoperiod or rainfall.

5.4.1.2 Male seasonality

Canid males at temperate latitudes also exhibit seasonality in reproductive parameters (see Figure 5.1), such as testosterone production, testis size, and spermatogenesis (red fox: Joffre 1977; arctic fox, *Alopex lagopus:* Smith et al. 1985; pampas fox, *Dusicyon gymnocercus,* and Andean fox, *D. culpaeus:* Crespo 1975; gray wolf: Asa et al. 1986, 1987). A notable exception is the male domestic dog, which is able to breed year-round (Taha, Noakes, & Allen 1981) and whose testosterone levels do not vary significantly throughout the year. Testosterone data are not available for crab-eating fox males but should either resemble those of the domestic dog or have a bimodal pattern coincident with the two cycles per year shown by the females.

5.4.1.3 Prolactin

Another hormone, prolactin, varies seasonally in both males and females, with a spring peak that occurs after the mating season, coinciding with the birth of pups (red fox: Maurel, Lacroix, & Boissin 1984; arctic fox: Mondain-Monval et al., 1985 and Smith et al. 1985; gray wolf: Kreeger et al., 1991). In the gray wolf, there is a seasonal prolactin peak in all pack members, males as well as females, subordinates as well as dominants, and even castrated and ovariectomized individuals (Figure 5.3). This also is a period of declining testosterone concentrations in male wolves (see Figure 5.1).

5.5 Phases of the Ovulatory Cycle

5.5.1 The Anestrous Phase

The anestrous phase, characterized by a quiescent reproductive system, is the time between breeding periods. It varies in length, depending primarily on (1) the number of reproductive cycles per year (species-typical), and (2) lactation (i.e., whether conception and live births occurred). Thus, anestrus is much more lengthy in species that breed only once per year and somewhat shorter, within a species, for individuals that conceive and experience a period of lactation.

5.5.2 Proestrous Phase

The proestrous period, as the name implies, is the transition period preceding estrus. During proestrus, as distinguished from estrus, the period when she is actually receptive to mating, the female is attractive to males. In females of some canid species, the onset of the breeding season is heralded by a sanguinous discharge from the vulva. Although the source of blood, the uterus, is similar (Mulligan 1942), canid proestrous bleeding differs from the menstrual phase of primates in the associated hormonal milieu. Menstruation occurs during a nadir of both estrogen and progesterone, following the luteal phase Canids, in contrast, show proestrous bleeding in response to gradually increasing concentrations of estrogen (domestic dog: Christie & Bell 1971a; gray wolf: Seal et al. 1979).

Proestrous bleeding has been reported in the literature only for species in the genus *Canis,* although a slight bloody discharge has been noticed from some female maned wolves (M. Rodden pers. comm. 1994). The pro-estrous

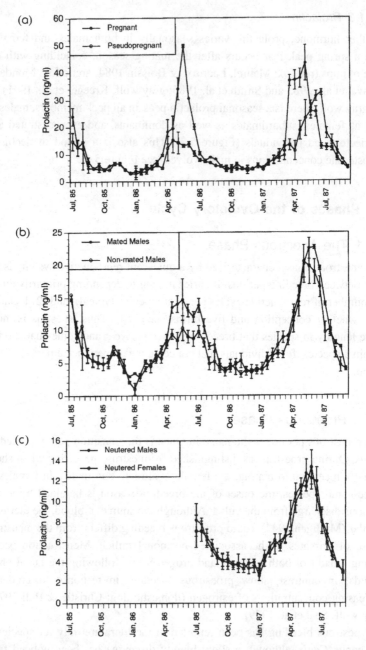

Figure 5.3. Mean (± SEM) prolactin levels for (a) pregnant and nonpregnant wolves, (b) intact males with or without mates, and (c) neutered males and females. From Kreeger et al. 1991.

phase, based on detection of sanguinous discharge, can be as long as two to three months for coyotes, *C. latrans* (Kennelly & Johns 1976; Stellflug et al. 1981), six weeks for gray wolves (Asa et al. 1986), and two to three weeks for dingos, *C. lupus dingo* (Kleiman 1968). Investigations of the domestic dog (Christie & Bell 1971a) and a feral subspecies of the domestic dog, the New Guinea singing dog, *C. lupus dingo* (Kleiman 1968), report about a one-week proestrus.

In the other canid species, proestrus generally lasts at least a week and can be detected by increased courtship activity and a swollen vulva, sometimes with a nonsanguinous discharge (e.g., raccoon dog, *Nyctereutes procyonoides:* Valtonen, Rajakoski, & Mäkelä 1977; Kleiman 1968). However, a proestrous phase of even one week is unusually long for mammals and is most comparable endocrinologically to the follicular phase of many primates, which is also a time when consortships may begin.

5.5.3 Estrous Phase

The estrous phase, when the female is receptive and mating occurs, is briefest in the fennec fox (one to two days: Gaultiers-Pilters 1967) and longest in the African hunting dog (up to 20 days: van Heerden & Kuhn 1985). However, for most canids, estrus lasts about one week (for an exhaustive list, see Hayssen, van Tienhoven, & van Tienhoven 1993), which is longer than for most mammalian species, in which one to three days is common. Exceptions include many primates and the equids, which like canids have an estrous period of about one week. These also are species in which males contribute to the well-being of infants, albeit usually indirectly by territory maintenance and guarding.

Although estrogen alone is sufficient to support the proestrous phase, the addition of progesterone appears necessary for full stimulation of estrous behavior, including receptivity to copulation (domestic dog: Beach & Merari 1968; Concannon, Hansel, & McEntee 1977a; gray wolf: Asa unpublished). An increase in progesterone just before the onset of copulatory behavior in the female arctic fox (Moller 1973) suggests that it is involved in full receptivity in this species as well. In addition, female dogs show a preference for particular males (Beach & LeBoeuf 1967; Beach 1970), suggesting a role for female choice of mate in this and perhaps other canid species.

Copulation in canids is distinguished by the copulatory lock or tie, when the swollen bulbus glandis at the base of the penis is held by the vaginal sphincter (Fuller & DuBois 1962) for varying lengths of time, ranging from a few minutes up to an hour or more. The African hunting dog is perhaps an exception, as observed copulations have ended without a tie (see Kleiman 1967). The function of the tie is likely a combination of mate guarding (i.e., the prevention of another copulation during the period of the tie) and of stimulation of sperm transport more directly into and through the uterus (e.g., see Adler 1968, 1969; Grandage 1972).

5.5.4 Pregnancy

The duration of pregnancy varies from about 50 to 70 days and is related primarily to body mass, with the smaller foxes having shorter and hunting dogs having longer gestations. Two exceptions are the bat-eared fox and the bush dog, two of the smaller species but with gestations of 65 to 70 days (Hayssen et al. 1993).

Both estradiol and progesterone are elevated during pregnancy (domestic dog: Smith & McDonald 1974; Hadley 1975; Concannon et al. 1977b; gray wolf: Seal et al. 1979; arctic fox: Moller 1973) and are apparently of ovarian origin. Both hormones peak during the first trimester and gradually decline throughout gestation.

5.5.5 Luteal Phase or Pseudopregnancy

The approximately two-month luteal phase (the period of elevated progesterone following ovulation) of the nonpregnant canid is significantly longer than that of other mammals, in which it averages two weeks. Because this prolonged luteal phase is roughly equivalent in length and in hormonal profile to pregnancy, it often is referred to as pseudopregnancy. Thus, if a female canid ovulates, she either becomes pregnant or pseudopregnant, based on reports from species for which data exist (Table 5.4).

Pseudopregnancy is recognized in domestic dogs by veterinarians only when there are overt symptoms such as a distended abdomen, swollen mammary glands that may secrete milk, or maternal behavior (Stabenfeldt & Shille 1977). Some females may even behave maternally toward, and try to care for, a toy, phantom pups, or surrogate pups (Fox & Bekoff 1975). However, although the expression of symptoms is variable by individual, the underlying endocrine basis for pseudopregnancy is present in all nonpregnant, postovulatory females (see Christie & Bell 1971a; Smith & McDonald 1974; Concannon, Mansel, & Visek 1975; Chakraborty 1987). This conclusion is supported by endocrine results from red and arctic foxes (Moller 1973; Mondain-Monval et al. 1977), gray wolves (Seal et al. 1979; Kreeger pers. comm. 1985; Asa et al. 1986), coyotes (Stellflug et al. 1981), and raccoon

Table 5.4. *Species in the family* Canidae *with obligate pseudopregnancy*

Common name	Scientific name	Reference
Red fox	*Vulpes vulpes*	Bonnin, Mondain-Monval, & Dutoume 1978
Arctic fox	*Alopex lagopus*	Moller 1973
Andean fox	*Dusicyon culpaeus*	Crespo 1975
Gray wolf	*Canis lupus*	Seal et al. 1979
Coyote	*C. latrans*	Gier 1975
Domestic dog	*C. familiaris*	Smith & McDonald 1974
Raccoon dog	*Nyctereutes procyonoides*	Valtonen et al. 1978

dogs (Valtonen et al. 1978). In fact, currently the only method for distinguishing pregnancy from pseudopregnancy in the dog and wolf (other than by palpation, ultrasound, or x-ray of fetuses) is the augmented prolactin increase in the pregnant female in response to exogenous administration of the opioid antagonist naloxone (Kreeger pers. comm. 1991).

5.6 The Endocrine Basis of Parental Behavior

Although the endocrine basis of maternal behavior has not been studied in any canid species, the general pattern that has emerged for other mammals likely applies to canids as well. Although the timing of release and the ratios of estrogen to progesterone vary by species, the hormones circulating during pregnancy appear to prepare the female to perform species-specific maternal behavior.

Numerous studies have focused on possible direct mediators of the onset of maternal behavior (see Numan 1988; Bridges 1990, & Pryce 1992 for reviews). Oxytocin, prostaglandin F2α, relaxin, endorphins, and lactogenic hormones from the pituitary (e.g., prolactin) and placenta have been considered for pivotal roles, as well as estrogen and progesterone. Perhaps the strongest case has been made for prolactin, preceded by estrogen and/or progesterone priming (see Bridges & Ronsheim, 1990). However, although most data are from rats and sheep, it is clear that, despite interspecies differences, the species-typical hormones of pregnancy prime females to behave maternally.

The case for support of paternal care is much less clear. First, paternal care is not common among mammals, and second, the results from studies of those that display paternal care are not conclusive (Brown 1985; Malcolm 1985).

However, as with females, prolactin may be important, because it has been implicated in the paternal behavior of marmosets *Callithrix jacchus* (Dixson & George 1982) and the California mouse, *Peromyscus californicus* (Gubernick & Nelson 1989).

5.7 The Interplay of Physiology and Social Organization

Although it is not possible to determine whether reproductive physiology has influenced the evolution of canid social organization, the unique combination of social and physiological features does favor pair or pack cohesion.

First, the prolonged proestrous phase affords the male–female pair extended time to establish or reaffirm the pair bond. The increasing attractiveness of the female encourages the male to remain in close proximity. In fact, Mech and Knick (1978) have documented that during the breeding season, the alpha pair of wolves can be distinguished from the other pack members by their closer sleeping or resting distances.

During proestrus, the rate of scent marking with urine commonly increases (for gray wolf, see Asa et al. 1990), with the male and female marking over each other's urine. The further observation that newly formed pairs of gray wolves (Rothman & Mech 1979) and bush dogs (Porton 1983) tandem mark more than do established pairs suggests a role for this behavior in pair formation. In addition, anosmia resulted in decreased tandem marking and failure to pair bond in a gray wolf pair (Asa et al. 1986).

It also is likely that the proestrous period is important for the female to evaluate the male's commitment, since the extent of his parental contribution may determine the survival of their pups. Although the data set is far from complete, there is a suggestion that, among canids, species in which the male makes the greatest contribution are those with the longest period of proestrus.

However, the relatively long proestrus may have been selected for primarily in response to monestrum. With only one estrus, and so only one opportunity for conception per year for the vast majority of canid females, ensuring access to a mate at that time is crucial. The relatively long period of estrus might be adaptive for the same reason.

From examination of models for other mammals, it is difficult to imagine an advantage to monestrum. Why would polyestrum, with its backup system of repeated cycles of estrus and ovulation, be selected against in canids? Possible explanations become apparent only when we consider the various elements of canid social organization.

To understand how monestrum might be advantageous in terms of breeding strategy, we must consider two other aspects of their reproductive systems:

(1) the role of postpubertal offspring in the social group and (2) the typical limitation of breeding to only the dominant pair.

The observation that the modal pattern for canids (see Table 5.2), with the exception of the maned wolf and perhaps some foxes, is to incorporate adult young into the social group suggests a common function or advantage. Data for jackals, particularly silverbacks, *Canis mesomelas* (Moehlman 1983), and African hunting dogs (Malcolm & Marten 1982) suggest that the presence of helpers, or alloparents, might provide the explanation. However, Harrington, Mech, & Fritts (1983) suggest that helping may be incidental and that adult subordinates stay with the group for more selfish reasons. In addition, all three studies (Malcolm & Marten 1982; Harrington et al. 1983; Moehlman 1983) found evidence that benefits of helping were related to ecological factors such as relative prey abundance that, of course, vary. The incorporation of additional members also would be advantageous for species that benefit from group hunting strategies.

So, although alloparenting may not be the primary motivation of young wolves that opt to remain in their natal groups, it might still provide a benefit to their parents and younger siblings. Perhaps the toleration by the dominant animals has contributed more to the selection of this type of social organization than has the motivation of their adult offspring.

In the expanded canid social group, the modal pattern is to limit reproduction to the dominant adult pair (e.g., Kleiman & Eisenburg 1973). The strategy apparently succeeds because direct fitness benefits accrue to the reproductive pair and indirect benefits accrue to the subordinates (e.g., increased fitness of kin, practice parenting, biding time until conditions improve for dispersal, etc.).

Thus, monestrum can be seen to contribute to the stability of the social unit by reducing the possible number of estrous periods among the postpubertal females of the group. Polyestry, with repeated estrous periods and attempts to mate, might wreak havoc on social stability in such groups. With monestrum, aided by the relative synchrony of estrus, intrapack aggression intended to suppress subordinate sexual behavior is limited to a brief period.

This interpretation, however, begs the questions, "Why only socially suppress reproduction? Why not suppress the subordinates physiologically so that they have no inclination to mate, eliminating competition and aggression entirely?" The answer may lie in pseudopregnancy. The ovarian events that result in the endocrine support of false pregnancies begin with follicular growth and include hormonal stimulation of estrus and ovulation. Because estrus is relatively synchronous within a pack, even the nonpregnant females are primed to behave maternally and perhaps even to lactate during the time the dominant female is pregnant. If this maternal response in subordinate females contributes to the survival of their parents' new offspring, their full siblings, the potential social disruption accompanying one estrous period may not be too high a price to pay.

The maternal behavior, apparently induced by the hormones of pseudo-pregnancy in subordinate females, is expressed in guarding and bringing food to their parents' pups. The long-term bond of the parents makes it very likely that a female's pups of successive years also share the same sire. Thus, a subordinate that is deferring reproduction shares on average as many genes with its siblings as it would with its own offspring. If the hormonal stimulation of pseudopregnancy is necessary to support the behavior, and the behavior does indeed increase the survival rate of pups, then the selective advantage of both monestrum and pseudopregnancy to subordinate females is more apparent.

This evidence can explain the parental care exhibited by subordinate females but not by the males. As shown in Table 5.1, some degree of paternal care is characteristic for all canid species that have been studied. Although the extent of the role of prolactin in mammalian parental behavior has not been completely elucidated, endogenous elevations or exogenous administration of prolactin has been associated not only with maternal (e.g., Bridges & Ronsheim 1990) but with paternal behavior as well (Dixson & George 1982; Gubernick & Nelson 1989).

Thus, the elevation of prolactin in males as well as females during the period young pups are present may support the parental behavior displayed by males and enhance that shown by females, perhaps also stimulating lactation (see Kreeger et al. 1991). However, further studies with canids are necessary to evaluate the possible role of the seasonal prolactin increase in support of parental behavior.

5.8 Conclusions

The advantages of cooperative hunting may have selected for the extended family organization seen in so many canid species. In turn, several unusual features of their reproductive biology seem to favor this form of sociality or perhaps are necessary for it to succeed. For example, synchronous monestrum may serve to minimize antagonistic interactions in the pack. The obligate pseudopregnancy in the absence of conception not only prevents further estrous cycles but also primes nonpregnant subordinate females to display maternal behavior toward the dominant female's pups and perhaps increase their chances of survival. Furthermore, seasonal prolactin elevations in males as well as females, coincident with the birth of pups, may stimulate parental behavior in the males and reinforce maternal behavior of the females.

Roughly similar social systems have been described for some primates without the concomitant alterations in reproductive cycles exhibited by canids. However, the level of accord and cooperation that may be necessary for coordinated hunting by these carnivores might not survive the social strife that accompanies polystrous reproductive cycles in other species.

References

Adler, N. T. (1968). Effects of the male's copulatory behavior in the initiation of pregnancy in the female rat. *Anat. Rec.* 160:305.

Adler, N. T. (1969). Effects of the male's copulatory behavior on successful pregnancy of the female rat. *J. Comp. Phys. Psych.* 69:613–622.

Asa, C. S., Mech, L. D., Seal, U. S., & Plotka, E. D. (1990). The influence of social and endocrine factors on urine-marking by captive wolves (*Canis lupus*). *Horm. Behav.* 24:497–509.

Asa, C. S., Seal, U. S., Letellier, M. A., & Plotka, E. D. (1987). Pinealectomy or superior cervical ganglionectomy do not alter reproduction in the wolf (*Canis lupus*). *Biol. Reprod.* 37:14–21.

Asa, C. S., Seal, U. S., Plotka, E. D., Letellier, M. A., & Mech, L.D. (1986). Effect of anosmia on reproduction in male and female wolves (*Canis lupus*). *Behav. Neur. Biol.* 46:272–284.

Bateson, P. (1983). Optimal outbreeding. In *Mate choice,* ed. P. Bateson, pp. 257–278. Cambridge: Cambridge University Press.

Beach, F. A. (1970). Coital behaviour in dogs. VIII. Social affinity, dominance and sexual preference in the bitch. *Behaviour* 36:131–148.

Beach, F. A., & LeBoeuf, B. J. (1967). Coital behaviour in dogs. I. Preferential mating in the bitch. *Anim. Behav.* 15:546–558.

Beach, F. A., & Merari, A. (1968). Coital behavior in dogs. IV. Effects of progesterone in the bitch. *Proc. Natl. Acad. Sci.* 61:442–446.

Bekoff, M., & Wells, M. C. (1982). Behavioral ecology of coyotes – Social organization, rearing patterns, space use and resource defense. *Z. Tierpsychol.* 60:281–305.

Biben, M. (1983). Comparative ontogeny of social behaviour in three South American canids, the maned wolf, crab-eating fox and bush dog: implications for sociality. *Anim. Behav.* 31:814–826.

Bonnin, M., Mondain-Monval, M., & Dutourne, B. (1978). Oestrogen and progesterone concentrations in peripheral blood in pregnant red foxes (*Vulpes vulpes*). *J. Reprod. Fert.* 54:37–41.

Brady, C. A. (1978). Reproduction, growth and parental care in crab-eating foxes, *Cerdocyon thous,* at the National Zoological Park, Washington. *Internat. Zoo Yearbook* 18:130–134.

Bridges, R. S. (1990). Endocrine regulation of parental behavior in rodents. In *Mammalian parenting: Biochemical, neurobiological, and behavioral determinants,* eds. N. A. Krasneger & R. S. Bridges, pp. 93–117. Oxford: Oxford University Press.

Bridges, R. S., & Ronsheim, P. M. (1990). Prolactin (PRL) regulation of maternal behavior in rats: Bromocriptine treatment delays and PRL promotes the rapid onset of behavior. *Endocrinology* 126:837–848.

Brown, R. E. (1985). Hormones and paternal behavior in vertebrates. *Amer. Zool.* 25:895–910.

Buehler, L. E. (1973). *Wild dogs of the world.* New York: Stein & Day.

Chakraborty, P. K. (1987). Reproductive hormone concentrations during estrus, pregnancy, and pseudopregnancy in the Labrador bitch. *Theriogenology* 27:827–840.

Christie, D. W., & Bell, E. T. (1971a). Endocrinology of the oestrous cycle of the bitch. *J. Small Anim. Pract.* 12:383–389.

Christie, D. W., & Bell, E. T. (1971b). Some observations of the seasonal incidence and frequency of oestrus in breeding bitches in Britain. *J. Small Anim. Pract.* 12:159–167.

Concannon, P. W., Hansel, W., & Visek, W. J. (1975). The ovarian cycle of the bitch: plasma estrogen, LH and progesterone. *Biol. Reprod.* 13:112–121.

Concannon, P. W., Hansel, W., & McEntee, K. (1977a). Changes in LH, progesterone, and sexual behavior associated with preovulatory luteinization in the bitch. *Biol. Reprod.* 17:604–613.

Concannon, P. W., Powers, M. E., Holder, W., & Hansel, W. (1977b). Pregnancy and parturition in the bitch. *Biol. Reprod.* 16:517–526.

Corbett, L., & Newsome, A. (1975). Dingo society and its maintenance: A preliminary analysis. In *The wild canids: Their systematics, behavioral ecology and evolution,* ed. M. W. Fox, pp. 369–379. New York: Van Nostrand Reinhold.

Crespo, J. A. (1975). Ecology of the pampas gray fox and the large fox (culpeo). In *The wild canids: Their systematics, behavioral ecology and evolution,* ed. M. W. Fox, pp. 179–191. New York: Van Nostrand Reinhold.

Davidar, E. R. C. (1975). Ecology and behavior of the dhole or Indian wild dog (*Cuon alpinus*). In *The wild canids: Their systematics, behavioral ecology and evolution,* ed. M. W. Fox, pp. 109–119. New York: Van Nostrand Reinhold.

Dixson, A. F., & George, L. (1982). Prolactin and parental behaviour in a male New World primate. *Nature* 299:551–553.

Egoscue, H. J. (1956). Preliminary studies of the kit fox in Utah. *J. Mammal.* 37:351–357.

Egoscue, H. J. (1962). Ecology and life history of the kit fox in Tooele County, Utah. *Ecology* 43:481–497.

Flint, A. P. F., Krzywinski, A., Sempéré, A. J., Mauget, R., & Lacroix, A. (1994). Luteal oxytocin and monoestry in the roe deer, *Capreolus capreolus. J. Reprod. Fert.* 101:651–656.

Fox, M. W., & Bekoff, M. (1975). The behaviour of dogs. In *The behaviour of domestic animals,* 3rd ed., ed. E. S. E. Hafez, pp. 370–409. Baltimore: Williams and Wilkins.

Frame, L. H., Malcolm, J. R., Frame, G. W., and Van Lawick, H. (1979). Social organization of African wild dogs (*Lycaon pictus*) on the Serengeti Plains, Tanzania 1967–1978. *Z. Tierpsychol.* 50:225–249.

Fuller, J. L., & DuBois, E. M. (1962). The behavior of dogs. In *The behavior of domestic animals,* ed. E. Hafez pp. 415–452. London: Bailliere, Tindall and Cox.

Gauthier-Pilters, H. (1967). The fennec. *Afr. Wildl.* 21:117–125.

Gier, H. T. (1975). Ecology and social behavior of the coyote. In *The wild canids: Their systematics, behavioral ecology and evolution,* ed. M. W. Fox, pp. 247–262. New York: Van Nostrand Reinhold.

Grandage, J. (1972). The erect dog penis: A paradox of flexible rigidity. *Vet. Rec.* 91:171–177.

Gubernick, D. J., & Nelson, R. J. (1989). Prolactin and paternal behavior in the biparental California mouse, *Peromyscus californicus. Horm. Behav.* 23:203–210.

Hadley, J. C. (1975). Total unconjugated oestrogen and progesterone concentrations in peripheral blood during pregnancy in the dog. *J. Reprod. Fert.* 44:453–460.

Harrington, F. H., Mech, L. D., & Firtts, S. H. (1983). Pack size and wolf pup survival: their relationship under varying ecological conditions. *Behav. Ecol. Sociobiol.* 13:19–26.

Hayssen, V., van Tienhoven, A., & van Tienhoven, A. (1993). *Asdell's patterns of mammalian reproduction: A compendium of species-specific data.* Ithaca, N.Y.: Cornell University Press.

Hersteinsson, P., & Macdonald, D. W. (1982). Some comparisons between red and arctic foxes, *Vulpes vulpes* and *Alopex lagopus,* as revealed by radio tracking. *Symp. Zool. Soc. Lond.* 49:259–289.

Joffre, M. (1977). Relationship between testicular blood flow, testosterone and spermatogenic activity in young and adult wild red foxes (*Vulpes vulpes*). *J. Reprod. Fert.* 51:35–40.

Kennelly, J. J., & Johns, B. E. (1976). The estrous cycle of coyotes. *J. Wildl. Manage.* 40:272–277.

Kilgore, D. L., Jr. (1969). An ecological study of the swift fox (*Vulpes velox*) in the Oklahoma Panhandle. *Amer. Midl. Nat.* 81:512–534.

Kleiman, D. G. (1967). Some aspects of social behavior in the Canidae. *Amer. Zool.* 7:365–372.

Kleiman, D. G. (1968). Reproduction in the Canidae. *Internat. Zoo Yearbook* 8:3–8.

Kleiman, D. G. (1977). Monogamy in mammals. *Quart. Rev. Biol.* 52:39–69.

Kleiman, D. G., & Eisenberg, J. F. (1973). Comparisons of canid and felid social systems from an evolutionary perspective. *Anim. Behav.* 21:637–659.

Kleiman, D. G., & Malcolm, J. R. (1981). The evolution of male parental investment in mammals. In *Parental care in mammals,* eds. D. J. Gubernick & P. H. Klopfer, pp. 347–387. New York: Plenum Press.

Kreeger, T. J., Seal, U. S., Cohen, Y., Plotka, E. D., & Asa, C. S. (1991). Characterization of prolactin secretion in gray wolves. *Can. J. Zool.* 69:1366–1374.

Kruuk, H., & Turner, M. (1967). Comparative notes on predation by lion, leopard, cheetah, and wild dog in the Serengeti area, East Africa. *Mammalia* 31:21–91.

Kuhme, W. (1965). Communal food distribution and division of labour in African hunting dogs. *Nature* 205:443–444.

Lamprecht, J. (1979). Field observations on the behaviour and social system of the bat-eared fox *Otocyon megalotis* Demarest. *Z. Tierpsychol.* 49:260–284.

Macdonald, D. W. (1979). "Helpers" in fox society. *Nature* 282:69–71.

Macdonald, D. W. (1983). The ecology of carnivore social behaviour. *Nature* 301:379–384.

Macdonald, D. W., & Moehlman, P. D. (1983). Cooperation, altruism, and restraint in the reproduction of carnivores. *Perspect. Ethol.* 5:433–467.

Malcolm, J. R. (1985). Paternal care in canids. *Amer. Zool.* 25:853–856.

Malcolm, J. R., & Marten, K. (1982). Natural selection and the communal rearing of pups in African wild dogs (*Lycaon pictus*). *Behav. Ecol. Sociobiol.* 10:1–13.

Maureal, D., Lacroix, A., & Boissin, J. (1984). Seasonal reproductive endocrine profiles in two wild animals: the red fox (*Vulpes vulpes*) and the European badger (*Meles meles* L.) considered as short day mammals. *Acta Endocr. Copenh.* 105:130–138.

Mech, L. D. (1970). *The wolf.* New York: Natural History Press.

Mech, L. D., & Knick, S. T. (1978). Sleeping distances in wolf pairs in relation to breeding season. *Behav. Biol.* 23:521–525.

Messire, F., & Barrette, C. (1982). The social system of the coyote (*Canis latrans*) in a forested habitat. *Can. J. Zool.* 60:1743–1753.

Moehlman, P. D. (1983). Socioecology of silverbacked and golden jackals (*Canis mesomelas* and *Canis aureus*). In *Recent advances in the study of mammalian behavior,* ed. J. F. Eisenberg & D. G. Kleiman, pp. 423–453. Lawrence, Kans.: American Society of Mammalogists.

Moehlman, P. D. (1989). Intraspecific variation in canid social systems. In *Carnivore behavior, ecology, and evolution,* ed. J. L. Gittleman, pp. 143–163. Ithaca, N.Y.: Cornell University Press.

Moller, O. M. (1973). Progesterone concentrations in the peripheral plasma of the blue fox (*Alopex lagopus*) during pregnancy and the oestrous cycle. *J. Endocr.* 59:429–438.

Mondain-Monval, M., Dutourne, B., Bonnin-Laffargue, M., Canivenc, R., & Scholler, R. (1977). Ovarian activity during the anoestrus and the reproductive season of the red fox (*Vulpes vulpes* L.). *J. Ster. Biochem.* 8:761–769.

Mondain-Monval, M., Moller, O. M., Smith, A. J., McNeilly, A. S., & Scholler, R. (1985). Seasonal variations of plasma prolactin and LH concentrations in the female blue fox (*Alopex lagopus*). *J. Reprod. Fert.* 74:439–448.

Morrell, S. (1972). Life history of the San Joaquin kit fox. *Calif. Fish Game* 58:162–174.

Mulligan, R. M. (1942). Histological studies on the canine female genital tract. *J. Morph.* 71:431–438.

Murie, A. (1944). *The wolves of Mount McKinley.* Seattle: University of Washington Press.

Numan, M. (1988). Maternal behavior. In *The physiology of reproduction,* Vol. 2, eds. E. Knobil & J. D. Neill, pp. 1569–1646. New York: Raven Press.

Packard, J. M., Mech, L. D., & Seal, U. S. (1983). Social influences on reproduction in wolves. In *Wolves in Canada and Alaska: Their status, biology and management,* ed. L. N. Carbyn, pp. 78–86. Canadian Wildlife Service Report, Series Number 45.

Packard, J. M., Seal, U. S., Mech, L. D., & Plotka, E. D. (1985). Causes of reproductive failure in two family groups of wolves (*Canis lupus*). *Z. Tierpsychol.* 68:24–50.

Porton, I. (1983). Bush dog urine-marking: Its role in pair-formation and maintenance. *Anim. Behav.* 31:1061–1069.

Porton, I. J., Kleiman, D. G., Rodden, M. (1987). Aseasonality of bush dog reproduction and the influence of social factors on the estrous cycle. *J. Mammal.* 68:867–871.

Pryce, C. R. (1992). A comparative systems model of the regulation of maternal motivation in mammals. *Anim. Behav.* 43:417–441.

Riley, G., & McBride, R. (1975). Status of the red wolf in the United States. In *The wild canids: Their sytematics, behavioral ecology and evolution,* ed. M. W. Fox, pp. 263–279. New York: Van Nostrand Reinhold.

Rothman, R. J., & Mech, L. D. (1979). Scent-marking in lone wolves and newly-formed pairs. *Anim. Behav.* 27:750–760.

Rowlands, I. W., & Hancock, J. L. (1949). The physiology of reproduction in the dog. *Vet. Rec.* 61:771–776.

Rowlands, I. W., & Parkes, A. S. (1935). The reproductive processes of certain mammals. VIII. Reproduction in foxes (*Vulpes* spp.). *Proc. Zool. Soc. Lond.* 4:823–841.

Schams, D., Barth, D., & Karg, H. (1980). LF, FSH, and progesterone concentrations in peripheral plasma of the female roe deer (*Capreolus capreolus*) during the rutting season. *J. Reprod. Fert.* 60:109–114.

Seal, U. S., Plotka, E. D., Mech, D., & Packard, J. M. (1987). Seasonal metabolic and reproductive cycles in wolves. In *Man and wolf,* ed. H. Frank, pp. 109–125. Dordrecht, The Netherlands: Junk.

Seal, U. S., Plotka, E. D., Packard, J. M., & Mech, L. D. (1979). Endocrine correlates of reproduction in the wolf. I. Serum progesterone, estradiol and LH during the estrous cycle. *Biol. Reprod.* 21:1057–1066.

Smith, A. J., Mondain-Monval, M., Moller, O. M., Scholler, R., & Hansson, V. (1985). Seasonal variations of LH, prolactin, androstenedione, testosterone and testicular FSH binding in the male blue fox (*Alopex lagopus*). *J. Reprod. Fert.* 74:449–458.

Smith, M. S., & McDonald, L. E. (1974). Serum levels of luteinizing hormone and progesterone during the estrous cycle, pseudopregnancy and pregnancy in the dog. *Endocrinology* 94:404–412.

Stabenfeldt, G. H., & Shille, V. M. (1977). Reproduction in the dog and cat. In *Reproduction in domestic animals,* 3rd edition, eds. H. H. Cole & P. T. Cupps, pp. 499–527. New York: Academic Press.

Stellflug, J. N., Muse, P. D. Everson, D. O., & Louis, T. M. (1981). Changes in serum progesterone and estrogen of the nonpregnant coyote during the breeding season. *Proc. Soc. Exp. Biol. Med.* 167:220–223.

Taha, M. B., Noakes D. E., & Allen, W. E. (1981). The effect of season of the year on the characteristics and composition of dog semen. *J. Small Anim. Pract.* 22:177–184.

Valtonen, M. H., Rajakoski, E. J., & Lähteenmäki, P. (1978). Levels of oestrogen and progesterone in the plasma of the raccoon dog (*Nyctereutes procyonoides*) during oestrus and pregnancy. *J. Endocr.* 76:549–550.

Valtonen, M. H., Rajakoski, E. J., & Mäkelä, J. I. (1977). Reproductive features in the female raccoon dog (*Nyctereutes procyonoides*). *J. Reprod. Fert.* 51:517–518.

Van Heerden, J., & Kuhn, F. (1985). Reproduction in captive hunting dogs *Lycaon pictus. S. Afr. J. Wildlf. Res.* 15:80–84.

Yalden, D. W., & Morris, P. A. (1975). *The lives of bats.* New York: New York Times Book Co.

Zabel, C. J., & Taggart, S. J. (1989). Shift in red fox, *Vulpes vulpes*, mating system associated with El Niño in the Bering Sea. *Anim. Behav.* 38:830–838.

6

Variation in Reproductive Suppression among Dwarf Mongooses: Interplay between Mechanisms and Evolution

SCOTT R. CREEL and PETER M. WASER

6.1 Introduction

Cooperative breeders are often divided into two types: plural breeders, in which most or all adults within a group produce young of their own, and singular breeders, in which subordinate group members do not produce young (Brown 1987; Stacey & Koenig 1990). Singular breeders form the majority among communal breeders of most taxa. For example, Brown (1987) classified 89 percent of 111 communally breeding bird species as singular breeders. Singular breeding also predominates among communally breeding carnivores (the focus of this chapter), among which subordinates do not normally raise offspring in 57 percent of 28 social species (Creel & Macdonald 1995).

6.1.1 Why Suppress Subordinates?

Reproductive suppression of subordinates probably arises from competition for resources that limit breeding opportunities within a group (Brown 1974; Stacey 1979; Emlen 1991). If the number of young that can be raised within a group is limited to fewer than a single female can produce, then selection should favor the ability to prevent group mates from breeding. (For dwarf mongooses, evidence that resources limit reproduction comes from supplementally fed groups in which subordinate females were more likely to become pregnant: $z = 1.89$, $P = 0.048$). For males, competition for low-cost mating opportunities might favor the ability to suppress others even when resources do not limit reproduction. In general, if reproduction by unrelated group mates reduces an individual's production of young, then selection should favor the ability to suppress group mates (provided that the actions

150

that impose suppression have little cost in themselves). If group mates are related, then the ability to suppress group mates should be favored when the direct fitness benefit outweighs the indirect fitness cost.

6.1.2 Why Accept Suppression?

But why should group mates tolerate suppression? If all group members compete on equal footing, it is difficult to imagine an evolutionary scenario in which suppression should be tolerated. Indeed, the few cooperatively breeding carnivores that lack dominance hierarchies usually do not show reproductive suppression (otters *Lutra lutra:* Kruuk 1991; lions *Panthera leo:* Packer et al. 1988; coatis *Nasua narica:* Russell 1983; some populations of Eurasian badgers *Meles meles:* Woodroffe & Macdonald in press). In contrast, subordinates in species with clear-cut dominance hierarchies do not normally breed successfully (dwarf mongooses *Helogale parvula:* Rasa 1973; Rood 1980; grey wolves *Canis lupus:* Packard et al. 1985; coyotes *Canis latrans:* Bekoff & Wells 1982; arctic foxes *Alopex lagopus:* Hersteinsson 1984; blackbacked and golden jackals *Canis mesomelas* and *Canis aureus:* Moehlman 1983, 1989; African wild dogs *Lycaon pictus:* Malcolm 1979; dholes *Cuon alpinus:* Johnsingh 1982; dingos *Canis f. dingo:* Corbett 1988).

The existence of a dominance hierarchy clearly favors the evolution of reproductive suppression, but in a sense this just rephrases the question from "why accept suppression?" to "why accept subordination?" The analyses of this chapter accept that dominance is normally a precondition for social reproductive suppression (following Vehrencamp 1983) and assume that dominance evolves as an honest indicator of fighting ability that reduces the costs of resolving conflicts of interest within a group (Grafen 1990). Our question becomes: "when dominance exists, why do subordinates accept suppression?"

Among carnivores, competition for breeding opportunities within a group often results in the death of subordinates' litters, either through direct infanticide (Rasa 1973; Malcolm 1979; Corbett 1988) or through biased provisioning (Malcolm 1979; Malcolm & Marten 1982). Infanticide by dominants and low survivorship of subordinates' young are probably the "levers" that favor the evolution of behavioral or endocrine mechanisms directly precluding reproduction among subordinates. Mating (for either sex), gestation and lactation (for females), and care of young (for either sex) are energetically costly and may involve mortality risks (Oftedal & Gittleman 1989; Clutton-Brock 1991; Creel & Creel 1991). Behavioral and endocrine mechanisms that preclude these costs and risks would be evolutionarily favored if subordinates' young have a sufficiently low probability of surviving to independence.

6.1.3 Variation in the Degree of Suppression

The dichotomy between singular and plural breeders is useful for comparisons across species, but it is an oversimplification. In all mammalian and avian cooperative breeders that have been studied in detail, reproductive suppression of subordinates is not absolute (Brown 1987; Emlen 1991). For example, reproductive suppression is the norm among subordinate dwarf mongooses (Rasa 1973; Rood 1980; Creel et al. 1992), but 12 percent of subordinates became pregnant, and subordinates accounted for 27 percent of 302 pregnancies in our Serengeti study population (Creel & Waser 1991). Fifteen percent of dwarf mongooses whose parentage was determined by DNA fingerprinting had subordinate mothers, and 25 percent had subordinate fathers (Keane et al. 1994).

Among other carnivores, African wild dogs normally have one breeding female per pack (e.g., 94% of packs in Kruger National Park: Fuller et al. 1992), but in some populations multiple litters are quite common (38% of packs in Masai Mara National Park: Fuller et al. 1992). Subordinates are suppressed in some populations of Eurasian badgers (Kruuk 1989) but not in others (Woodroffe & Macdonald in press). Reproductive suppression has been noted to fail under some conditions among red foxes (*Vulpes vulpes:* Macdonald 1980), arctic foxes (Hersteinsson 1984), coyotes (Camenzind 1978), and wolves (Packard et al. 1985). Suppression even fails occasionally in the naked mole-rat, though it is considered eusocial (Jarvis 1991; see also Faulks & Abbott, this volume).

Clearly, failures of reproductive suppression are common and widespread in species for which suppression is typical. This suggests that reproductive suppression should be addressed as a matter of degree, rather than as a trait that is present or absent. Taking this view, we can ask if patterns of reproduction among subordinates are the product of selection, rather than regarding subordinate reproduction as a simple failure of the "normal" mechanisms that suppress reproduction. At the population level, the degree of reproductive suppression can be measured as the percentage of subordinates that do not breed, either annually (Creel & Waser 1991) or in their lifetime (Lacey & Sherman 1991). At the individual level, the degree of suppression can be measured as the age- or sex-specific probability of breeding. At the behavioral and endocrine levels, the degree of suppression can be quantified by comparison of subordinates' mating rates and reproductive hormone levels with those of dominants (French et al. 1984; French, Inglett, & Dethlefs 1989; Faulkes, Abbott, & Jarvis 1990; Creel et al. 1992; Creel, Wildt, & Monfort 1993).

Vehrencamp (1983) provides an evolutionary model of the degree of reproductive suppression within a social group. In her model, the degree of reproductive suppression within a group varies continuously from "egalitarian" groups, in which reproductive success is evenly shared, to "despotic" groups with complete suppression of subordinates. The model assumes that dominants will bias fitness within the group in their own favor, up to the point at which subordinates' expected fitness would be higher outside the group (leading them to emigrate).

In this chapter, we use Vehrencamp's model to predict the direct reproductive success that subordinate dwarf mongooses (of given sex, age, and pack size) must receive in order to make the fitness payoff to nondispersal exceed that of dispersal. We then compare these predictions to actual patterns of reproduction by subordinates. This comparison allows us to ask if reproduction by subordinates is simply an accidental failure of the normal mechanisms of reproductive suppression, or if it represents an adaptive attempt by subordinate mongooses to obtain a more equal distribution of reproductive success.

This approach simply assumes that reproductive suppression is something that dominants can impose if it is in their interest to do so (Vehrencamp 1983), without consideration of the mechanisms involved. But suppression can be the result of widely divergent mechanisms. In some species, behavioral mechanisms preclude mating (e.g., interference by dominants in wolves: Seal et al. 1979; Packard et al. 1985). In others, endocrine mechanisms prevent reproduction even among subordinates that mate (e.g., hormonally suppressed ovulation in common marmosets *Callithrix jacchus* and cotton-top tamarins *Saguinus oedipus*: Abbott 1984, 1987; French et al. 1984; see also French this volume). Because males and females differ in many aspects of reproductive function, differences between the sexes in mechanisms of suppression are likely. In particular, lower energetic costs of reproduction among males suggest that males should be more likely to "resist" suppression (Kleiman 1980; Creel et al. 1992).

The pattern of subordinate reproduction across ages, sexes, and group sizes of subordinates might well be influenced by the mechanisms involved. Having predicted patterns of subordinate reproduction in dwarf mongooses with Vehrencamp's evolutionary model, we ask if lack of fit between predicted and observed patterns of subordinate reproduction is explained by the behavioral and endocrine mechanisms underlying suppression. First, we give a general description of communal breeding in dwarf mongooses.

6.2 Dwarf Mongoose Social Organization

Dwarf mongooses are the smallest members of the carnivore family Herpestidae and are widely distributed across woodland and wooded savanna in sub–Saharan Africa (Kingdon 1977). Most mongoose species are solitary, and sociality in dwarf mongooses is probably due to two factors. First, they prey primarily on invertebrates, which are sufficiently abundant and rapidly renewing to allow a territory to be shared with conspecifics at little cost (Waser 1981; Waser & Waser 1985). Second, small size and diurnality combine to make dwarf mongooses highly vulnerable to predation (Rasa 1986; Rood 1986, 1990), so that coordinated vigilance provides benefits to life in groups.

6.2.1 The Study Population

Our data come from a population of 12 to 20 packs living on 25 square kilometers along the Sangere River in Serengeti National Park, Tanzania (2°20'S, 30°50'E). The population was under study from 1977 to 1991 (by Jon Rood from 1977 until 1987). The mongooses were marked (713 total) and observed to record age-specific survivorship, reproduction, immigration, and emigration. From these data we constructed a data base of individual life-histories.

6.2.2 Social Organization

In our population, packs average nine adults and yearlings (9.0 ± 0.3, mean $\pm$ SEM, $n = 208$ pack-years: Rood 1990; Creel & Waser 1991), plus zero to 15 dependent young of the year. Although dispersal is weakly male biased, packs are made up of natal and immigrant individuals of both sexes (Rood 1986, 1990). Sex differences in the frequency of dispersal and its age distribution reflect mortality risks (Waser, Creel, & Lucas 1994) and inclusive fitness payoffs (Creel & Waser 1994).

Within a pack, only the dominant individual of each sex is assured of breeding (Rasa 1973; Rood 1980). A dominant female produces up to four litters of 1 to 7 young (mean $\pm$ SEM $= 2.5 \pm 0.2$ litters, $n = 302$) in a breeding season that coincides with rains between November and May. Dwarf mongooses are born blind, hairless, and incapable of moving with a foraging pack for six weeks. Subordinates of both sexes guard, groom, carry, and feed the young at rates comparable to those of the dominants (Rood 1980). Subordinate females that become pregnant nurse a joint litter with the alpha female, although the young are predominantly the alpha's (Rood 1980; Creel & Waser 1991; Keane et al. 1994). Spontaneous lactation by nonbreeders, perhaps the most extreme form of helping in mammals, occurs among 4 percent of subordinates annually (Creel et al. 1991). The help of subordinates of both sexes is effective in increasing breeders' reproductive success (Creel & Waser 1994).

Dominance is tightly correlated with age among both natal and immigrant mongooses (Rood 1980; Creel et al. 1992). Subordinates may wait to attain dominance in their natal pack or disperse in search of breeding opportunities elsewhere (Rood 1990). Successful dispersers are more likely than nondispersers to attain breeding positions (Rood 1990), but the mortality risk of dispersal is great (51% among males and 78% among females: Waser et al. 1994). Primarily because mongooses face poor odds of dispersing successfully, nondispersers of both sexes accrue greater direct fitness than do dispersers at most ages (Creel & Waser 1994).

Interestingly, dwarf mongooses can obtain indirect fitness after dispersal, either by immigration into groups where relatives are already established as breeders or by taking over another pack with a group of related dispersers. (Rood 1990; Creel & Waser 1994). Nonetheless, nondispersers accrue greater indirect fitness than dispersers do because nondispersers generally provide help to closer relatives (Creel & Waser 1994; Lucas, Waser, & Creel this volume).

The final option open to subordinates is reproduction despite the presence of a same-sexed dominant. The analyses that follow use Vehrencamp's (1983) model of bias in reproductive success to predict which subordinate mongooses should adopt this strategy.

6.3 Evolutionary Modeling of Subordinate Reproduction

We have previously applied Vehrencamp's model to predict subordinate pregnancies among female mongooses (Creel & Waser 1991). In that analysis we assumed that subordinates did not accrue indirect fitness after dispersing. Subsequent analysis revealed that indirect fitness can be a surprisingly large proportion of dispersers' inclusive fitness (Creel & Waser 1994): During the year of dispersal, indirect fitness comprised up to 15 percent of inclusive fitness for dispersing females and up to 26 percent for dispersing males. Here, we have modified the model to account for indirect fitness among dispersers and have applied it to both males and females.

The model compares two measures for subordinates of each age and group size: (1) the indirect fitness that a subordinate obtains by remaining in its group and helping its group mates, and (2) the inclusive fitness that the subordinate could expect if it dispersed to compete in a new group. When the first measure exceeds the second, nondispersal is favored, even if it entails complete reproductive suppression. When the second measure exceeds the first, then the indirect fitness benefits of helping are insufficient to make helping (nondispersal) preferable to dispersal, and nondispersing subordinates should

Table 6.1. *Definitions of parameters in the evolutionary model*

Model element	Definition
Parameters	
$W_{w(k,t)}$	Reproductive success of a subordinate of age t in a group of size k.
$\Delta W_{(k)}$	Change in total reproductive success within group as group size increases from $k-1$ to k.
W_a	Reproductive success of an inexperienced alpha, averaged across all group sizes.
d	Probability that a mongoose who disperses will successfully immigrate into a new group.
$b_{(t)}$	Probability of obtaining an alpha position within a group, as a function of age.
$r_{N(t)}, r_{D(t)}$	Coefficient of relatedness between a helper and the breeders in its group, as a function of the helper's age, N = nondisperser, D = disperser.
Subscripts	
w	Subordinate
a	Alpha
k	Group size
t	Age in years
D	Disperser
N	Nondisperser

require direct reproductive success to make up the difference (Vehrencamp 1983).

The inclusive fitness of a nondispersing subordinate is given (Eq. 6.1):

$$IF_{w(k,t)} = W_{w(k)} + r_{N(t)} \cdot [\Delta W_{(k)} - W_{w(k)}] - p \cdot \overline{\Delta W}_{(k)} \tag{6.1}$$

The inclusive fitness that a subordinate can expect if it disperses is given (Eq. 6.2):

$$IF_{D(t)} = d \cdot [(b_{(t)} \cdot W_a) + (1 - b_{(t)}) \cdot r_{D(t)} \cdot \Delta W_{(k)}] - p \cdot \overline{\Delta W}_{(k)} \tag{6.2}$$

Definitions of parameters are in Table 6.1. The equations are explained in detail in Creel & Waser (1991): equation 6.2 has been modified to incorporate the indirect fitness of dispersers via the second term within square brackets. The two measures can be set equal and solved to yield the direct reproductive success that a subordinate should require to make nondispersal yield greater (or equal) inclusive fitness than dispersal (Eq. 6.3).

$$W_{w(k,t)} = \frac{d \cdot [(b_{(t)} \cdot W_a) + (1 - b_{(t)}) \cdot r_{D(t)} \cdot \Delta W_{(k)}] - r_{N(t)} \cdot \Delta W_{(k)}}{1 - r_{N(t)}} \tag{6.3}$$

The reproductive success expected for subordinates of a given age and group size, $W_{w(k,t)}$, depends on two sets of parameters. One set affects the indirect fitness payoff to nondispersers ($r_{N(t)}$, $\Delta W_{(k)}$). The other set affects the direct and indirect payoffs to dispersers (d, $b_{(t)}$, W_a, $r_{D(t)}$, $\Delta W_{(k)}$). In turn, most of these parameters are functions of group size (k) or age (t).

6.4 Evolutionary Predictions for Subordinate Reproduction

6.4.1 Estimating Parameter Values

6.4.1.1 Relatedness

Relatedness between subordinates and dominants was measured for dispersers and nondispersers, using 381 subordinate–dominant dyads from the genealogies of five packs ($n = 126$ mongooses: Creel 1991). The genealogies were based on the assumption that all offspring in a pack were produced by the pack's dominants, but molecular data on parentage (Keane et al. 1994) allowed post-hoc correction of r values from the genealogies to account for subordinate reproduction. True r values between subordinates and dominants are only 86 percent of those that assume reproduction exclusively by dominants (Creel & Waser 1994; Keane et al. 1994). Therefore, we have multiplied all r values from the genealogies by 0.86 as a correction factor. Case by case correction of r values using molecular data was not possible because the genealogies extended back to before the advent of DNA fingerprinting (an interesting commentary on rates of methodological change in molecular biology and evolutionary ecology).

As expected, subordinate nondispersers were more closely related to dominants than were dispersers. More surprisingly, subordinate dispersers of some ages were on average quite closely related to the dominants in their new pack (Creel & Waser 1994). This pattern arose from group dispersal and repeated dispersal between given pairs of packs (Rood 1990; Waser et al. 1994). For nondispersing males of one to six years, $r_{N(t)} = [0.39, 0.30, 0.26, 0.22, 0.20, 0.17]$. For dispersing males, $r_{D(t)} = [0.22, 0.15, 0.10, 0.08, 0.05, 0.03]$. Among females, relatedness of nondispersers to breeders was slightly lower than among males; $r_{N(t)} = [0.35, 0.26, 0.20, 0.16, 0.13, 0.10]$. For dispersing females, relatedness was initially higher than among males but fell to similar or lower values at age 3 and beyond; $r_{D(t)} = [0.46, 0.26, 0.14, 0.05, 0.00, 0.00]$. Note that average relatedness of dispersing females to the breeders in the new pack was roughly that of full-sibs at one year and that of half-sibs at

two years. These values are quite high even when compared with values for
nondispersers in many species.

6.4.1.2 Effects of subordinates on breeders' reproduction

To estimate the effect of subordinates' help on breeders' reproduction, we
regressed annual changes in reproductive success (measured in yearlings
raised) on annual changes in group size. This methodology is borrowed from
econometrics, where it is used as a test of causality (Kennedy 1992). In effect,
this method asks whether increases and decreases in group size are associated
with subsequent increases or decreases in reproductive success. The regression indicated that each subordinate accounted for an increase of 0.27 ± 0.06
yearlings raised ($t = 4.47$, $P < 0.001$, $R^2 = 0.13$). All tests for associations
between environmental variables and reproductive success were not significant (Waser et al., 1995), further suggesting that the relationship between
group size and reproduction success is causal. There was not a significant difference in the effectiveness of male versus female helpers (Creel & Waser
1994).

6.4.1.3 Dispersal risk

Waser et al. (1994) developed a graphical method for estimating the mortality
risk of dispersal and applied the method to dwarf mongooses. The probability
that a mongoose that dispersed would survive and join a new pack differed
significantly between the sexes. The best estimate of the probability that an
interpack transfer would succeed if attempted was 0.22 for females (bootstrap
95% C.I. 0.12 to 0.33) and 0.49 for males (bootstrap 95% C.I. 0.37 to 0.62).
The lower success of females suggests that females are more choosy in evaluating packs to join, or that females encounter greater resistance to immigration (Waser et al. 1994).

6.4.1.4 Probabilities of obtaining an alpha position

For both dispersers and nondispersers, the probability of attaining an alpha
position increased with age (Creel et al. 1992; Creel & Waser 1994).
Successful dispersal also increased the probability that a mongoose of given
age would attain a breeding position (Rood 1990; Waser et al. 1994).

For dispersing subordinates, we directly measured the proportion $p_{(t)}$ of
mongooses of each sex that obtained an alpha position upon immigration,
pooling age classes as necessary to keep $n > 10$ (Creel & Waser 1994). For

nondispersers, we measured the age-specific probability of becoming a breeder as the product of three probabilities: (1) death of the same-sexed breeder, (2) replacement of the breeder by an existing pack member, and (3) a given subordinate winning the replacement contest.

For nondispersing females, the annual probability of attaining a breeding position steadily increased from 2 percent at one year to 17 percent at seven years. Among successfully dispersing females, a much higher percentage attained breeding slots: 23 percent among 1-year-olds and 43 percent thereafter. For nondispersing males, age-specific probabilities of attaining dominance ranged from 1 percent among 1-year-olds to 21 percent at seven years, similar to the pattern among females. No dispersing males immigrated into a breeding position at age one, but the proportion increased to 28 percent among 2- and 3-year-olds and to 43 percent thereafter. The only pronounced difference between the sexes is among 1-year-old dispersers, where males fare poorly.

6.4.1.5 Postdispersal reproductive success

If an emigrant mongoose does successfully join a new pack and establish dominance, it still must produce young and raise them in order to accrue direct fitness. Among males, there is no relationship between reproductive success and experience as a breeder, so that inexperienced immigrant breeders attain good reproductive success (2.34 ± 0.42 yearlings raised). Among females, reproductive success is correlated with prior experience as a breeder ($b = 0.42 \pm 0.12$, $t = 3.45$, $P < 0.001$), so that the expected reproductive success of an inexperienced immigrant is somewhat lower, 1.86 ± 0.43.

6.4.2 Predicted Patterns of Subordinate Reproduction

The foregoing data can be used with equation (6.3) to produce a distribution (across ages and group sizes) of the reproductive success a subordinate should require to remain in its pack and help, rather than dispersing. The predicted distribution of subordinate reproduction is given in Figure 6.1 for females and in Figure 6.2 for males. The two distributions are qualitatively similar. For both males and females, there is no association predicted between group size and subordinate reproduction (males: partial $b = -0.01 \pm 0.01$, $t_{34} = 1.05$, $P = 0.31$; females: partial $b = -0.01 \pm 0.004$, $t_{34} = 1.08$, $P = 0.29$). For both males and females, a positive association is predicted between age and subordinate reproduction (males: partial $b = 0.06 \pm 0.01$, $t_{34} = 4.66$, $P < 0.0001$; females: partial $b = 0.11 \pm 0.01$, $t_{34} = 17.35$, $P < 0.0001$). Although

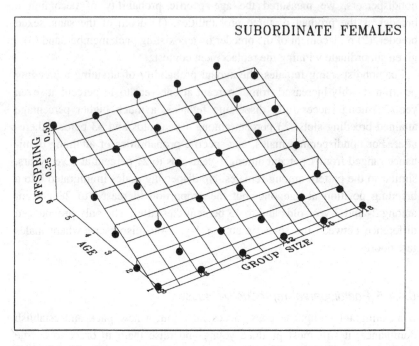

Figure 6.1. The direct reproductive success (measured in units of mature yearlings raised) that nondispersing subordinate females are predicted to require, $W_{w(k,t)}$, across ages and group sizes. Females gaining less than this reproductive success are predicted to disperse and pursue alternative reproductive options. Multiple regression plane is shown with data points connected by droplines.

the predictions for the two sexes are qualitatively similar, two quantitative differences exist. First, the predicted relationship between age and subordinate reproduction is almost twice as steep among females as among males, as can be seen by comparing the slopes of the planes in Figures 6.1 and 6.2. Second, there is less variability in the predictions for females ($R^2 = 0.90$, compared with $R^2 = 0.37$ for males). In Figures 6.1 and 6.2, this can be seen as more pronounced scatter of points away from the plane for males.

6.4.3 Observed Patterns of Subordinate Reproduction

Actual patterns of subordinate reproduction are shown in Figure 6.3 for females and in Figure 6.4 for males. For females, subordinate reproduction was measured by the percentage of females that became pregnant. For males, subordinate reproduction was measured by the mating rate.

Among females, the observed distribution of subordinate reproduction is

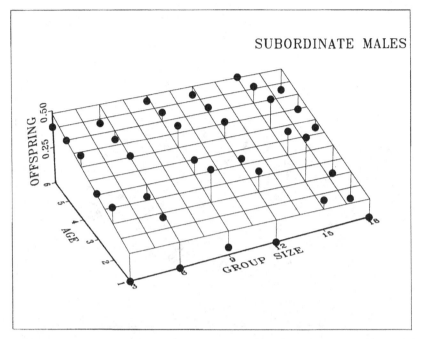

Figure 6.2. The direct reproductive success (measured in units of mature yearlings raised) that nondispersing subordinate males are predicted to require, $W_{w(k,t)}$, across ages and group sizes. Males gaining less than this reproductive success are predicted to disperse and pursue alternative reproductive options. The multiple regression plane is shown with data points connected by droplines.

quite similar to the predicted distribution. As predicted, there is a significant association between age and subordinate pregnancy (partial $b = 0.08 \pm 0.01$, $t_{34} = 7.12$, $P < 0.0001$). Also as predicted, there is no relationship between group size and subordinate pregnancy (partial $b = 0.01 \pm 0.01$, $t_{34} = 1.34$, $P = 0.18$). The regression of observed frequency of subordinate pregnancy on predicted subordinate reproduction (with one point for each age–group-size class) gives a strong test of how well the model fits the data. The model is a good predictor of subordinate pregnancies ($r = 0.67$, $F_{1,35} = 21.8$, $P < 0.001$), explaining 45 percent of the variance in the observed distribution of subordinate pregnancy. Subordinates that become pregnant are often those for whom nondispersal and helping would otherwise not yield a sufficient fitness payoff to be the favored strategy. Subordinate pregnancies appear to be a form of adaptive power sharing by the alpha female that is used to retain older helpers who have relatively good odds of attaining dominance in another pack.

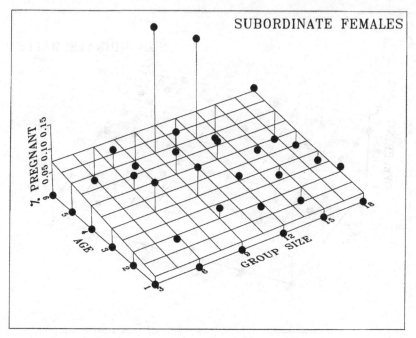

Figure 6.3. The observed distribution of subordinate pregnancies, across ages and group sizes; compare slopes with the predicted distribution of Figure 6.1. The multiple regression plane is shown with data points connected by droplines.

Among males, the picture is quite different. The predicted association between age and subordinate reproduction does not emerge in the regression of subordinates' mating rates on age (partial $b = 0.05 \pm 0.12$, $t_{34} = 0.40$, $P = 0.70$). Furthermore, no association between group size and subordinate reproduction was predicted, but the partial regression between mating rate and group size is highly significant (partial $b = 0.24 \pm 0.08$, $t_{34} = 2.80$, $P = 0.009$). Subordinate males mate more frequently in large packs than in small packs. Testing the overall fit of the model to the data, there is actually a *negative* correlation between subordinates' mating rates and reproduction predicted by the model ($r = 0.38$, $t_{32} = 2.35$, $P = 0.025$). For males, the model is an abject failure at predicting mating rates, and subordinate reproduction cannot necessarily be interpreted as adaptive power sharing by the alpha male. Matings are not shared preferentially with those that are predicted to require direct reproductive success in order to remain in the pack. Another explanation for patterns of mating among male subordinates must be sought.

Digressing briefly, it would be desirable to use a more direct measure of subordinate reproduction among males – for instance, genetic data on pater-

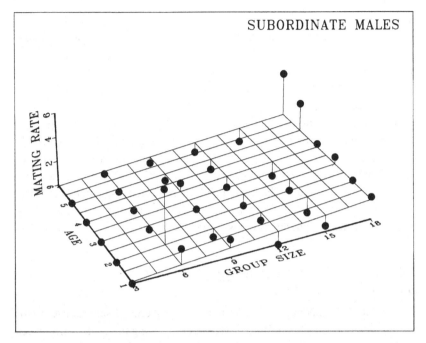

Figure 6.4. The observed distribution of mating by subordinate males, across ages and group sizes; compare slopes with the predicted distribution of Figure 6.2. The multiple regression plane is shown with data points connected by droplines.

nity. Insufficient genetic data precluded this analysis but several lines of evidence suggest that mating rate is a reasonable currency for measuring dwarf mongoose males' reproduction. First, neither age nor rank affects androgen levels (Creel et al. 1992, 1993), so that subordinate males are likely to be fertile. Second, DNA fingerprints confirm that subordinate males' matings can produce young (Keane et al. 1994). Finally, dominance rank is tightly correlated with mating rate (Creel et al. 1992), and dominance rank is the only significant predictor of genetic paternity among demographic variables (Keane et al. 1994). Together, these patterns suggest (but do not guarantee) that mating rates are likely to yield reasonable estimates of subordinate males' reproductive success. Factors other than mating rate may also play a role in paternity. For instance, testis size is positively correlated with rank (Creel et al. 1992), possibly affecting sperm competition.

6.4.3.1 Mean degree of suppression

For both sexes, the partial regression of expected subordinate reproductive success on age is statistically significant. For males, expected subordinate reproductive success is equal to 0.09 + 0.06 · age (in years). For females, expected subordinate reproductive success is equal to −0.17 + 0.11 · age (in years). These partial regression equations and the frequency distribution of individuals across ages can be combined to give weighted means for the expected reproductive success of subordinates. For subordinate males, this method predicts a mean reproductive success of 0.23 yearlings raised annually. For subordinate females, the value is considerably lower, 0.10 yearlings per year.

DNA fingerprinting has shown that subordinate females on average produce 0.10 yearlings per year (95% C.I.: 6 to 30%; Keane et al. 1994), exactly matching prediction. DNA fingerprinting showed subordinate males' reproductive success to be 0.16 yearlings per year (95% C.I.: 12 to 40%), a slightly higher-point estimate than that of females (as predicted) but somewhat lower than the predicted value. Subordinate males produce more offspring than subordinate females do, but not as many as predicted.

6.4.4 Mechanistic Influences on Patterns of Subordinate Reproduction

The evolutionary model is sufficient to explain patterns of subordinate reproduction among female dwarf mongooses but not among males. Do differences in the proximate mechanisms underlying reproductive suppression help to explain sex differences in patterns of subordinate reproduction?

Suppression of subordinate females has both endocrine and behavioral components (Creel et al. 1992). Baseline estrogen levels are correlated with dominance, and when a mating period begins (females within a pack come into estrus synchronously), mating rate is also correlated with dominance. Low baseline estrogens and low mating rates probably lead to poor positive feedback between the two, with the result that peak estrogen levels during estrus are also correlated with rank. The correlation between rank and peak estrogen levels underlies the correlation between age and the probability of becoming pregnant seen in Figure 6.3. Older, higher-ranking subordinates have endocrine function closely resembling that of dominant females, and they are more likely to become pregnant and produce offspring. Younger, lower-ranking subordinates are less likely to establish pregnancy, probably because their peak estrogen levels are insufficient to trigger the surge of luteinizing hormone that causes ovulation.

These mechanistic patterns accord well with the patterns of subordinate reproduction predicted by inclusive fitness theory. Because age is highly correlated with dominance, even following dispersal (Creel et al. 1992), older subordinates are more likely to attain breeding positions elsewhere. This gives older subordinates greater power to demand concessions from the alpha female. If the alpha female is to retain older subordinates as helpers (and thus increase her own reproductive success), she must concede her monopoly on reproduction.

In contrast, male dwarf mongooses are suppressed primarily by behavioral mechanisms (Creel et al. 1992; Creel et al. 1993). There is no correlation between dominance and androgen levels. Mating rates correlate with dominance primarily because subordinate mongooses are likely to have their mating attempts thwarted by aggression from dominants. Excluding alpha males, mating rate does not correlate significantly with age among subordinates. Subordinate reproduction among males is generally opportunistic: Most males appear to be fertile, and suppression results primarily from failing to mate with fertile females, due to aggression from the alpha male.

This proximate mechanism of suppression sheds light on the failure of the model to predict patterns of mating among subordinate males. The lack of association between mating rate and age reflects the fact that even young, low-ranking males have normal androgen levels. The positive relationship between group size and subordinates' mating rate probably reflects the increasing difficulty that alpha males have in controlling the mating activity of larger groups. For instance, alpha males monopolize mating less effectively when a pack is scattered (Creel et al. 1992). The degree of scattering and the chances for sneaked matings would be expected to increase as group size increased.

The lack of fit between subordinate males' predicted reproductive success and mating rates might also arise because mating rate does not accurately reveal paternity. We consider it unlikely (based on behavioral and endocrine data) that broad patterns in mating rate do not reflect broad patterns in dwarf mongooses' paternity, but this assumption could be tested with a larger sample of genetic and behavioral data.

Two questions remain: Why has the proximate mechanism of reproductive suppression evolved relatively good adjustment to the optimal degree (for the alpha) of reproductive suppression among females, whereas among males it has not? Why does suppression have a primarily endocrine basis among females but not males? The fundamental difference may arise with confidence of parentage. Infanticide is the final "veto" by which dominants can enforce the degree of reproductive suppression that is optimal for them. To use this power of veto, dominants must be able to distinguish their own offspring.

Although females normally give birth within several hours of one another when subordinates become pregnant, an alpha female probably has considerable confidence in distinguishing which offspring are her own, through odor cues or simply by keeping track of individual newborns. Males can share paternity within single litters (Keane et al., 1994) so that the alpha male would have far greater difficulty in establishing which young were his own. Genetic self-matching cues (e.g., MHC-based odor signatures) would be the only feasible means of identification. Even genetic cues would be problematic, because relatedness is usually high within packs (though this also mitigates the cost of sharing paternity). A dominant male, having the information that he has had a disproportionately large share of matings but lacking direct information on which particular offspring are his own, is poorly positioned to kill the young of other males. This denies him the power to enforce a greater degree of suppression than prior mechanisms (e.g., mating-rate biases) have yielded. A dominant female, confident about the parentage of offspring, can accurately enforce the degree of suppression that is optimal for her. This sex difference in veto power means that subordinate males are selected to attempt to sneak offspring whenever possible, whereas subordinate females are constrained to accept the degree of suppression optimal for the dominant.

A prediction of this argument for future work is that in species where confidence of paternity is higher, male suppression will have a stronger endocrine basis. Interestingly, multiple paternity within litters is emerging as a common pattern among communally breeding carnivores (e.g., lions: C. Packer, pers. comm. 1993; Ethiopian wolves: Gotelli et al. 1994; Eurasian badgers: Evans, Macdonald, & Cheeseman 1989; African wild dogs: R. Wayne pers. comm. 1995). When more is known about variation in paternity, quantitative predictions can be made for the degree of reproductive suppression among males of various species.

6.5 Summary

Evolutionary and mechanistic approaches to reproductive suppression in dwarf mongooses shed interesting light on one another. Information on the proximate mechanisms of reproductive suppression helps to explain why inclusive fitness modeling has good power to explain variation in the degree of suppression among females but little power to do so among males. In turn, evolutionary arguments suggest why sex differences in the mechanisms of suppression have arisen. Both approaches suggest that it is an oversimplication to regard reproductive suppression as a dichotomous trait: Suppression occurs to varying degrees, and this variation is an important element of the social organization of communal breeders. Our analyses raise many open questions. Dominant males father 75 percent of all offspring. Why, lacking the leverage of infanticide, can males enforce this degree of suppression?

Infanticidal leverage allows dominant females to enforce a degree of suppression roughly optimal for them. Given the obvious costs of subordination, why is age-dependent dominance evolutionarily stable? In the small sample of mammals in which mechanisms of suppression are known for both sexes, females are suppressed by more rigid endocrine mechanisms than are males. Will this prove to be a broad pattern? Further work integrating evolutionary mechanistic approaches will, it is hoped, answer these questions.

Acknowledgments

Our thanks to Nancy Marusha Creel, Steve Monfort, Brian Keane, Lee Elliott, David Wildt, and Dennis Minchella for their work in the lab and field. Our particular thanks to Jon Rood for allowing us to continue his study and for access to prior demographic data. For helpful comments or discussions, we thank Nancy Marusha Creel, David Macdonald, Sandra Vehrencamp, and Rosie Woodroffe. We are grateful to Tanzania National Parks, the Serengeti Wildlife Research Institute, and the Tanzania Commission for Science and Technology for permission to work at the SWRC. Our research was supported by grants from the National Geographic Society, the Scholarly Studies Program of the Smithsonian Institution, the National Science Foundation (BSR–88180040), and a David Ross fellowship from Purdue University. This chapter was prepared under support from an NSF–NATO postdoctoral fellowship and the Frankfurt Zoological Society, Help for Threatened Animals Project 1112/90.

References

Abbott, D. H. (1984). Behavioural and physiological suppression of fertility in female marmoset monkeys. *Am. J. Primatol.* 6:169–186.

Abbott, D. H. (1987). Behaviourally mediated suppression of reproduction in female primates. *J. Zoo. (Lond.)* 213:455–470.

Bekoff, M., & Wells, M. C. (1982). The behavioral ecology of coyotes: Social organization, rearing patterns, space use, and resource defense. *Z. Tierpsychol.* 60:281–305.

Brown, J. L. (1974). Alternate routes to sociality in jays – with a theory for the evolution of altruism and communal breeding. *Am. Zool.* 14:63–80.

Brown, J. L. (1987). *Helping and communal breeding in birds.* Princeton: Princeton University Press.

Camenzind, F. J. (1978). Behavioral ecology of coyotes on the National Elk Refuge, Jackson, Wyoming. In *Coyotes: Biology, behavior and management,* ed. M. Bekoff, pp. 267–294. New York: Academic Press.

Clutton-Brock, T. H. (1991). *The evolution of parental care.* Princeton: Princeton University Press.

Corbett, L. K. (1988). Social dynamics of a captive dingo pack: Population regulation by dominant female infanticide. *Ethology* 78:177–198.

Creel, S. R. (1991). Reproductive suppression and communal breeding in dwarf mongooses, *Helogale parvula*. PhD dissertation, Purdue University, West Lafayette, Indiana.

Creel, S. R., & Creel, N. M. (1991). Energetics, reproductive suppression and obligate communal breeding in carnivores. *Behav. Ecol. Sociobiol.* 28:263–270.

Creel, S. R., & Macdonald, D. W. (1995). Sociality, group size and reproductive suppression among carnivores. *Adv. Stud. Behav.* 24:203–257.

Creel, S. R., & Waser, P. M. (1991). Failures of reproductive suppression in dwarf mongooses (*Helogale parvula*): Accident or adaptation? *Behav. Ecol.* 2:7–15.

Creel, S. R., & Waser, P. M. (1994). Inclusive fitness and reproductive strategies in dwarf mongooses. *Behav. Ecol.* 5:339–348.

Creel, S. R., Monfort, S. L., Wildt, D. E., & Waser, P. M. (1991). Spontaneous lactation is an adaptive result of pseudopregnancy. *Nature* 351:660–662.

Creel, S. R., Creel, N. M., Wildt, D. E., & Monfort, S. L. (1992). Behavioral and endocrine mechanisms of reproductive suppression in Serengeti dwarf mongooses. *Anim. Behav.* 43:231–245.

Creel, S. R., Wildt, D. E., & Monfort, S. L. (1993). Androgens, aggression and reproduction in wild dwarf mongooses: A test of the challenge hypothesis. *Am. Nat.* 141:816–825.

Emlen, S. T. (1991). Evolution of cooperative breeding in birds and mammals. In *Behavioural ecology: An evolutionary approach,* 3rd ed., ed. J. R. Krebs & N. B. Davies, pp. 301–337. London: Blackwell.

Evans, P. G. H., Macdonald, D. W., & Cheeseman, C. L. (1989). Social structure of the Eurasian badger (*Meles meles*): Genetic evidence. *J. Zool. Lond.* 216:587–595.

Faulkes, C. G., Abbott, D. H., & Jarvis, J. U. M. (1990). Social suppression of ovarian cyclicity in captive and wild colonies of naked mole-rats, *Heterocephalus glaber*. *J. Reprod. Fert.* 88:559–568.

French, J. A., Abbott, D. H., Scheffler, G., Robinson, J., & Goy, R. W. (1984). Cyclic excretion of urinary oestrogens in female tamarins (*Saguinus oedipus*). *J. Reprod. Fert.* 68:177–184.

French, J. A., Inglett, B. J., & Dethlefs, T. M. (1989). The reproductive status of non-breeding group members in captive golden lion tamarin social groups. *Am. J. Primatol.* 18:73–86.

Fuller, T. K., Kat, P. W., Bulger, J. B., Maddock, A. H., Ginsberg, J. R., Burrows, R., McNutt, J. W., & Mills, M. G. L. (1992). Population dynamics of African wild dogs. In *Wildlife 2001: Populations,* ed. D. R. McCullough & R. H. Barrett, pp. 1125–1139. London: Elsevier.

Gotelli, D., Sillero-Zubiri, C., Applebaum, G. D., Roy, M. S., Girman, D. J., Garcia-Moreno, J., Ostrander, E. A., & Wayne, R. K. (1994). Molecular genetics of the most endangered canid: The Ethiopian wolf *Canis simensis*. *Molec. Ecol.* 3:301–312.

Grafen, A. (1990). Biological signals as handicaps. *J. Theor. Biol.* 144:517–546.

Hersteinsson, P. (1984). The behavioural ecology of the arctic fox (*Alopex lagopus*) in Iceland. PhD dissertation, Oxford University, Oxford.

Jarvis, J. U. M. (1991). Reproduction of naked mole-rats. In *The biology of the naked mole-rat,* eds. P. W. Sherman, J. U. M. Jarvis, & R. D. Alexander, pp. 384–425. Princeton: Princeton University Press.

Johnsingh, A. J. T. (1982). Reproductive and social behaviour of the dhole, *Cuon alpinus* (Canidae). *J. Zool. (Lond.)* 198:443–463.

Keane, B., Waser, P. M., Creel, S. R., Creel, N. M., Elliott, L. F., & Minchella, D. J. (1994). Subordinate reproduction in dwarf mongooses. *Anim. Behav.* 47:65–75.

Kennedy, P. (1992). *A guide to econometrics,* 3rd ed. Oxford: Blackwell.

Kingdon, J. (1977). *East African mammals. Vol. 3: Carnivores.* London: Academic Press.

Kleiman, D. G. (1980). The socioecology of captive propagation. In *Conservation biology: An evolutionary-ecological perspective,* ed. M. E. Soule & B. A. Wilcox, pp. 243–261. Sunderland, Mass.: Sinauer.

Kruuk, H. (1989). *The social badger.* Oxford: Oxford University Press.

Kruuk, H. (1991). The spatial organization of otters (*Lutra lutra*) in Shetland. *J. Zool. (Lond.)* 224:41–57.

Lacey, E. A., & Sherman, P. W. (1991). Social organization of naked mole-rat colonies: Evidence for division of labor. In *The biology of the naked mole-rat,* ed. P. W. Sherman, J. U. M. Jarvis, & R. D. Alexander, pp. 275–336. Princeton: Princeton University Press.

Macdonald, D. W. (1980). Social factors affecting reproduction amongst red foxes. In *The red fox: Symposium on behavior and ecology,* ed. E. Zimen, pp. 123–175. The Hague: Junk.

Malcolm, J. R. (1979). Social organization and the communal rearing of pups in African wild dogs (*Lycaon pictus*). PhD dissertation, Harvard University, Cambridge, Mass.

Malcolm, J. R., & Marten, K. (1982). Natural selection and the communal rearing of pups in African wild dogs (*Lycaon pictus*). *Behav. Ecol. Sociobiol.* 10:1–13.

Moehlman, P. D. (1983). Socioecology of silverbacked and golden jackals (*Canis mesomelas* and *Canis aureus*). In *Advances in the study of mammalian behavior,* ed. J. F. Eisenberg, & D. G. Kleiman. Lawrence, Kans.: American Society of Mammalogists.

Moehlman, P. D. (1989). Intraspecific variation in canid social systems. In *Carnivore behavior, ecology and evolution,* ed. J. L. Gittleman, pp. 143–163. Ithaca: Cornell University Press.

Oftedal, O. T., & Gittleman, J. L. (1989). Patterns of energy output during reproduction in Carnivores. In *Carnivore behavior, ecology and evolution,* ed. J. L. Gittleman, pp. 355–378. Ithaca: Cornell University Press.

Packard, J. M., Seal, U. S., Mech, L. D., & Plotka, E. D. (1985). Causes of reproductive failure in two family groups of wolves. *Z. Tierpsychol.* 68:24–40.

Packer, C., Herbst, L., Pusey, A. E., Bygott, J. D., Hanby, J. P., Cairns, S. J., & Borgerhoff Mulder, M. (1988). Reproductive success in lions. In *Reproductive success: Studies of individual variation in contrasting breeding systems,* ed. T. H. Clutton-Brock, pp. 363–383. Chicago: University of Chicago Press.

Rasa, O. A. E. (1973). Intra-familial sexual repression in the dwarf mongoose, *Helogale parvula. Naturwissenschaften* 60:303–304.

Rasa, O. A. E. (1986). Coordinated vigilance in dwarf mongoose family groups: the "watchman's song" hypothesis and the costs of guarding. *Z. Tierpsychol.* 71:340–344.

Rood, J. P. (1980). Mating relationships and breeding suppression in the dwarf mongoose. *Anim. Behav.* 28:143–150.

Rood, J. P. (1986). Ecology and social evolution in the mongooses. In *Ecological aspects of social evolution,* ed. D. I. Rubenstein & R. W. Wrangham, pp. 131–152. Princeton: Princeton University Press.

Rood, J. P. (1990). Group size, survival, reproduction and routes to breeding in dwarf mongooses. *Anim. Behav.* 39:566–572.

Russell, J. K. (1983). Altruism in coati bands: Nepotism or reciprocity? In *Social behavior of female vertebrates,* ed. S. K. Wasser. New York: Academic Press.

Seal, U. S., Plotka, E. D., Packard, J. M., & Mech, L. D. (1979). Endocrine correlates of reproduction in the wolf. I. Serum progesterone, estradiol and LH during the estrous cycle. *Biol. Reprod.* 21:1057–1066.

Stacey, P. B. (1979). Habitat saturation and communal breeding in the acorn woodpecker. *Anim. Behav.* 27:1153–1166.

Stacey, P. B., & Koenig, W. D. (1990). *Cooperative breeding in birds: Longterm studies of ecology and behavior.* Cambridge: Cambridge University Press.

Vehrencamp, S. L. (1983). A model for the evolution of "despotic" versus "egalitarian" species. *Anim. Behav.* 31:667–682.

Waser, P. M. (1981). Sociality or territorial defense? The influence of resource renewal. *Behav. Ecol. Sociobiol.* 8:231–237.

Waser, P. M., & Waser, M. S. (1985). *Ichneumia albicauda* and the evolution of viverrid gregariousness. *Z. Tierpyschol.* 68:137–151.

Waser, P. M., Creel, S. R., & Lucas, J. R. (1994). Death and disappearance: Estimating mortality risks associated with philopatry and dispersal. *Behav. Ecol.* 5:135–141.

Waser, P. M., Elliot, L., Creel, N. M., & Creel, S. (1995). Habitat variation and mongoose demography. In *Serengeti II: Dynamics, management and conservation of an ecosystem,* ed. A. R. E. Sinclair & P. Arcese, pp. 421–450. Chicago: University of Chicago Press.

Woodroffe, R., & Macdonald, D. W. (in press). Female/female competition in European badgers, *Meles meles:* Effects on breeding success. *J. Anim. Ecol.*

7

Dynamic Optimization and Cooperative Breeding: An Evaluation of Future Fitness Effects

**JEFFREY R. LUCAS, SCOTT R. CREEL,
and PETER M. WASER**

7.1 Introduction

Why do some reproductively mature animals stay at home, delay breeding, and help to rear the offspring of others? Over the past 25 years, this question has been addressed by authors studying a wide variety of animals, from insects (Alexander, Noonan, & Crespi 1991) to mammals (Brown 1987; Emlen 1991). On one level, the answer is simple: An individual should stay at home and help when its inclusive fitness is higher than it would be if it dispersed and attempted to breed (Hamilton 1964; Brown 1987; note: we will refer to allo-parental behavior as "help"). Unfortunately, for several reasons, quantifying the inclusive fitness consequences of these alternatives is not a simple matter. For example, the sequence of behavioral decisions that lead to helping may differ among species, as we discuss later. Also, there are four components of inclusive fitness that need to be measured (Brown 1987): current direct fitness, current indirect fitness, future direct fitness, and future indirect fitness. The calculation of some of these components is quite complicated.

To quantify the fitness payoffs to individuals that make these decisions, we first need to understand clearly what decisions they are making. As Brown (1987) has suggested, cooperative breeding often involves three distinct decisions: to remain in the natal territory (as opposed to dispersing), to delay breeding, and to help. In empirical studies, these decisions are often simplified to a dichotomy: philopatry/help versus disperse/breed. For example, in black-backed and golden jackals, virtually all individuals that delay breeding and help do so on their natal territory, and conversely, virtually all dispersal events are associated with breeding attempts (Moehlman 1983).

In other cooperative breeders, however, the three decisions can be more independent. In dwarf mongooses (Creel & Waser 1994) and wild dogs (Malcolm & Marten 1982), dispersers often enter nonnatal groups as helpers.

Thus, in these species, the decision to help is decoupled from the decision to disperse. Similarly, decisions to help may be decoupled from decisions to attempt or to delay breeding. Reproductive suppression of subordinates is found in the majority of mammalian (and avian) cooperative breeders (Creel & Macdonald 1995; French this volume; Moehlman & Hofer this volume; Solomon & Getz this volume), but it is often incomplete, and occasional reproduction by subordinates can make a substantial contribution to fitness (e.g., dwarf mongooses: Keane et al. 1994).

Our approach here is to examine the consequences of what we view as a critical decision virtually all cooperative breeders make: whether to disperse instead of remaining at home. The consequences of the dispersal decision may dictate (or at least impact) the degree of alloparental care offered by an individual, and may also influence potential delays in breeding.

Species differ markedly in the ability of individuals to gather information about dispersal options. In many avian cooperative breeders, philopatric animals can remain in their natal groups and vie for breeding opportunities that arise from the deaths of dominant animals in nearby territories as well as in their own (Wiley & Rabenold 1984; Woolfenden & Fitzpatrick 1984; Zack & Stutchbury 1992). In dwarf mongooses, naked mole-rats (Lacey & Sherman this volume), and probably most mammalian cooperative breeders, animals vying for breeding opportunities cannot simultaneously assess options in more than one group. Subordinate animals may queue for a dominant position in their natal group or in a nonnatal group, but not both. The fundamental question is thus: Where to queue? The animal's subsequent decisions as to when to help and when to attempt breeding should depend on the breeding opportunities inherent in the queue it has joined. Our task is to calculate the four components of inclusive fitness associated with joining a natal or a non-natal queue.

The direct components of fitness have been measured in several ways. The simplest estimate of a breeder's current direct fitness is the number of surviving young it produces during the present breeding season. Unfortunately, this method double counts offspring, because young produced due to the action of helpers will be attributed to both the breeder and to the helper (Grafen 1982). A common solution to the problem (which appears to follow from a verbal description of direct fitness in Hamilton 1964) has been to equate current direct fitness with the number of offspring the breeder would have in the absence of help (e.g., Grafen 1984). In short, this method strips the number of young produced as a result of aid by helpers from each breeder's reproductive success. Creel (1990a) has shown that this estimate also is incorrect, as becomes clear if one thinks of species like dwarf mongooses or African wild dogs in which unaided breeders raise no offspring. Following the traditional verbal definition of direct fitness, these breeders have no fitness at all. In fact, an extension of Hamilton's algebra shows that the direct component of fitness is the number of offspring produced by a breeder, decreased by "the average effect of one individual on others' reproductive success" (Creel 1990a, p. 220). Thus, breeders' reproductive success should be decremented by the increase in reproductive success attributable to the average individual in the population, including all helpers and nonhelpers. The calculation of direct fitness is discussed further when we outline our model.

Estimation of the current direct fitness of subordinates is also more complex than it might appear. The recent discovery in cooperative breeders of subordinate parentage (detected by DNA fingerprinting in bicolored and stripebacked wrens: Rabenold et al. 1990; Haydock et al. in review; and in dwarf mongooses: Keane et al. 1994; see also Mumme et al. 1985; Brooker et al. 1990; Faaborg et al. 1995) shows that the most common method used to estimate direct fitness effects – behavioral observations of mating and dominance – is imperfect. Some of the current direct fitness traditionally attributed to breeders in fact belongs to helpers.

The current indirect component of fitness is typically taken to be the contribution to the production of offspring from the marginal helper, devalued by the genetic relatedness between the helper and those offspring. (The contribution of the marginal helper would be the net effect on group reproductive success of the removal of a single helper.) Creel (1990a) has shown that a full accounting of the inclusive fitness accrued by helpers requires that this number, like the direct fitness of breeders, be devalued by the mean effect of an individual on others' reproductive success. When helpers are not completely suppressed, the breeders also may accrue indirect fitness. We resolve the resulting terminological morass by considering breeding and helping to be two roles that both dominant and subordinate mongooses can play; subordinates are helpers for most, but not all, of a group's offspring.

Some authors have estimated the inclusive fitness of reproductive alternatives based on the payoffs summed over a single year (e.g., Emlen & Wrege 1991; Creel & Waser 1994), but there is evidence that the future consequences of reproductive decisions can be substantial (Reyer 1984; Wiley & Rabenold 1984; Mumme, Koenig, & Ratnieks 1989; Creel 1990b; Emlen & Wrege 1994). Methods used to calculate future inclusive fitness consider the effect of helping on the survivorship and fecundity of the helper (e.g., a reduction in survivorship may represent one cost of helping) and the breeder (e.g., an increase in survivorship may increase future reproductive success and increase indirect fitness for the helper). In some cases, a helper's future inclusive fitness may also be enhanced by offspring that result from its assistance (Creel 1990b; Solomon 1994).

The dispersal decision can influence future inclusive fitness through several mechanisms. One possibility is that dispersal can shorten the queue occupied by an individual, in which case the disperser will breed earlier than will individuals that remain philopatric (Rood 1990; Creel & Waser 1994). In some species, experience can affect reproductive success; experience gained as a helper may increase the reproductive output of that individual when it finally breeds (Brown 1987). The same may be true for a breeder: Breeders may experience low reproductive success in their first year but may have enhanced success in future years as a result of this experience (Creel & Waser 1991). In both cases, these delayed effects can contribute to the fitness payoffs associated with dispersal decisions.

Delayed direct benefits of helping also include the possibility that the helper will ascend into the breeding position of the territory in which it helps (Woolfenden & Fitzpatrick 1984). The benefits are compounded through reciprocity if the offspring that the helper provides care for themselves assist the helper when it ascends (Wiley & Rabenold 1984). There can be several future indirect fitness effects, including an increase in the probability that a breeder will survive subsequent seasons and an increase in future production of offspring that arise as a result of help given by offspring the helper is currently aiding (Mumme et al. 1989; Creel 1990b). We assume that the act of helping increases the immediate reproductive output of breeders, or that it may increase survivorship of breeders. Solomon (1994) has recently shown that helpers also may increase the future reproductive success of young that they care for. This will be particularly important in species in which juvenile growth correlates with adult size and fecundity.

Complete accounting of future fitness effects has resisted treatment in part because of the number and complexity of the possible fitness components (e.g., Emlen & Wrege 1994). But in addition, there is a knotty problem that cannot be addressed by traditional methods of fitness accounting. This is that the fitness consequences of dispersal decisions to be made in the future feed back into current dispersal decisions.

Many of the papers published on helping behavior assume that a decision (e.g., to disperse or stay) is made only once in the animal's lifetime (Creel & Waser 1991; Emlen & Wrege 1991; Walters, Doerr, & Carter 1992), or if the current decision is to help, that the animal disperses the next year (e.g., Zack

& Stutchbury 1992). However, if there are delayed fitness effects or probabilistic changes in future reproductive options (e.g., the death of the breeder in the natal territory caused by some chance event), the consequences of current reproductive decisions will be influenced by decisions potentially made in the future. For example, if a subordinate is likely to ascend to dominance status in the near future, its best option might be to remain in its natal territory in order to capitalize on this possible event. On the other hand, if there is a chance that conditions will change such that the animal would do better to disperse the next year, it might be better to disperse immediately if experience gained in the nonnatal group in the current year increases reproductive success in the subsequent years. In particular, if the disperser found a breeding position, its reproductive success might be low in its first year in that position, but reproductive success in subsequent years might increase because of experience gained in that initial year. Thus, to calculate the payoff to current decisions, we need to estimate what future decisions the animal will make and how they will be impacted by current decisions.

The idea that future decisions should impact current decisions is well founded in the foraging literature (Houston & McNamara 1982; Lucas 1983) and has recently been used in the study of some aspects of life-history theory (Clark 1993; Lucas, Howard, & Palmer 1996). The best tool available to incorporate future fitness effects into the analysis of current reproductive decisions is dynamic programming (Mangel & Clark 1988).

Dynamic programming has not yet been applied to the analysis of cooperative breeding; our primary goal here is to show how it can be used to evaluate the inclusive fitness consequences, including future fitness components, for an animal faced with joining a natal versus a nonnatal queue. The rationale is as follows: Inclusive fitness payoffs to the subordinate will depend on relatedness to the breeders in the group. In the natal territory, stochastic mortality events will cause relatedness to be a random variable. For example, subordinates of the same age in different groups may vary in their relatedness to the dominants in their group as a result of random differences in dominants' mortality. We show later that dynamic programming can be used to address this problem. The model specifically deals with dwarf mongooses, although it can be generalized to other cooperative breeders. The motivation for the model is twofold. First, Creel and Waser (1991, 1994) have modeled helping and dispersal decisions in dwarf mongooses, considering only benefits that accrue over the course of a single breeding season. However, subordinates derive ·future fitness benefits from ascending into the breeding position in their territory, and the probability of ascending is higher in nonnatal territories (Creel &

Waser 1994). Also, breeding success increases with experience. These delayed effects could potentially influence dispersal tactics. We ask here whether the magnitude of delayed effects is sufficient to alter predictions about dispersal derived from Creel and Waser's analysis. Second, Creel and Waser (1994) found that there was individual variation among same-aged individuals in dispersal decisions (i.e., not all individuals of any given age dispersed), yet their fitness calculations considered only age-dependent effects. Dynamic programming allows dispersal decisions to be analyzed with more than one conditionality (e.g., age, group size, relatedness). Here we ask whether differences among subordinates in the genetic relatedness to dominants can account for some of the age-independent individual variation in dispersal observed in mongooses. Dynamic programming offers an evolutionary perspective of this problem by providing a means of calculating the lifetime inclusive fitness consequences of dispersal decisions.

We address one additional issue: How important are indirect fitness effects in the expression of dispersal decisions? This issue can be addressed by partitioning indirect and direct fitness effects derived from the model.

7.2 The Animals

Parameter values for the model come from demographic data recorded between 1974 and 1990 from 12 to 20 dwarf mongoose packs in the vicinity of the Serengeti Wildlife Research Institute, Tanzania (Rood 1990; Creel & Waser 1994). Mongooses were marked individually and censused at the beginning of each breeding season. Additional data came from behavioral observations during most breeding seasons and DNA fingerprinting of the members of 10 central packs from 1987 to 1990 (Keane et al. 1994; Creel & Waser, this volume).

7.3 The Model

The model focuses on dispersal decisions made by subordinate dwarf mongooses on their natal territory. These individuals are assumed to have two available alternatives: to remain in the natal queue or to disperse and enter the breeding queue in another territory. We assume that an individual will choose the alternative that maximizes its lifetime inclusive fitness. The lifetime reproductive success of either alternative includes inclusive fitness accrued as a subordinate in the current reproductive year and future reproductive fitness. Future reproductive fitness includes fitness accrued as a dominant weighted by the probability that dominant status will be attained, as well as fitness accrued as a subordinate weighted by the probability that dominant status will not be attained, each option devalued by its own mortality risk. Because both subordinates and dominants can contribute offspring to the litter (Keane et al. 1994), we estimate direct and indirect fitness for both classes of individuals each year they live. Males and females are treated separately because their survival and fecundity schedules differ, and age-related differences in fecundity and survival are explicitly considered.

We characterize each individual on the basis of three states: age, dominance status, and, if the dominance status is natal subordinate (i.e., subordinate individual in the territory in which it was born), relatedness to the dominants. Dominance status is divided into four categories: natal subordinate, natal dominant, nonnatal subordinate, and nonnatal dominant. Once individuals ascend to dominant status, we assume that they remain dominants on that territory until they die. Dominance is tightly related to age in dwarf mongooses, and the oldest dominant male and female are the parents of most offspring (Creel & Waser 1991).

At any given age, subordinates aid in the rearing of the dominant's offspring by guarding them from predators, carrying them, feeding them, grooming them, and, for some females, by nursing them (Rood 1978; Creel et al. 1991). By the same mechanisms, dominants increase the survival of those few young whose parents are subordinates. The probability that any given subordinate ascends to dominant status in its current group depends on the joint probability that the same-sex dominant dies, that there is an ascension from within the group, and that the subordinate is the oldest subordinate in its group (i.e., is first in the queue) (Creel & Waser 1994).

In our model, the genetic relationship between a subordinate and any one dominant falls into one of three categories: first-order relatives (dominant is the father or mother of the subordinate), second-order relatives (dominant is a group member that ascended to dominant status), and nonrelatives (dominant is an individual not born in the group). The coefficient of relatedness between subordinates and individuals in these three categories are 0.54, 0.26, and 0.0, respectively. These values are approximations derived from pedigree analyses (Creel & Waser 1991; and unpubl. data).

Our model does not include the effect of group size on mongoose dispersal decisions. In dwarf mongooses, mortality rates are higher in smaller groups, and the length of the queue is shorter (Rood 1990). However, the effect of a subordinate on group reproductive success is independent of group size (Creel & Waser 1991). Furthermore, group size is not temporally autocorrelated (Creel & Waser 1994), so if a subordinate is in a large group in any given year, this does not necessarily mean that it will breed in a large group if it ascends to the dominant breeding position. For simplicity and because relatedness between the subordinate and the dominant has historically been the condition most discussed in the cooperative breeding literature, we model the system assuming that subordinate–breeder relatedness is the most relevant state variable determining the expression of dispersal decisions.

7.3.1 The Dynamic Program

We use the standard dynamic programming algorithm (Mangel & Clark 1988) to calculate lifetime inclusive fitness of individuals occupying each of the four categories of breeding status. That is, we start at the maximum possible age (here operationally defined as 10 years – we assume that no animals live past this age; Waser et al. 1995) and work backward in time to age 1. For each breeding status, we calculate current inclusive fitness and expected future fitness. All individuals in the pack, both dominant and subordinate, derive inclusive fitness benefits directly from descendent offspring and indirectly from helping the offspring of others. However, since dominants produce more offspring than do subordinates, the weighting of these different avenues of fitness differs with status.

For dominant mongooses, production of offspring can increase with experience (i.e., with the number of years the individual has bred). For dominants in their natal territory, inclusive fitness (see Tables 7.1, 7.2, and 7.3 for definitions of terms and parameter values) is as follows:

$$I_{DN}[age, experience] = (r_{breed} \times (\beta_{0N} + \beta_{1N} \times experience) \times (1 - P_{subrep}))$$

$$+ (r_{DS} \times \Delta_{help} \times P_{subrep}) - e^{\circ}$$

$$+ (P_{surv,D} \times I_{DN}[age + 1, experience + 1]) \qquad (7.1)$$

(Throughout, square brackets in these equations denote terms that are functions of the bracketed variables.) There are four parts to this relationship. The first is the current direct fitness, which is the number of offspring produced by the pack $(\beta_{0N} + \beta_{1N} \times experience)$, multiplied by the fraction of these that are attributable to the dominant mongoose $(1 - P_{subrep})$, devalued by the relatedness between the dominant and its offspring (r_{breed}). The second is the indirect fitness derived from helping the offspring of subordinates $(\Delta_{help} \times P_{subrep})$, devalued by the relatedness between the dominant and the subordinates' offspring (r_{DS}). The third part is Hamilton's (1964) social factor (e°). The last part accounts for all future fitness effects $(I_{DN}[age + 1, experience + 1])$, discounted by the probability of survival $(P_{surv,D})$.

Table 7.1. *Age-independent parameter values used in the dwarf mongoose dynamic program*

Parameter	Female	Male	Definition
β_{0N}	1.78	2.00	Intercept of function relating offspring production and age of natal dominant
β_{1N}	0.73	0.00	Slope of function relating offspring production and age of natal dominant
β_{0O}	1.55	2.52	Intercept of function relating offspring production and age of nonnatal dominant
β_{1O}	0.28	0.00	Slope of function relating offspring production and age of nonnatal dominant
r_{breed}	0.5	0.50	Relatedness between dominant breeder and its offspring
r_{DS}	0.33	0.33	Genetic relatedness between the dominant and offspring produced by the subordinates
P_{subrep}	0.15	0.24	Proportion of young for which subordinates are actual parents
Δ_{help}	0.21	0.25	Number of additional offspring each helper contributes to group reproduction
$P_{surv,D}$	0.70	0.69	Probability that breeder survives 1 year
RS_{sub}	0.10	0.16	Direct fitness (number of offspring produced) of average subordinate
P_{inpack}	0.77	0.68	Probability that a dead breeder is replaced from within the group

Sources: Creel & Waser 1991, 1994; Waser et al. 1994.

As discussed in the introduction, Hamilton's social factor is the mean effect of nonbreeders on breeders' reproductive output. Following Creel (1990a), e° is subtracted from all individuals in the population. In essence, subtracting e° balances the indirect fitness added to helper fitness therefore eliminating double accounting. For the dwarf mongooses, Hamilton's social factor is calculated as follows:

$$e^{\circ} = (\Delta_{help} \times \rho_{dom} \times P_{subrep}) + \left(\Delta_{help} \times \rho_{sub} \times \left(1 - P_{subrep} + \frac{GS - 2}{GS - 1} \times P_{subrep} \right) \right) \quad (7.2)$$

This represents the mean contribution of help given by dominants ($\Delta_{help} \times \rho_{dom} \times P_{subrep}$), plus the mean contribution of help given by subordinates to dominants ($\Delta_{help} \times \rho_{sub} \times (1 - P_{subrep})$), plus the mean contribution of help given by subordinates to other subordinates in the group ($\Delta_{help} \times \rho_{sub} \times (GS - 2)/(GS - 1) \times P_{subrep}$), where $GS = (1 - \rho_{sub})^{-1}$. Note that our defini-

Table 7.2. *Age-dependent parameter values used in dwarf mongoose dynamic program*[a]

Parameter	Sex	AGE									
		1	2	3	4	5	6	7	8	9	10
r_{SO}	Female	0.46	0.26	0.14	0.05	0	0	0	0	0	0
	Male	0.22	0.15	0.10	0.08	0.05	0.03	0.01	0	0	0
$P_{surv,SO}$	Male[b]	0.60	0.77	0.77	0.63	0.63	0.63	0.63	0.63	0.63	0.63
	Male[b]	0.73	0.90	0.90	0.67	0.67	0.67	0.67	0.67	0.67	0.67
	Female[b]	0.74	0.74	0.74	0.74	0.74	0.74	0.74	0.74	0.74	0.74
	Female[b]	0.80	0.80	0.80	0.80	0.80	0.80	0.80	0.80	0.80	0.80
$P_{surv,SN}$	Male[b]	0.62	0.79	0.79	0.63	0.63	0.63	0.63	0.63	0.63	0.63
	Male[b]	0.73	0.90	0.90	0.67	0.67	0.67	0.67	0.67	0.67	0.67
	Female[b]	0.74	0.74	0.74	0.74	0.74	0.74	0.74	0.74	0.74	0.74
	Female[b]	0.77	0.77	0.77	0.77	0.77	0.77	0.77	0.77	0.77	0.77
$P_{surv,disp}$	Male[b]	0.56	0.67	0.67	0.55	0.55	0.55	0.55	0.55	0.55	0.55
	Male[b]	0.40	0.48	0.48	0.40	0.40	0.40	0.40	0.40	0.40	0.40
	Female[b]	0.50	0.50	0.50	0.50	0.50	0.50	0.50	0.50	0.50	0.50
	Female[b]	0.28	0.28	0.28	0.28	0.28	0.28	0.28	0.28	0.28	0.28
P_{breed}	Male	0	0.28	0.28	0.43	0.43	0.43	0.43	0.43	0.43	0.43
	Female	0.23	0.46	0.46	0.46	0.46	0.46	0.46	0.46	0.46	0.46
$P_{maxage,O}$	Male	0.14	0.28	0.51	0.77	1	1	1	1	1	1
	Female	0.50	0.50	0.62	1	1	1	1	1	1	1
$P_{maxage,N}$	Male	0.09	0.14	0.38	0.83	1	1	1	1	1	1
	Female	0.09	0.12	0.24	0.38	0.60	0.67	0.80			

Sources: Creel & Waser 1991, 1994; Waser et al. 1994.

[a] Discontinuities are caused by the combining of age classes where sample sizes were low (see Creel & Waser 1994).

[b] Two sets of parameter values represent the upper and lower bounds of our estimates of these mortality risks.

Table 7.3. *Definitions of terms not defined in Table 7.1*

Term	Definition
age	Chronological age of the dominant breeder (measured in years).
experience	Number of years the dominant breeder has been reproducing.
$e°$	Hamilton's social factor, which is the mean effect of an individual on breeder reproductive output.
GS	Number of same-sex individuals in a group.
$P_{surv,SO}[age]$	Age-dependent survivorship of the nonnatal subordinate.
$P_{surv,SN}[age]$	Age-dependent survivorship of the natal subordinate.
$P_{maxage}[age]$	Age-dependent probability that the helper is the oldest in the group and therefore takes over the breeding position.
$P_{surv,disp}[age]$	Probability that the dispersing mongoose survives until it finds a territory (this is assumed to take about 1 month).
$P_{dom}[age]$	Probability that the individual will enter the new group as a dominant.
$P_{ascend,O}[age]$	Probability that the nonnatal subordinate will ascend to the dominant breeding position in the next year.
$\pi_{PS}[parent\ state]$	Transition probabilities from the current breeder status (*parent state*) to the breeder status next year (*PS*).
$r_{SO}[age]$	Relatedness between a nonnatal helper and the offspring it helps to rear.
ρ_{sub}	Percentage of subordinates in the population (estimated from forward iteration).
ρ_{dom}	Percentage of dominants in the population (estimated from forward iteration).

tion of group size, $GS = (1 - \rho_{sub})^{-1}$, can be thought of as the number of individuals per each dominant individual in the group. The contribution of subordinates to other subordinates' reproduction assumes that the number of subordinates of a given sex is $GS - 1$, and that the number of subordinates a given subordinate will help is $(GS - 1) - 1$. We estimate ρ_{sub} and ρ_{dom} by using forward iteration, which is discussed later.

The inclusive fitness of dominants in nonnatal territories is similar to that of dominants in their natal territories (only offspring production, $\beta_{00} + \beta_{10}$, and future fitness, $I_{DO}[age + 1, experience + 1]$, are different):

$$I_{DO}[age, experience] = (r_{breed} \times (\beta_{00} + \beta_{10} \times experience) \times (1 - P_{subrep}))$$

$$+ (r_{DS} \times \Delta_{help} \times P_{subrep}) - e°$$

$$+ (P_{surv,D} \times I_{DO}[age + 1, experience + 1]) \tag{7.3}$$

We assume that in any given year nonnatal subordinates remain in the subordinate position that year. They then either ascend to breed the next year if the dominant dies and if they are first in the queue, or they continue as subordinates for at least one more year:

$$I_{SO}[age] = (RS_{sub} \times r_{breed})$$

$$+ (r_{SO}[age] \times \Delta_{help} \times (1 - P_{subrep}) + r_{SO}[age] \times 0.26)$$

$$\times \Delta_{help} \times \frac{GS - 2}{GS - 1} \times P_{subrep})$$

$$- e^{\circ} + (P_{surv,SO}[age] \times ((P_{ascend,O}[age]$$

$$\times I_{DO}[age + 1, experience = 0])$$

$$+ ((1 - P_{ascend,O}[age]) \times I_{SO}[age + 1]))) \qquad (7.4)$$

Equation (7.4) sums the current direct ($RS_{sub} \times r_{breed}$) and indirect fitness accrued by the nonnatal subordinate, devalued by Hamilton's social factor (e°). There are two components to the indirect fitness: that derived from caring for the dominant's young ($r_{SO}[age] \times \Delta_{help} \times (1 - P_{subrep})$) and that derived from caring for the young of other subordinates (($r_{SO}[age] \times 0.26) \times \Delta_{help} \times ((GS - 2)/(GS - 1)) \times P_{subrep}$). We assume that the mean relatedness of a subordinate to another breeding subordinate is its relatedness to the dominant multiplied by an approximation (0.26) of the relatedness of dominants to breeding subordinates, $r_{SO}[age] \times 0.26$. The increase in litter size, Δ_{help}, is attributed to the presence of a subordinate. Some fraction of this increase (P_{subrep}) is caused by subordinate reproduction, and a fraction of this fraction (($GS - 2)/(GS - 1)$) is attributable to subordinates other than the focal individual being modeled. All future effects are weighted by the probability that the subordinate will survive at least one more year ($P_{surv,SO}[age]$). The future fitness of a nonnatal subordinate is the fitness of a dominant ($I_{DO}[age + 1, experience = 0]$) if it ascends (with probability $P_{ascend,O}[age]$), or the fitness of a subordinate ($I_{SO}[age + 1]$) if it does not ascend (with probability ($1 - P_{ascend,O}[age]$)). The probability that a subordinate will ascend in a nonnatal group is as follows:

$$P_{ascend,O}[age] = (1 - P_{surv,D}) \times P_{inpack} \times P_{maxage,N}[age] \qquad (7.5)$$

Thus the probability of ascending is the probability that the dominant will die ($1 - P_{surv,D}$) times the probability that the dominant will be replaced from within the pack (P_{inpack}) times the probability that the subordinate will be first in the queue ($P_{maxage,N}[age]$).

Subordinates on their natal territories have two alternatives: They can remain there at least one more year or disperse and attempt to join another group. The inclusive fitness of an individual choosing the dispersal option is as follows:

$$I_{disp}[age] = P_{surv,disp}[age]$$

$$\times (P_{dom}[age] \times ((r_{breed} \times \beta_{00} \times (1 - P_{subrep})) + (r_{DS} \times \Delta_{help} \times P_{subrep})$$

$$+ (P_{surv,D}[age]^{11/12} \times I_{DO}[age + 1, experience = 1])$$

$$+ (1 - P_{dom}[age]) \times ((r_{SO}[age] \times \Delta_{help} \times (1 - P_{subrep})$$

$$+ (r_{SO}[age] \times 0.26) \times \Delta_{help} \times \frac{GS - 2}{Gs - 1} \times P_{subrep}) + (RS_{sub} \times r_{breed})$$

$$+ P_{surv,SO}[age]^{11/12} \times ((P_{ascend,o}[age] \times I_{DO}[age + 1, experience = 0])$$

$$+ ((1 - P_{ascend,o}[age]) \times I_{SO}[age + 1])))) - e^{\circ} \qquad (7.6)$$

Here we assume that the disperser suffers some mortality risk ($P_{surv,disp}[age]$ is the probability that it survives) during the dispersal event that we assume to take about 1 month. If it survives, it enters a nonnatal pack. If it enters as a dominant (with probability $P_{dom}[age]$), its expected current fitness is ($r_{breed} \times \beta_{00} \times (1 - P_{subrep}) + (r_{DS} \times \Delta_{help} \times P_{subrep})$. If it survives the following 11 months (with probability $P_{surv,D}[age]^{11/12}$), then its future fitness is $I_{DO}[age + 1, experience = 1]$. Finally, if the mongoose enters as a subordinate (with probability $(1 - P_{dom}[age])$), then its current fitness will be $(r_{SO}[age] \times \Delta_{help} \times (1 - P_{subrep}) + (r_{SO}[age] \times 0.26) \times \Delta_{help} \times ((GS - 2)/(GS - 1)) \times P_{subrep}) + (RS_{sub} \times r_{breed})$. If it survives the following 11 months (with probability $P_{surv,SO}[age]^{11/12}$), it will ascend the next year with probability $P_{ascend,o}[age]$, in which case its future fitness is $I_{DO}[age + 1, experience = 0]$, or it will fail to ascend the next year with probability $(1 - P_{ascend,o}[age])$, in which case its future fitness is $I_{SO}[age + 1]$.

The inclusive fitness of the individual that remains as a subordinate in its natal pack is dependent, in part, on its relatedness to the current dominants and on the expected change in relatedness in the future. We define *parent state* as the current relationship between a single dominant (i.e., either the male or female) and a subordinate. There are three different levels of *parent state*, first-order, second-order, and nonrelative, and the level will change in any year that at least one of the dominants dies. The following transition matrix was used to estimate the transition probabilities, $\pi_{PS}[parent state]$, from the current degree of relatedness (*parent state*) to the relatedness for the next year (*PS*):

Parent State	PS = First Order	PS = Second Order	PS = Nonrelative
First order	$P_{surv,D}$	$(1 - P_{surv,D})$ $\times P_{inpack}$	$(1 - P_{surv,D})$ $\times (1 - P_{inpack})$
Second order	0	$(1 - P_{surv,D})$ $\times P_{inpack} + P_{surv,D}$	$(1 - P_{surv,D})$ $\times (1 - P_{inpack})$
Nonrelative	0	$(1 - P_{surv,D})$ $\times P_{inpack}$	$(1 - P_{surv,D})$ $\times (1 - P_{inpack}) + P_{surv,D}$

These transition probabilities are used to calculate the inclusive fitness of the mongoose that remains as a subordinate on its natal territory, as follows:

$I_{SN}[age, parent\ state] =$

$(r_{SN}[parent\ state] \times \Delta_{help} \times (1 - P_{subrep})$

$$+ (0.26 \times \Delta_{help} \times \frac{GS - 2}{GS - 1} \times P_{subrep})$$

$$+ (RS_{sub} \times r_{breed}) - e^\circ$$

$$+ P_{surv,SN}[age] \times (P_{ascend,N}[age] \times I_{DN}[age + 1, experience = 0]$$

$$+ (1 - P_{ascend,N}[age]) \times \sum_{PS=1}^{9} (\pi_{PS}[parent\ state] \times I_{SN}[age + 1, PS])) \quad (7.7)$$

Here current indirect $(r_{SN}[parent\ state] \times \Delta_{help} \times (1 - P_{subrep}) + (0.26 \times \Delta_{help} \times ((GS - 2)/(GS - 1)) \times P_{subrep}))$ and direct $(RS_{sub} \times r_{breed})$ inclusive fitness are devalued by Hamilton's social factor (e°). Future fitness is accrued as a dominant $(I_{DN}[age + 1, experience = 0])$ if the subordinate survives (with probability $P_{surv,SN}[age]$) and ascends (with probability $P_{ascend,N}[age]$), and future fitness as a subordinate $(\sum_{PS=1}^{9} (\pi_{PS}[parent\ state] \times I_{SN}[age + 1, PS]))$ in the next year if it survives but fails to ascend then (with probability $(1 - P_{ascend,N}[age])$).

If $I_{Sn}[age, parent\ state] \geq I_{disp}[age]$ (i.e., if the fitness from philopatry exceeds the fitness from dispersal), we assume that the individual will remain on its natal territory, and that otherwise it will disperse.

7.4 Forward Iteration

The dynamic program will generate the optimal age-specific dispersal decisions for natal subordinates. We can use these age-specific dispersal decisions in a population simulation (i.e., a forward iteration). The forward iteration can be used to estimate several demographic characteristics of the population, including the expected age structure of the population, the expected proportion of the population that are subordinates, and the mean genetic relationship between dominants and subordinates. The forward iteration starts with a population of juveniles that are resident on their natal territories. Given the survivorship values for parents and offspring, the state-specific dispersal decisions and the transition equations from the dynamic program (see previous section), we can then trace the expected fate of these juveniles.

Forward iteration, as we have used it here, is a standard approach to solving a second type of problem: Predicted decisions may depend on certain variables that in turn are affected by decisions made by the animals (e.g., Green 1980; Houston & McNamara 1987). For example, in game theory models the payoff on any given decision is affected by the frequencies with which alternative decisions are expressed by individuals of the population, but these frequencies are in turn affected by the payoff matrix (e.g., Maynard Smith 1982).

In our model the proportion of individuals in the population that are subordinates will covary with the state-dependent decisions made by those subordinates. This is because Hamilton's social factor (e°) is a function of the proportion of subordinates in the population (see Eq. 7.2) and will affect the fitness consequences of the dispersal decision. But the proportion of subordinates in the population will vary with the dispersal decisions made by the mongooses. Thus e° will affect the dispersal decisions, but its value is dependent in part on the dispersal decisions.

There are two alternatives for incorporating this feedback into the model. One is essentially to ignore it by fixing the proportion of subordinates at the level observed in our mongoose populations and estimating the dispersal decisions based on this fixed number. The other is to use a forward iteration process to estimate both the decision matrix and the proportion of subordinates. We have taken the latter approach: We start with an estimate of the proportion of subordinates, then use dynamic programming to determine the optimal decision matrix, and then use forward iteration to estimate the expected proportion of subordinates that would result from this decision matrix. We then rerun the dynamic program with this new estimate of the proportion of subordinates. This process is repeated until the proportion of subordinates generated from the forward iteration is the same as the proportion of subordinates used in the dynamic program.

This iteration process solves for the equilibrium set of state- and age-dependent behaviors, which will in turn determine the equilibrium proportion of subordinates in the population (i.e., group size). Again, the calculations are necessary because behavior affects group size and group size affects behav-

Table 7.4. *Equilibrium proportion of helpers predicted by the dynamic program*

Model	Sex	$P_{surv,disp}[age]^a$		Proportion helpers
Current fitness only	Female	low		0.869
		high		0.748
	Male	low		0.836
		high		0.807
Lifetime fitness	Female	low		0.872
		high		0.765
	Male	low		0.836
		high		0.807
Breeder reprod. only	Female	low		0.872
		high		0.765
	Male	low		0.849
		high		0.807

[a]See Table 7.1 for low and high parameter values.

ior. If group size were fixed, the model would not be internally consistent, because the predicted behavior matrix for a given group size may generate a population characterized by a different group size. Table 7.4 shows that the equilibrium proportion of helpers determined by forward iteration using a wide range of parameter values ranges from 0.748 to 0.872, in close agreement with the true proportion of helpers in mongoose groups.

Note that although the iteration procedure is similar to the algorithm typically used to solve dynamic games (see Houston & McNamara 1987), this is not a dynamic game because none of the payoffs are assumed to be frequency dependent. Also, with some dynamic games the iteration process never reaches a steady state (e.g., Houston & McNamara 1987; Lucas et al. 1996); however the results reported here always reached a steady state within three iterations.

7.5 Decisions Based on Immediate Payoffs

To determine the importance of future fitness effects for reproductive decisions by subordinate mongooses, we ran our dynamic program both by using equations (7.1) to (7.7) and by using them minus all the future fitness terms. The results of the latter procedure should resemble those obtained by alternative models based solely on current fitness. Creel and Waser (1991, 1994) developed two different sets of equations estimating age-specific current fitness for philopatric and dispersing mongooses. In Creel and Waser (1991), the subordinate mongoose was assumed to remain subordinate on its natal territory if it chose to stay. Alternatively, it could enter a new group as either a subordinate or as a dominant if it chose to disperse, and if it survived the dispersal event. This is an oversimplification because approximately 30 percent of dominants turn over every year, and some of this turnover occurs during the breeding season. Thus, there is some (relatively low) probability that a subordinate will ascend during the current breeding season.

Creel and Waser (1994) allowed for the additional possibility that the natal subordinate may ascend on its territory during the current year by assuming that all turnover occurs before the breeding season starts. Unfortunately, this latter formulation overestimates the probability that a subordinate will ascend into dominant breeder status. Implicitly, it assumes that all dominants that die do so before breeding begins, so that all the direct fitness they would have acquired that year goes to the ascending subordinate.

The better of the two estimates of inclusive fitness comes from Creel and Waser (1991), and we use this estimate. In essence, we assume that the year begins at the beginning of the breeding season and that dominants that die do so after the year's breeding is complete.

7.6 Lifetime Inclusive Fitness, Fitness Gradients, and Parameter Estimates

The results of our model were calculated as the net fitness consequences associated with choosing to stay (i.e., inclusive fitness of a philopatric subordinate minus the inclusive fitness of a dispersing subordinate), taken as a fraction of the expected lifetime inclusive fitness (essentially its reproductive value v_x; Fisher 1958). This ratio is the fitness gradient; it gives the net consequences to lifetime fitness for each of the options available to the natal subordinate (see McNamara & Houston 1986 for a somewhat different approach). The consequences of all decisions are given as a function of two variables: the subordinate's age and its genetic relatedness to the current dominants.

We have direct estimates from field observations of all parameter values except mortality rates (see Tables 7.1, 7.2). We estimated reasonable bounds on mortality rates that were based on known disappearance rates, immigration rates, and mortality rates of individuals unlikely to disperse (i.e., dominants). The technique is described in detail in Waser, Creel, and Lucas (1994). In short, from annual census data, we know how many individuals of any given age and sex survive in a group and how many immigrate into a group. From these data, we can estimate an average mortality risk in the population, but we do not know the circumstances of these deaths. More specifically, some unknown fraction occur during dispersal, some unknown fraction occur in the natal group, and some unknown fraction occur after a subordinate immigrates into a nonnatal group. We can place absolute bounds on the three sources of mortality because they must sum to the overall mortality rate in the population. We have narrowed these bounds by assuming that mortality risks are the same in the natal and nonnatal groups and by assuming that the mortality risk experienced during dispersal is no less than the risk experienced in the group. One result of our method of estimating mortality risk is that there is a negative correlation between dispersal mortality risk and the mortality risk experienced by philopatric individuals. As a result, maximum estimates of dispersal risk are used with minimum estimates of mortality risk on the territory.

7.7 Results

Our results strongly suggest that future fitness effects cannot be ignored in modeling the evolution of cooperative breeding. Moreover, they indicate that differences among individuals in relatedness to the dominants can result in individual differences in dispersal tactics, even after accounting for the effects of age (which covaries strongly with relatedness). Finally, they suggest that predicted dispersal tendencies are strongly influenced by assumptions about mortality risk. This last finding points up a critical need for better data from field studies, because dispersal mortality has been notoriously difficult to measure with accuracy.

7.8 Future Effects

Creel and Waser (1994) showed that dispersal in female mongooses occurs primarily at 2 years of age; females that fail to emigrate at this age tend to remain on their natal territories. In contrast, males disperse at all ages, although most leave before they are 4 years old. The version of our model in which dispersal decisions are based solely on consequences in the current year correctly predicts a more extensive dispersal in males than in females but does not correctly predict the age distribution of dispersers, especially in females. The model generally predicts that dispersal decisions should be approximately age independent for all individuals older than 1 year of age (Figure 7.1); the prediction holds for females irrespective of dispersal mortality, and it holds for males if dispersal mortality is low. If the dispersal mortality is high for males, then all dispersal is predicted to occur before individuals are 4 years old. Neither of these patterns was observed. The discrepancy between the model results and the observed dispersal patterns suggests that

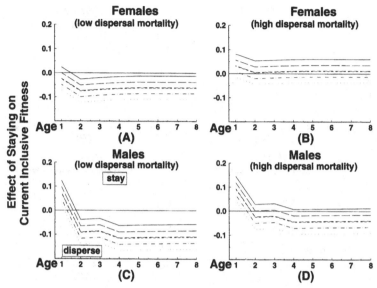

Figure 7.1. Net effect of nondispersal on current inclusive fitness as a function of age for subordinate mongooses on their natal territory. The net effect of nondispersal is the inclusive fitness accrued in the current year by staying minus the inclusive fitness accrued in the current year by dispersing females (A, B) and males (C, D). Figures 7.1A and B assume low dispersal mortality and high mortality on the territory; B and D assume high dispersal mortality and low mortality on the territory. Different lines represent different levels of relatedness to dominant breeders: solid line: $r = 0.54$, long dash: $r = 0.40$, medium dash: $r = 0.27$, short dash: $r = 0.26$, dot dash: $r = 0.13$, dots: $r = 0.00$. Parameter values are given in Tables 7.1 and 7.2. For values >0, the mongoose should remain on the territory; for values <0, the mongoose should disperse.

dispersal decisions are governed by more than just immediate fitness consequences.

The predictions differ in several respects when future reproductive success is added to the model (Figure 7.2). First, the mongooses should be more philopatric if future effects are considered. These effects are primarily a result of the low contribution of current reproductive success to lifetime fitness. Current inclusive fitness payoffs are only about 10 percent of the total lifetime fitness of young individuals (Figure 7.3). Because future fitness effects represent a majority of the lifetime fitness of subordinates, mongooses should be more sensitive to mortality risk if they base dispersal decisions on total lifetime reproductive success as opposed to basing their decision on just the payoff accrued during the current year. A longer time horizon to the dispersal

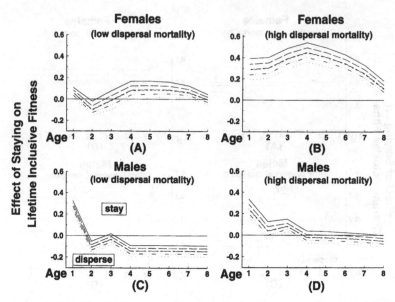

Figure 7.2. Net effect of staying on lifetime inclusive fitness, as a function of age, for subordinate mongooses on their natal territory. These results consider current and future and both indirect and direct fitness effects. See Figure 7.1 for details.

decision should favor philopatry, because the mortality risk during dispersal is relatively high (see Tables 7.1 and 7.2).

Second, female dispersal decisions are potentially more strongly age dependent than those of males (see Figure 7.2). Assuming that mortality rates during dispersal are near our lower estimated rates, the patterns predicted by the model incorporating lifetime consequences of dispersal decisions are consistent with the observed dispersal behavior: Some females leave at age 2, but those that stay should remain in the natal territory for their lifetime. The high cost of leaving for older females is not seen in males (see Figure 7.2).

However, a major caveat to these conclusions is that the results are strongly affected by the assumptions made concerning mortality rates. If the dispersal mortality rates are high, then no dispersal is predicted, except by older males. In addition, if the dispersal mortality rates are high, the fitness consequences of dispersal are quite severe for females; they would lose more than 50 percent of their age-specific lifetime inclusive fitness (i.e., current plus future fitness) if they dispersed at any age. Thus, if the dispersal mortality rates were high, there would be very strong selection against dispersal. Given that about 20 percent of 2-year-old females disperse (Creel & Waser 1994), these results

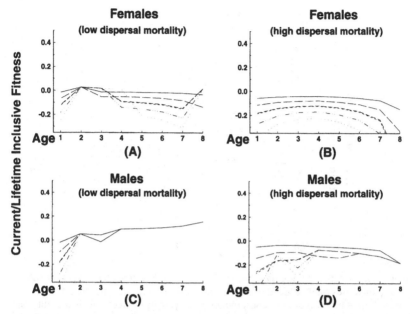

Figure 7.3. Ratio of current to lifetime inclusive fitness for subordinate mongooses on their natal territory, as a function of age. Results assume that dispersal decisions are based on lifetime inclusive fitness, considering both indirect and direct effects: (A,B) females, (C,D) males. Figures 7.3A and B assume low dispersal mortality and high mortality on the territory; B and D assume high dispersal mortality and low mortality on the territory.

indicate that the true dispersal risk faced by female mongooses is closer to our lower estimate. At this point, nothing more can be said about this conclusion, except that it is a testable hypothesis.

7.9 Indirect Effects

We can evaluate the relative impact of indirect fitness effects on dispersal decisions by assuming that the dispersal decisions are based solely on dispersal's current and future consequences for mortality and production of offspring. If indirect fitness effects are important determinants of dispersal decisions in dwarf mongooses, then the predicted decisions should change when indirect effects are dropped from the model.

When we compare model predictions with and without indirect fitness effects, the results suggest a reason why controversy over this topic has been so persistent. On the one hand, the general pattern of age-dependence in the

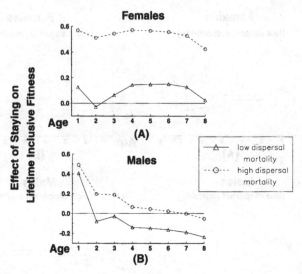

Figure 7.4. Effect of staying on lifetime direct fitness, as a function of age, for subordinate mongooses on their natal territory. These results measure only the effects of the dispersal decision on mortality and production of young; indirect effects are not considered.

fitness gradient is similar with or without indirect fitness included (compare Figure 7.2 and 7.4). On the other hand, Figure 7.2 demonstrates that indirect fitness effects can tip the balance in favor of staying in the natal group. Higher relatedness between subordinates and dominants substantially shifts the fitness gradient toward philopatry. Indirect effects have a particularly marked effect on the dispersal decision if dispersal mortality is low (see Fig. 7.2).

The results of the model reflect the impact of a number of age-related tradeoffs that potentially affect mongoose dispersal decisions. The genetic relatedness between subordinate and dominant declines with subordinate age. The decrease is greater for males than for females and occurs on both natal and nonnatal territories (Creel & Waser 1994). The decline in relatedness on the natal territory is caused by mortality of the subordinate's parents and by the potential ascension of less closely related dominants. The high relatedness in young nonnatal subordinates is primarily a result of group dispersal of related individuals into a territory and to the ascension of co-dispersing kin into the breeding position (Creel & Waser 1994). As these kin die, relatedness declines (see Table 7.2). One result of the decrease in mean relatedness between helper and breeder is a decrease with age in the expected contribution of indirect fitness to total lifetime inclusive fitness. This also means that indirect fitness consequences will have greater effects on decisions of young animals.

An important factor that causes inclusive fitness to be low for subordinates is the relatively high value of $e°$, Hamilton's social factor. This is subtracted from the inclusive fitness of all individuals in the population. Subordinates accrue some direct fitness benefits, but these are not very large (fewer than 0.2 offspring/subordinate/year; see Table 7.1). The number of offspring equivalents added by each subordinate as a result of its aid of pack mates is somewhat larger (see Table 7.1). However, both types of fitness are generally less than Hamilton's social factor for nondispersing subordinates, so the inclusive fitness of this class of individual tends to be negative (see Figure 7.3). The primary reason is the high proportion of subordinates in the population. Hamilton's social factor is the mean amount of assistance each individual in the population gives to its neighbors. To simplify the discussion, we consider a system in which there is a perfect dichotomy between helpers and breeders. In such a system, $e°$ is equivalent to the effect on breeder reproductive success of the helper multiplied by the proportion of helpers in the population. Indirect fitness is the effect on breeder reproductive success of a helper, devalued by its relatedness to the breeder. Thus, if the proportion of helpers is greater than the relatedness between helper and breeder, helpers will realize a negative inclusive fitness. Because both subordinates and dominants help, the calculations are somewhat more complicated for mongooses, but the general problem still holds.

Given the negative inclusive fitness values, the decision made by older female natal subordinates to stay in their natal territories (see Figure 7.2) is more a reflection of expected future breeding opportunities than of the indirect fitness derived from helping the dominants rear offspring. In this respect, a number of factors favor philopatry in females compared to males: (1) breeder survivorship is lower in females and therefore the dominant position is more likely to open up (see Table 7.1); (2) given that a dominant position has become vacant, females from within the pack (as opposed to immigrants) are more likely than are resident subordinate males to ascend into the dominant position (see Table 7.1); and (3) the mortality risk of dispersal appears to be higher for females than for males (see Table 7.2).

7.10 Discussion

The relative importance of indirect inclusive fitness on the evolution of helping behavior is controversial. Some authors feel that indirect effects are likely to be relatively unimportant (Zahavi 1974; Ligon 1981; Walters et al. 1992), whereas others argue that indirect effects cannot be overlooked (Mumme et al. 1989; Emlen & Wrege 1991; Creel & Waser 1994). Our results from the dwarf mongooses add an additional complication to this controversy. Indirect effects may account for a small fraction of the total lifetime inclusive fitness on average, yet they potentially have a major impact on dispersal decisions in cases where the fitness consequences of dispersal are similar to the fitness consequences of philopatry. We have shown that this condition is met when the mortality risk associated with dispersal is low. Thus, for mongooses of some age–sex classes, indirect fitness can be critical to dispersal decisions.

This raises a second complication: The predicted behavior is sensitive to assumptions about mortality risks. Unfortunately, mortality risks are exceedingly difficult to measure directly (e.g., Lima & Dill 1990). We show elsewhere that certain types of demographic data can be used to estimate bounds on mortality risks (Waser et al. 1994), and we have employed this method here. Our bounds are based on >1,600 mongoose years of demographic data, yet they are still too wide to allow unequivocal predictions from the model! Perhaps the strongest comment we can make is that the lack of good data on dispersal mortality may be a major cause of the persistence of controversy over the "causes" of cooperative breeding. Perhaps part of the effort currently devoted to obtaining more precise estimates of parentage would be better spent using telemetry to get better estimates of dispersal mortality.

An additional point is worth mentioning about the magnitude of indirect fitness effects. In species where virtually all individuals are either helpers or breeders, and in particular where the proportion of helpers in the population is large, the inclusive fitness to helpers may be negative for all helpers in the population. This follows from the formulation of Hamilton's social factor derived by Creel (1990a), who showed that Hamilton's social factor was the mean amount of help given by the average individual in a population (or alternatively, the mean amount of help received). If the fraction of helpers in a population is large (as is the case in dwarf mongooses) and each contributes a substantial amount of help in rearing young, then Hamilton's social factor could be a relatively large number. Creel (1990a) also showed that this number should be subtracted from each individual in the population. Thus, when the indirect fitness accrued by helpers is less than Hamilton's social factor, then the current inclusive fitness derived by helpers will be negative.

Creel (1990a; also see Creel & Waser 1994) suggested that the magnitude of $e°$ should not affect behavioral decisions because it is a constant subtracted equally from all individuals in the population. The implication is that relative payoff of different dispersal tactics can be correctly evaluated even if we miscalculate $e°$. This is important because $e°$ has rarely been calculated correctly.

However, when future effects are considered, the magnitude of $e°$ can alter the predictions about dispersal behavior if the mortality risks associated with the alternative behaviors differ. To show this, we consider a simple example. Assume that dispersing subordinates always die in the current year and that nondispersing subordinates survive the current year but always die the following year. The net impact of $e°$ on dispersers will be $1 \times e°$, and on nondispersers it will be $2 \times e°$. In this case, the magnitude of $e°$ could be quite important in the net benefit of dispersal. Given that mortality risks of behavioral alternatives will generally differ, an accurate evaluation of $e°$ seems warranted.

Dwarf mongooses exhibit a great deal of individual variation in their dispersal behavior. For example, some individuals can remain as natal subordinates after 9 years of age, whereas others leave at age 1 (Rood 1987). A full understanding of the evolution of dispersal requires that we refine our analyses by identifying variables that account for a significant amount of variation among individuals. Much of the individual variation in the mongooses can be accounted for by variables that change with age. These include mortality risks and the probability of ascending into the dominant breeding position. These age-related effects were the basis of our previous modeling efforts (Creel & Waser 1991, 1994). However, there are additional variables that can potentially account for some of this variation that is not age-related per se. We have shown here that variation in the relatedness between subordinate and dominants can account for some of the variation in dispersal tendencies in same-aged individuals; helpers closely related to the current dominants should be more likely to stay than subordinates that are only distantly related.

7.11 Conclusions

Models of the inclusive fitness of helping have been developed in an incremental fashion. Early models (e.g., Grafen 1984) considered only the immediate effects of helping on inclusive fitness. Future effects were added by a number of authors, including Brown (1987), Mumme et al. (1989), and Creel (1990b). Hamilton's $e°$, the social factor that should be stripped from breeders' and helpers' fitness, has only recently been rigorously defined (Creel 1990a). Our models contribute to this iterative process by showing how dispersal can be modeled as a series of dynamic decisions (Houston, Clark, & Mangel 1988; Mangel & Clark 1988), in which both state (here relatedness) and the effect of future decisions are considered in the evaluation of a behavioral trait. Our model by necessity leaves out some additional states. Perhaps the most important is group size. Our original models (Creel & Waser 1991,

1994) ignored group size as a factor affecting dispersal decisions, and we have done so here as well. Several factors that influence the inclusive fitness of subordinates are independent of group size. These include the contribution each subordinate makes to group reproductive output (Creel & Waser 1991) and the temporal autocorrelation in group size (Creel & Waser 1994). However, survival rates can increase with group size, at least for breeders (Rood 1990). Two other factors that we have ignored are individual differences in size or nutrition and in the opportunity for dispersal. These additional factors may account for some variation between individuals in dispersal decisions, in addition to the differences in relatedness considered in our model.

Acknowledgments

Computer time was provided by Purdue University. Recent field work and demographic analysis by S. R. Creel and P. M. Waser was supported by the National Geographic Society, the Frankfurt Zoological Society, and the National Science Foundation. Modeling by J. R. Lucas was supported by NSF grant IBN–9222313.

References

Alexander, R. D., Noonan, K. M., & Crespi, B. J. (1991). The evolution of eusociality. In *The biology of the naked mole-rat,* ed. P. W. Sherman, J. U. M. Jarvis, & R. D. Alexander, pp. 3–44. Princeton: Princeton University Press.
Brooker, M. G., Rowley, I., Adams, M., & Baverstock, P. R. (1990). Promiscuity: An inbreeding avoidance mechanism in a socially monogamous species? *Behav. Ecol. Sociobiol.* 26:191–200.
Brown, J. L. (1987). *Helping and communal breeding in birds.* Princeton: Princeton University Press.
Clark, C. W. (1993). Dynamic models of behavior: an extension of life history theory. *Trends Ecol. Evol.* 8:205–209.
Creel, S. (1990a). How to measure inclusive fitness. *Proc. Royal Soc. Lond.* B241:229–231.
Creel, S. (1990b). The future components of inclusive fitness: Accounting for interactions between members of overlapping generations. *Anim. Behav.* 40:127–134.
Creel, S. R., & Macdonald, D. W. (1995). Sociality, group size, and reproductive suppression in carnivores. *Adv. Study Behav.* 24:203–257.
Creel, S. R. & Waser, P. M. (1991). Failure of reproductive suppression in dwarf mongooses (*Helogale parvula*): Accident or adaptation? *Behav. Ecol.* 2:7–15.
Creel, S. R., & Waser, P. M. (1994). Inclusive fitness and reproductive strategies in dwarf mongooses. *Behav. Ecol.* 5:339–348.
Creel, S. R., Monfort, S. C., Wildt, D. E., & Waser, P. M. (1991). Spontaneous lactation is an adaptive result of pseudopregnancy. *Nature, Lond.* 351:660–662.
Emlen, S. T. (1991). Evolution of cooperative breeding in birds and mammals. In *Behavioural ecology: An evolutionary approach,* 3rd ed., ed. J. R. Krebs & N. B. Davies, pp. 301–335. Oxford, U.K.: Blackwell.

Emlen, S. T., & Wrege, P. H. (1991). Breeding biology of white-fronted bee-eaters at Nakuru: The influence of helpers on breeder fitness. *J. Anim. Ecol.* 60:309–326.

Emlen, S. T., & Wrege, P. H. (1994). Gender, status and family fortunes in the white-fronted bee-eater. *Nature, Lond.* 367:129–132.

Faaborg, J., Parker, P. G., Delay, L., deVries, T., Bednarz, J. C., Paz, S. M., Naranjo, J., & Waite, T. A. (1995). Confirmation of cooperative polyandry in the Galapagos Hawk (*Buteo galapagoensis*). *Behav. Ecol. Sociobiol.* 36:83–90.

Fisher, R. A. (1958). *The genetical theory of natural selection.* New York: Dover.

Grafen, A. (1982). How not to measure inclusive fitness. *Nature, Lond.* 298:425.

Grafen, A. (1984). Natural selection, kin selection and group selection. In *Behavioural ecology: An evolutionary approach,* 2nd ed., ed. J. R. Krebs & N. B. Davies, pp. 62–84. Oxford, U.K.: Blackwell.

Green, R. F. (1980). Bayesian birds: A simple example of Oaten's stochastic model of optimal foraging. *Theo. Pop. Biol.* 18:244–256.

Hamilton, W. D. (1964). The genetical evolution of social behaviour. *J. Theo. Biol.* 7:1–16.

Houston, A. I., & McNamara, J. M. (1982). A sequential approach to risk-taking. *Anim. Behav.* 30:1260–1261.

Houston, A. I., & McNamara, J. M. (1987). Singing to attract a mate: A stochastic dynamic game. *J. Theo. Biol.* 129:57–68.

Houston, A., Clark, C., McNamara, J., & Mangel, M. (1988). Dynamic models in behavioural and evolutionary ecology. *Nature, Lond.* 332:29–34.

Keane, B., Waser, P. M., Creel, S. R., Creel, N. M., Elliot, L. F., & Minchella, D. J. (1994). Subordinate reproduction in dwarf mongooses. *Anim. Behav.* 47:65–75.

Ligon, J. D. (1981). Demographic patterns and communal breeding in the green woodhoopoe, *Phoeniculus purpuratus.* In *Natural selection and social behavior: Recent research and new theory,* ed. R. D. Alexander & D. W. Tinkle, pp. 231–243. New York: Chiron Press.

Lima, S. L., & Dill, L. M. (1990). Behavioral decisions made under the risk of predation: a review and prospectus. *Can. J. Zool.* 68:619–640.

Lucas, J. R. (1983). The role of foraging time constraints and variable prey encounter in optimal diet choice. *Am. Nat.* 122:191–209.

Lucas, J. R., Howard, R. D., & Palmer, J. G. (1996). Callers and satellites: Chorus behaviour in anurans as a stochastic dynamic game. *Anim. Behav.* 51:501–518.

Malcolm, J. R., & Marten, K. (1982). Natural selection and the communal rearing of pups in African wild dogs (*Lycaon pictus*). *Behav. Ecol. Sociobiol.* 10:1–13.

Mangel, M., & Clark, C. W. (1988). *Dynamic modeling in behavioral ecology.* Princeton: Princeton University Press.

Maynard Smith, J. (1982). *Evolution and the theory of games.* Cambridge: Cambridge University Press.

McNamara, J., & Houston, A. I. (1986). The common currency for behavioral decisions. *Am. Nat.* 127:358–378.

Moehlman, P. (1983). Socioecology of silverbacked and golden jackals, *Canis mesomelas* and *C. aureus.* In *Advances in the study of mammalian behavior,* ed. J. F. Eisenberg & D. L. Kleiman, pp. 423–453. Lawrence, Kans.: American Society of Mammalogists Special Publication #7.

Mumme, R. L., Koenig, W. D., & Ratnieks, F. L. W. (1989). Helping behaviour, reproductive value, and the future component of indirect fitness. *Anim. Behav.* 38:331–343.

Mumme, R. L., Koenig, W. D., Zink, R. M., & Martin, J. A. (1985). Genetic variation and parentage in a California population of acorn woodpeckers. *Auk* 102:305–312.

198   *J. R. Lucas, S. R. Creel, and P. M. Waser*

Rabenold, P. P., Rabenold, K. N., Piper, W. H., Haydock, J., & Zack, S. W. (1990). Shared paternity revealed by genetic analysis in cooperatively breeding tropical wrens. *Nature, Lond.* 348:538–540.

Reyer, H.-U. (1984) Investment and relatedness: a cost/benefit analysis of breeding and helping in the pied kingfisher (*Ceryle rudis*). *Anim. Behav.* 32:1163–1178.

Rood, J. P. (1978). Dwarf mongoose helpers at the den. *Z. Tierpsychol.* 48:277–287.

Rood, J. P. (1987). Dispersal and intergroup transfer in the dwarf mongoose. In *Mammalian dispersal patterns: The effects of social structure on population genetics*, ed. B. D. Chepko-Sade & Z. Halpin, pp. 85–102. Chicago: Chicago University Press.

Rood, J. P. (1990). Group size, survival, reproduction, and routes to breeding in dwarf mongooses. *Anim. Behav.* 39:566–572.

Solomon, N. G. (1994). Effect of the pre-weaning environment on subsequent reproduction in prairie voles, *Microtus ochrogaster*. *Anim. Behav.* 48:331–341.

Walters, J. R., Doerr, P. D., & Carter, J. H. III. (1992). Delayed dispersal and reproduction as a life-history tactic in cooperative breeders: Fitness calculations from red-cockaded woodpeckers. *Am. Nat.* 139:623–643.

Waser, P. M., Creel, S. R., & Lucas, J. R. (1994). Death and disappearance: estimating mortality risks associated with philopatry and dispersal. *Behav. Ecol.* 5:135–141.

Waser, P. M., Elliott, L. F., Creel, N. M., & Creel, S. R. (1995). Habitat variation and mongoose demography. In *Serengeti II: Dynamics, management, and conservation of an ecosystem*, ed. A. R. E. Sinclair & P. Arcese, pp. 421–450. Chicago: Chicago University Press.

Wiley, R. H., & Rabenold, K. N. (1984). The evolution of cooperative breeding by delayed reciprocity and queuing for favorable social positions. *Evolution* 38:609–621.

Woolfenden, G., & Fitzpatrick, J. (1984). *The Florida scrub jay: Demography of a cooperative-breeding bird.* Princeton: Princeton University Press.

Zack, S., & Stutchbury, B. J. (1992). Delayed breeding in avian social systems: The role of territory quality and "floater" tactics. *Behaviour* 123:194–219.

Zahavi, A. (1974). Communal nesting by the Arabian babbler. *Ibis* 116:84–87.

8

Examination of Alternative Hypotheses for Cooperative Breeding in Rodents

NANCY G. SOLOMON and LOWELL L. GETZ

8.1 Introduction

Within the order Rodentia, cooperative or communal nesting and care of young have been reported for 35 species and from 9 of 30 (30%) families. Table 8.1 is based on a review of the literature and lists rodents in which cooperative breeding has been documented. We are not trying to suggest that species can always be discretely classified as cooperative or noncooperative breeders. For example, not all deer mice (*Peromyscus maniculatus*) breed cooperatively although cooperative breeding has been observed. It may be more useful to envision cooperative breeding as a continuum rather than a unitary phenomenon (Sherman et al. 1995; Lewis & Pusey this volume). At one end of the continuum are groups with more than one reproductively active female. Approximately 57 percent of the 35 species of cooperatively breeding rodents contain more than one breeding female per social group; this system is referred to as "plural breeding." In groups with plural breeders, all breeding females provide care for young, which typically are reared in a single nest (Brown 1987). At the other end of the continuum of cooperative breeding systems are social groups in which only one female produces offspring and the young that remain at the natal nest typically are reproductively suppressed. These are termed "singular breeders" or are referred to as social systems with "helpers-at-the-nest" (Brown 1987). In the latter type of social system, at least some nonbreeding group members assist in care of the young born to the breeding female. Assistance can be in the form of direct care (huddling with young, licking or grooming pups, or retrieving young that stray out of the nest) or indirect care (caching food, building nests, or maintaining nests or runways).

In compiling Table 8.1, we defined plural breeders as species in which multiple breeding females were observed in a nest or, in precocial rodents, communally nursing young. A rodent was categorized as a singular breeder if

Table 8.1. *Rodent species in which cooperative breeding has been observed*

Species	Family	Type of data	Reference
Singular breeders			
Prairie vole (*Microtus ochrogaster*)	Cricetidae	L	Solomon 1991a, b
		F	Getz et al. 1993
		L	Wang & Novak 1992
Pine vole (*Microtus pinetorum*)	Cricetidae	F	Fitzgerald & Madison 1983
		L	Powell & Fried 1992
		L	Solomon & Vandenbergh 1994
Montane vole (*Microtus montanus*)	Cricetidae	F	Jannett 1978
Gudaur vole (*Microtus gud*)	Cricetidae	F	Ognev 1964
Water vole (*Arvicola terrestris*)	Cricetidae	L	Meylan 1977
Mongolian gerbil (*Meriones unguiculatus*)	Cricetidae	F	Agren et al. 1989
		L	Ostermeyer & Elwood 1984
		L	Salo & French 1989
California mouse (*Peromyscus californicus*)	Cricetidae	L/F	Gubernick pers. comm. 1992
Australian native mouse (*Pseudomys albocinereus*)	Muridae	F	Happold 1976
Four-striped grass mouse (*Rhabdomys pumilio*)	Muridae	L/F	Choate 1972
Beaver (*Castor canadensis*)	Castoridae	F	Brady & Svendsen 1981
		F	Patenaude 1983
Naked mole-rat (*Heterocephalus glaber*)	Bathyergidae	F	Jarvis 1981
		L	Lacey & Sherman 1991
Damaraland mole-rat (*Cryptomys damarensis*)	Bathyergidae	L/F	Bennett & Jarvis 1988
		F	Jarvis & Bennett 1993
Zambian mole-rat (*Cryptomys hottentotus*)	Bathyergidae	F	Burda 1990
Giant mole rat (*Cryptomys mechowi*)	Bathyergidae	L/F	Burda 1993
Brush-tailed porcupine (*Atherurus africanus*)	Hystricidae	F	Kingdon 1974
Plural breeders			

Plural breeders

Common vole (*Microtus arvalis*)	Cricetidae	F	Boyce & Boyce 1988
Meadow vole (*Microtus pennsylvanicus*)	Cricetidae	F	McShea & Madison 1984
Townsend's vole (*Microtus townsendii*)	Cricetidae	F	Lambin & Krebs 1991
Grey red-backed vole (*Clethrionomys rufocannus bedfordiae*)	Cricetidae	F	Saitoh 1989
Sagebrush vole (*Lemmiscus curtatus*)	Cricetidae	L/F	Nowak 1991
Sagebrush vole (*Lagurus lagurus*)	Cricetidae	F	Ognev 1964
Deer mouse (*Peromyscus maniculatus*)	Cricetidae	F	Hansen 1957
		F	Millar & Derrickson 1992
		F	Wolff 1994
White-footed mouse (*Peromyscus leucopus*)	Cricetidae	F	Jacquot & Vessey 1994
		F	Wolff 1994
Pygmy mouse (*Baiomys taylori*)	Cricetidae	F	Pitts & Garner 1988
House mouse (*Mus musculus*)	Muridae	L	Saylor & Salmon 1971
		L	König 1994
		F	Wilkinson & Baker 1988
		S	Manning et al. 1992
Norway rat (*Rattus norvegicus*)	Muridae	F	Calhoun 1963
		L	Mennella et al. 1990
		L	Nováková & Babicky 1977
		L	Solomon 1984
Spiny mouse (*Acomys cahirinus*)	Muridae	L	Porter & Doane 1978
Black-tailed prairie dog (*Cynomys ludovicianus*)	Sciuridae	F	Hoogland et al. 1989
Columbia ground squirrel (*Spermophilus columbianus*)	Sciuridae	F	Waterman, cited in Packer et al. 1992
Dwarf cavy (*Microcavia australis*)	Cavidae	L	Rood 1972
Cui (*Galea musteloides*)	Cavidae	L	Rood 1972
Cavy (*Cavia aperea*)	Cavidae	Unknown	Rood 1972
			Kunkele & Hoeck, cited in Packer et al. 1992
Domestic guinea pig (*Cavia porcellus*)	Cavidae	L	Fullerton et al. 1974
Mara (*Dolichotis patagonum*)	Cavidae	F	Taber & Macdonald 1992
Capybara (*Hydrochoerus hydrochaeris*)	Hydrochoeridae	F	Macdonald 1981

L = lab: F = field: S = seminatural

there was a report in the literature of groups including at least the breeding pair and more than one litter of young. The definitions of plural and singular breeding used to compile Table 8.1 are conservative and exclude many potential cooperative breeders who may have been described in the literature as gregarious, socially tolerant, or characterized by formation of large groups.

Although no formal phylogenetic analysis has been conducted, it appears that there is no clear-cut taxonomic division between species that typically display singular or plural breeding within the order Rodentia. Plural breeding rodents are found in species as disparate as house mice (*Mus musculus*) and capybaras (*Hydrochoerus hydrochaeris*) (Saylor & Salmon 1971; Macdonald 1981; Wilkinson & Baker 1988). Singular breeding is found in naked mole-rats (*Heterocephalus glaber*), Mongolian gerbils (*Meriones unguiculatus*), and beavers (*Castor canadensis*) (Brady & Svendsen 1981; Jarvis 1981; Patenaude 1983; Agren, Zhou, & Zhong 1989; Jarvis 1991). Even within a genus, both types of cooperative breeders can be found; for example, within the genus *Peromyscus,* deer mice tend to be plural (Hansen 1957; Millar & Derrickson 1992; Wolff 1994) and California mice (*P. californicus*) are singular breeders (Gubernick pers. comm. 1992). Within the genus *Microtus,* common voles (*M. arvalis*) are plural (Boyce & Boyce 1988), and pine voles (*M. pinetorum*) are typically singular breeders (FitzGerald & Madison 1983). Communal groups in the white-footed mouse (*P. leucopus*) (Wolff 1994) appear to be intermediate; sometimes groups have multiple breeding females, whereas at other times they include only a single breeding female and non-breeding alloparents.

Species cannot always be neatly categorized on the basis of the number of reproductive females. It is likely that in a number of species there are both intraspecific and interspecific variations in the number of breeding females per group (see also Mumme this volume). Trying to define the boundaries of cooperative breeding is difficult. Therefore, in compiling Table 8.1, we had to create boundaries that may be somewhat arbitrary.

In this review, we will examine species that typically display singular breeding with helpers-at-the-nest. Since naked mole-rats will be discussed in detail by Lacey and Sherman (this volume), we will primarily focus on data from muroid rodents but use mole-rats for comparison whenever feasible. First, we will describe criteria by which to recognize cooperatively breeding rodents, and then we will discuss hypotheses pertaining to the three hallmarks of cooperative breeding: philopatry, reproductive suppression, and alloparental care. This division reflects the three questions typically posed to investigate the functional hypotheses for cooperative breeding (see also Solomon & French this volume, French this volume).

8.2 Identification of Cooperative Breeding in Rodents

What criteria can we use to verify cooperative breeding (singular breeding and care of young by helpers) in a secretive small mammal? We need to show that groups occur in free-living populations, that only one female breeds per group, and that nonbreeders display aspects of parental care typically associated with reproductive individuals. In singular cooperative breeders, at least some offspring display philopatry, that is, remain at the natal nest beyond the age of maturity or when individuals typically disperse from their natal nest. The presence of philopatric offspring results in age-structured groups containing related, nonreproducing individuals.

The presence of social groups, as well as the size and composition of these groups, is typically determined by intensive live trapping, radiotelemetry (e.g., FitzGerald & Madison 1983; Solomon, Vandenbergh, & Sullivan in prep) , use of fluorescent or ultraviolet reflective powder (Lemen & Freeman 1985), or finding excavations, as in naked mole-rats (Brett 1991a). From these data, a social group can be defined as consisting of individuals (1) that have extensive overlapping home ranges that are spatially separate from the overlapping home ranges of other such groups and (2) that share a single nest within the home range (Figure 8.1).

There is considerable diversity in the size of social groups, both among and within species (Table 8.2). Naked mole-rats form the largest social groups, averaging 70 to 80 individuals (Brett 1991b). Two other species of mole-rats, Damaraland mole-rats (*Cryptomys damarensis*) and giant mole-rats (*C. mechowi*), are found in groups that can range up to 40 to 60 individuals; some other rodents occur in yet smaller groups (e.g., Zambian mole-rats (*Cryptomys hottentotus*), prairie voles (*Microtus ochrogaster*), pine voles, Mongolian gerbils, and beavers; see Table 8.2). It is not clear what effect group size has on the options for future reproduction, dispersal, or alloparental behavior of nonbreeders living at the nest (see Braude 1991; Lacey & Sherman this volume).

Field data show that groups are age-structured. Groups containing one or more adult males and females plus offspring from different litters have been reported for a number of species. In naked mole-rats and pine voles, offspring first trapped at the natal nest remained there beyond the age of sexual maturity, during which time the original breeders produced successive litters (FitzGerald 1984; Braude 1991; Solomon & Vandenbergh unpubl. data) (Figure 8.2). McGuire et al. (1993) reported a similar pattern for prairie voles; long-term studies reveal that only 22.5 percent of offspring disperse from natal nests.

It has been assumed that social groups are composed primarily of parents and one or more litters of offspring, but this assumption has been tested in

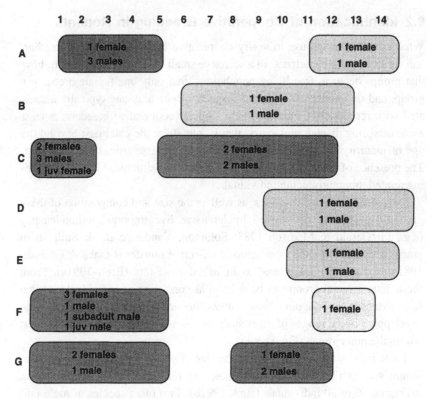

Figure 8.1. Pine vole social groups in an apple orchard in Henderson County, North Carolina (Solomon & Vandenbergh unpubl. data). The numbers on the horizontal axis represent apple trees, and the letters on the vertical axis represent rows of trees. Individuals were followed by using standard mark-recapture methods for approximately two weeks. Social groups could then be defined as consisting of individuals showing extensive overlap in space but no overlap with other such groups. The areas in which individuals were trapped are shaded in this figure.

few species. DNA fingerprinting has verified that naked mole-rats and California mice live in extended families (Reeve et al. 1990; Ribble 1991). Only one of four litters of prairie voles examined so far was sired by two males (Bouzat pers. comm. 1993); the male residing at the nest sired the majority of litters. But it is not clear that resident males are sires of young trapped at the nest in all species of cooperatively breeding rodents. For example, preliminary data suggest that the degree of relatedness between alloparents and young may not be consistently high in Mongolian gerbils. Female Mongolian gerbils copulate with neighboring males (Agren et al. 1989), which could result in multiple paternity of litters. In addition, although we are

Table 8.2. *Group sizes in cooperatively breeding rodents*

Species	Range of group sizes	Reference
Damaraland mole-rats	8–41	Bennett & Jarvis 1988
		Jarvis & Bennett 1993
Giant mole-rats	8–60	Burda 1993
Naked mole-rats	25–295	Brett 1991b
Zambian mole-rats	2–25	Burda 1990
Four-striped grass mice	12–30	Choate 1972
Prairie voles	2–19	Getz unpubl.
Pine voles	2–11	FitzGerald & Madison 1983;
		Gourley 1983;
		Solomon et al. in prep.
Mongolian gerbils	2–17	Agren et al. 1989
Beavers	4–9	Patenaude 1983

focusing on singular-breeding rodents, turnover of breeding males between litters would decrease within-group relatedness, making alloparents less than full siblings of the subsequent young born at the nest.

Finally, what evidence do we have for the existence of helping or alloparental behavior by nonbreeders? In avian studies, the strongest evidence of helping is increased survival of offspring (Brown 1987 and references therein). The lack of such evidence in species like prairie dogs (*Cynomys ludovocianus*) has resulted in some heated discussions (Hoogland 1983; Michener & Murie 1983), but the original definition of alloparental behavior did not specify that benefits would accrue to the alloparents (see Solomon & French this volume). By necessity, observations of alloparental behavior in rodents have been made from animals in laboratory settings. The majority of these rodents occupy surface or subsurface nests or burrow systems, making observations of their behavior in a natural setting impossible. Laboratory studies demonstrate that nonbreeders engage in all parental behaviors except nursing (Jarvis 1981; Fried 1987; Lacey & Sherman 1991; Solomon 1991a; Powell & Fried 1992; Wang & Novak 1992). Alloparents huddle over and groom pups as well as engage in indirect parental care (e.g., nest building and, in some species, food caching). Not all family members necessarily contribute equally to care of young. In prairie voles, for example, male and female young from the previous litter spend as much time in the nest with pups as do parents. However, not all family members engage in equivalent amounts of pup licking (Solomon unpubl.). Mothers lick pups significantly more than do fathers, who lick pups significantly more than do juveniles (Figure 8.3). There

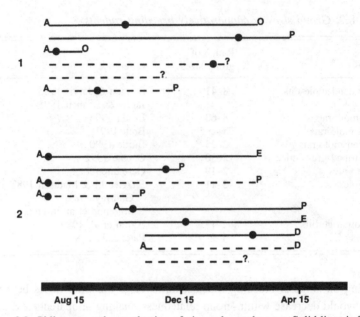

Figure 8.2. Philopatry and reproduction of pine voles at the nest. Solid lines indicate males and dashed lines indicate females. Lines beginning with an "A" indicate animals that were adults (>20 g) at first capture. Unmarked lines indicate animals that were less than 20 g at first capture and were considered to have been born in that group. Dots indicate the time when males first exhibited scrotal testes or females exhibited pregnancy. Symbols at the end of each line indicate the fate of each individual: anesthesia (O), predation (P), disease (D), emigration (E) or unknown cause (?). Numbers on the left indicate social groups. Philopatric individuals are those that were first caught when less than 20 g and remained beyond sexual maturity (about two months). Adapted from FitzGerald & Madison 1983.

is no difference in the amount of care displayed by male versus female alloparents.

8.3 Evolution and Maintenance of Philopatry

The decision of whether to remain at the natal nest (i.e., remain philopatric) or to disperse and breed can be examined by using a framework that focuses on the demographic and ecological factors that affect an individual's decision. In their delayed dispersal threshold model, Koenig et al. (1992) examine some of the factors that might influence this decision: (1) risks associated with dispersal, (2) the probability of successful establishment in a suitable territory, (3) the probability of securing a mate, and (4) the probability of successful independent reproduction (see also Emlen 1982). These factors can in turn be

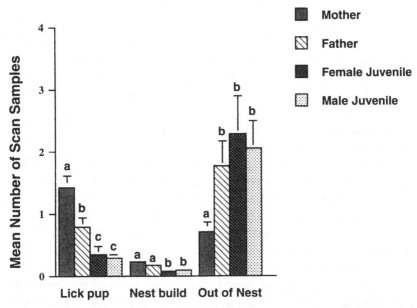

Figure 8.3. Comparison of behaviors of prairie vole family members. Figure shows the mean number of scan samples spent by each family member in each behavior. The sample period lasted for 15 minutes every other day from days 1 to 2 through days 19 to 20. Scans were taken at 30-second intervals. n = 10 family groups.

influenced by demographic and ecological variables such as population density and distribution of territories of differing quality. Although most of the predictions from the delayed dispersal threshold model have not been directly addressed in rodent studies, there are some suggestive data.

One factor that may be involved in the maintenance of philopatry is the risk or cost associated with dispersal, which would include the mortality risk through predation and the need to find food during dispersal. Lucas, Creel, and Waser, (this volume) have shown that data relating to the mortality risk from dispersal are critical in evaluating an individual's decision to remain philopatric versus disperse from the natal nest. For risk during dispersal to be an important constraint, the risks incurred during dispersal must be greater in cooperatively breeding species than in noncooperatively breeding species.

Although there are no data available from which to compare dispersal risks in species with different social organizations, there are data relating to the energetic risk of dispersal for mole-rats. According to Lovegrove's (1991) model, the risk of starvation of a single individual attempting to disperse (in an arid region with a low density of food resources) is 89 to 98 percent, suggesting that solitary dispersers would have only a slight chance of success. In contrast, in mesic regions, where the density of food resources is higher and solitary dispersers should be successful in establishing territories, mole-rats are solitary instead of living in groups. Support for this prediction is suggested by the observation that Zambian mole-rats, which live in habitats with a higher density of food resources, live in smaller groups than do other social mole-rats.

Habitat quality can affect philopatry in various ways, depending on the scale at which it is examined (see Mumme this volume). An individual's decision about whether to remain philopatric or disperse can be influenced by variation in territory quality within a habitat. Not all territories have to be occupied for extrinsic constraints to be a major factor in philopatry (Stacey & Ligon 1987; Koenig et al. 1992). Within a habitat, individuals in higher-quality territories may gain more benefits by remaining on their natal territories than by dispersing to lower-quality available territories. These high-quality territories would tend to be continuously occupied, and therefore the opportunity for independent reproduction would be restricted on them. Conversely, some low-quality territories could be vacant because individuals in them may gain greater benefits by dispersing than by remaining at home. Koenig et al.'s (1992) delayed dispersal threshold model predicts that, within a population, some individuals would remain philopatric and others would disperse at independence. Within-habitat variability has not been examined in most studies. Agren et al. (1989), in their study of gerbils in the Inner Mongolian steppes, found considerable variation in the abundance of the main food plant (an annual herb, *Artemsia sieversiana*) within the study area but did not examine the relationship between habitat quality and philopatry.

Limited suitable breeding habitat also suggests that there may be competition for territorial vacancies. Due perhaps to the difficulty in observing these animals, competition for territorial vacancies has not been reported. We can indirectly assess the probability of competition by examining turnover in territory occupancy and movement of individuals within the population. Survival of adult pine voles is greater in orchards than in natural habitats (Gentry 1968; Goertz 1971; FitzGerald 1984); the latter are presumably areas with less food. This suggests that turnover of territory vacancies would be lower in orchards than in natural habitats and that competition should be more intense in the former.

Although remaining philopatric, individuals may engage in forays that allow them to monitor vacancies in nearby groups. Field data from pine voles in an orchard in New York provide support for this hypothesis. Of 24 observed dispersal events, 69 percent of the male dispersers moved into adjacent burrow systems (FitzGerald 1984). Furthermore, 90 percent of these dispersing males secured a mate through this move. A similar pattern is seen in female pine voles. In 16 dispersal events within a season, 69 percent of the females moved from the adjacent tree line in the orchard (Solomon & Vandenberg unpubl. data).

In some avian species, it has been suggested that one potential benefit to philopatric individuals is inheritance of a portion of the territory after the death of the breeders. Few studies of cooperatively breeding rodents have been conducted for a long enough time to address this prediction. Data from prairie voles show that offspring rarely stay and become reproductive at the natal nest following loss of their parents (McGuire et al. 1993). Inheritance of territory is not an important selective pressure for remaining as alloparents in this species. In contrast, naked mole-rats always remain philopatric; succession of group members to breeder status has been documented in the laboratory as well as in the field (Braude 1991; Lacey & Sherman this volume).

It also is possible that the total abundance or quality of food resources is not the primary limiting factor; rather, the abundance of one particular resource may constrain the formation of new social groups. Alexander, Noonan, & Crespi (1991) and Seger (1991) have hypothesized that an important factor in the evolution of insect eusociality is the availability of a permanent nest. A similar argument may be made for the existence of a safe, expandable subterranean nest or burrow system in some rodents (Alexander et al. 1991). This may be especially true for subterranean rodents like mole-rats that have to dig to forage and disperse. Although fossorial rodents like pine voles do not have to dig to disperse, building a new burrow system in which to rear young may still be energetically expensive (Powell & Fried 1992). Long-distance above-ground dispersal to find a vacant burrow system would expose individuals to predators that are not able to enter a burrow system. Dispersal through neighboring burrow systems would not be feasible because members of pine vole social groups are aggressive toward intruders (Horsfall 1964; Boyette 1965). The availability of underground nests or burrows also could be a critical resource for Mongolian gerbils. In this species as well, group members are aggressive toward intruders (Agren et al. 1989). Although burrow systems or underground nests should be critical resources in both low and high food habitats, they may be even more limiting in high food habitats because of higher population densities (Gentry 1968; Miller & Getz 1969; Paul 1970; Agren et al. 1989). Although the cost of constructing an underground nest or burrow may be high in some cooperatively breeding rodents, there are also many solitary rodents that inhabit burrows, such as bannertailed kangaroo rats (*Dipodomys spectabilis*) and plains pocket gophers (*Geomys bursarius*) (Nevo 1979; Waser & Jones 1983), so use of a burrow cannot be sufficient to predispose individuals to remain philopatric.

At a somewhat larger scale, there may be variation in the degree of cooperative breeding among habitats even within a single species. Most data on cooperatively breeding rodents come from populations living in food-rich habitats (i.e., prairie voles in alfalfa and pine voles in apple orchards). Perhaps there are differences in social organization of populations in food-rich habitats and those living in more typical low food habitats (tallgrass prairie for prairie voles and deciduous forest edges for pine voles). Rich habitats are relatively stable in terms of food resources; this may result in higher population densities and in turn a shortage of suitable breeding territories. However, Getz, Gudermuth, and Benson (1992) observed prairie voles living in groups in high food (alfalfa), medium food (bluegrass), and low food (tallgrass prairie) in the fall and winter. Experimental examination of the effect of food resource quality on social organization during the primary breeding season is now in progress by G. R. Cochran and N. G. Solomon at Miami University in Oxford, Ohio.

There is evidence that population densities are higher in rich- as compared to low-quality habitats. For example, in the Mongolian gerbil, population densities in agricultural cropland were more than twice those in nearby bush plantations with little herbaceous growth (Agren et al. 1989). In prairie voles, peak densities were greatest in alfalfa (high food), lowest in tallgrass prairie (low food), and intermediate in bluegrass (medium food) habitats (Getz et al. 1987). The same relationship between density and food availability has been found in pine voles (Cornblower & Kirkland 1983) and naked mole-rats (Brett 1991b). The largest naked mole-rat colony was found in an area of good food productivity, and colonies were smaller in areas with poor food supply (Brett 1991b).

Increased densities may constrain alloparents because of a lack of suitable territory vacancies. Support for this possibility stems from trapping data on prairie voles. Even though groups form when population density is low (<100/ha), a much larger percentage of prairie vole breeding units are groups at high population density (>100/ha) (Table 8.3). Male–female pairs and single females with young are the most prevalent types of breeding units at low densities (Getz et al. 1993). Population density appears to influence dispersal of prairie voles in multiple ways. Significantly more individuals dispersed from their natal nests at low (39.4%) than at high (27.0%) density (McGuire et al. 1993). Dispersal also occurred at an earlier age at low than high densities (38.8 ± 3.3 days and 53.7 ± 2.5 days, respectively). Dispersers settled significantly farther from the natal nest at low (39.6 ± 3.1 m) than at high (21.1 ± 2.4 m) density (Getz et al. 1994). Observational data make it difficult

Table 8.3 *Types of prairie vole social groups at different densities*

	Percentage	
	High density (>100 voles/ha)	Low density (<100 voles/ha)
Group	53.5	14.7
Male–female pair	30.9	47.6
Single female	15.6	37.7

Groups are defined as consisting of more than two resident adults of the same sex and can also contain one or more adult opposite-sex conspecifics and young. Single-female breeding units can contain either only a single adult female or an adult female and young.

to untangle the causal relationships among juvenile survival, density, and group formation. Experimental manipulations would help to clarify the relationship between density and dispersal.

Between-habitat variability in resource quality may promote philopatry in some species but not in others. Jarvis and Bennett (1993) have suggested that social mole-rats are found in arid regions characterized by patchy distributions of geophytes, a primary food resource. Sociality may increase the probability of locating such dispersed food resources in an animal that must burrow through hard soil, an energetically expensive proposition, to find food. Jarvis et al. (1994) suggest that the erratic rainfall may impose severe constraints on dispersal and solitary living because, in the limited time when a mole-rat can move through the soil with reasonable ease, it may not find enough food to sustain it until the next rain. In contrast, the more solitary species of mole-rats live in mesic environments where food resources are more evenly dispersed.

There are some data available that address the third factor postulated to be important in the maintenance of philopatry, that is, the probability of securing a mate. One way that finding a mate could be a problem would be if the adult sex ratio were skewed. For example, if the adult sex ratio were male-biased, then females would be a limiting factor. Males that could not acquire mates could benefit by remaining philopatric until a mating opportunity arose. Only data from naked mole-rats support this hypothesis. Adult sex ratios in naked mole-rats, common mole-rats, and Mongolian gerbils are male-biased (Genelly 1965; Agren et al. 1989; Braude 1991). We therefore would expect that, because the adult sex ratios in Mongolian gerbils and mole-rats are male-biased, the sex ratio of philopatric juveniles also would be male-biased. Data are not available on gerbils and common mole-rats, but male-biased juvenile sex ratios in naked mole-rats (Braude 1991) support the hypothesis that philopatry occurs because of limitations in finding a mate. In contrast, the adult sex ratio among prairie voles does not deviate significantly from 1:1 (range: 48–52% males; Getz unpubl. data). Sex ratios of adult pine voles range from 1:1 to slight female or male biases over time (FitzGerald 1984; Anthony et al. 1986). In the pine vole, FitzGerald (1984) found that the juvenile sex ratio was male-biased; there were almost twice as many male as female juveniles. In another population, Solomon et al. (in prep.) found no difference in the number of male and female juveniles recruited into the population. Again, in this species, there does not seem to be consistent support for the lack-of-mate hypothesis. Furthermore, it is important to keep in mind that field data regarding sex ratios may be subject to error; differences in trappability may bias results. Finally, there does not seem to be any a priori reason that a biased sex ratio alone should result in philopatry, for, at least in birds, biased sex ratios also are found in noncooperative breeders (Koenig & Pitelka 1981).

In cooperatively breeding rodents, one way to acquire a mate would be to remain at a nest where the opposite-sex breeder is not related (e.g., a stepparent) and mate with that individual. Again, few data exist to test the predictions that alloparents are (1) unrelated to the opposite-sex breeder and (2) mate with that individual – either by formation of a pair bond when the same-sex breeder (parent) dies or by extra-pair matings. In prairie voles, 48 percent of the male alloparents and 33 percent of the female alloparents occur in social groups that also include unrelated opposite-sex adults (Getz unpubl. data). If philopatry is a way to acquire a mate, then dispersal should be affected by the amount of competition for mates within groups. Data collected by McGuire et al. (1993) show that natal dispersal was not affected by the within-group level of competition for mates. There was no significant difference in the percentage of dispersers from groups with differing numbers of potential competitors or mates. Furthermore, if individuals disperse from their natal groups, then we would expect that they would move to a group with less competition for mates. Most dispersers in either the primary breeding season (spring to early autumn) or during late autumn to winter increased their opportunity to mate by joining other groups (McGuire & Getz 1995). In contrast, in naked mole-rats, DNA fingerprinting suggests that 85 percent of the matings are between parents and offspring (Reeve et al. 1990). Data from DNA fingerprinting are needed to determine if alloparents and adults mate in other species; in naked mole-rats the hypothesis that philopatry results in an opportunity to secure an unrelated mate does not seem to apply.

The final factor posited to affect philopatry is the cost of independent reproduction (Emlen 1982). Benefits to philopatry may exceed the cost of missed breeding opportunities if the probability of successful independent reproduction is low. Under these circumstances, we would expect to see that groups would have greater reproductive success than would pairs alone. Unfortunately, there are no data from singularly breeding rodents. If assistance from alloparents increases reproductive success, then remaining philopatric may be more beneficial than dispersing and breeding independently. In this case, the presence of potential alloparents becomes a critical determinant of territory quality. Therefore, offspring should disperse unless there is a shortage of vacancies with potential alloparents.

It also is possible that the cost of independent reproduction is higher for younger than for older animals. One way that the cost of reproduction could be decreased in older animals is that by remaining philopatric and helping care for younger siblings, alloparents gain skills critical for successful rearing of young. If there is a decreasing cost as individuals mature, then we would expect that older individuals would display less philopatry than younger siblings would.

8.4 Evolution and Maintenance of Reproductive Suppression

In cooperatively breeding rodents, social influences on reproduction are strong and may delay or suppress reproduction for extended periods in females born into the social group: beavers (Novak 1977); prairie voles (Getz, Dluzen, & McDermott 1983; Carter & Getz 1985); pine voles (Schadler 1983, 1990; Lepri & Vandenbergh 1986); Mongolian gerbils (Payman & Swanson 1980; French 1994); Damaraland mole-rats (Jarvis & Bennett 1993); and naked mole-rats (Faulkes et al. 1991a). In these species, there is usually only one reproductive female per social group (Wasser & Barash 1983). For example, in naked mole-rats and pine voles respectively, 90 and 80 percent of the free-living groups contained only one breeding female (Braude 1991; Solomon et al. submitted). Presence of a parent may prevent offspring from reproducing; the mother has been shown to exert a suppressive effect on her daughters' reproduction (Getz et al. 1983; Carter, Getz, & Cohen-Parsons 1986; Schadler 1990; Faulkes et al. 1991a; Solomon & Vandenbergh 1993). The father also may exert some suppressive effect. In the presence of their fathers, female prairie voles exposed to experienced males were less likely to become reproductive (McGuire & Getz 1991). Effects of the father's presence currently are being investigated in the pine vole by C. L. Brant and N. G. Solomon at Miami University, Oxford, Ohio.

The functional hypotheses proposed to explain reproductive suppression posit benefits to the suppressor or to the individual that is suppressed. Reproductive suppression has been hypothesized to (1) decrease the number of reproductive females at a specific site, (2) prevent inbreeding, or (3) delay pregnancy until female offspring are older and more likely to be successful in rearing young (Haigh 1987). According to the first hypothesis, mothers would benefit by exerting a suppressive effect on the reproduction of other females living at the nest. Successful reproduction by daughters may indirectly increase the mother's fitness but only to the extent that the daughter's reproduction would not adversely affect successful reproduction by the mother. Competition for limiting resources may be such that reproductive success of the mother would be adversely affected by reproductive daughters. Also, granddaughters are "worth" only half as much to the mother as daughters in terms of gene equivalents. The second hypothesis, inbreeding avoidance, suggests that reproductive suppression of daughters living at the natal nest is adaptive to the daughter and her male relatives if the only available mates are the father or brothers. Death or emigration of the brothers or father, but not the mother, from the natal group should alter the daughter's reproductive status. Once the potential for inbreeding is removed, the daughter should breed with a new male that immigrates into the group.

Long-term studies of naked mole-rats have shown that aggression by the breeding female suppresses reproduction in other females, primarily daughters (Faulkes et al. 1991a). Removal of the breeding female results in reproduction by one or two formerly nonbreeding females in the colony (Braude 1991; Jarvis 1991). Furthermore, Reeve et al. (1990) found a high degree of band sharing in multilocus DNA fingerprints, which suggests a high level of inbreeding within naked mole-rat colonies. These data suggest that in naked mole-rats reproductive suppression is the outcome of female–female competition, not inbreeding avoidance, and that it controls the number of reproductive females per colony.

In pine voles as well, the presence of an adult female decreases the probability that daughters will produce litters (Schadler 1990; Solomon & Vandenbergh 1993). The effect of the father's presence is not yet known, but Schadler (1983) has shown that male siblings suppress reproduction. These data suggest that hypotheses one and two are not mutually exclusive and that both reproductive competition and inbreeding avoidance may play roles in female reproductive suppression in female pine voles.

The third hypothesis posits that it may be adaptive for females to forgo reproduction until after dispersal and establishment of their own nest sites. It would not be advantageous for younger philopatric females to produce litters if they could not rear the pups to weaning. French (1994) has shown that in

Table 8.4. *Reproductive variables in female pine voles paired at two months versus six months of age*

	Age of pairing	
Measure	2 months	6 months
Percentage of females producing litters	81 (n = 32)	83 (n = 29)
Median time to first litter (days)	46.5	30.5*
Litter size (mean ± SEM)	2.2 ± 0.1	2.2 ± 0.1
Mean interlitter interval (days)	30.3 (n = 26)	29.5 (n = 24)
Weaning weight of pups (g)	11.7 ± 0.5	13.1 ± 0.5*
Expected number of offspring within 18 months	22.2 (n = 32)	17.5 (n = 29)

*Significant differences between age groups.
Source: Modified from Hellkamp 1991.

Mongolian gerbils, litters born to alloparents two weeks or more after the mother's delivery suffered significantly higher mortality than did litters born immediately after the mother gave birth. Production of litters under these conditions could incur a significant energetic expense but no benefit.

We would predict that females that delay breeding should show increased lifetime reproductive success compared to nondelayed females. Reproductive success did not increase in laboratory-housed female pine voles that delayed reproduction. Most components of reproductive success did not differ between female pine voles paired at two months of age or six months of age (Table 8.4; Hellkamp 1991). Although female pine voles that did not breed until they were six months of age produced their first litter in a shorter period than did females that were paired at two months of age, this difference did not compensate for the time they lost before reproducing. Furthermore, under field conditions, the proportion of females that successfully disperse and reproduce is unknown. It is likely that this proportion is quite low, which makes one wonder why females would delay breeding. Additionally, care must be taken in interpreting results of breeding studies conducted under optimal laboratory conditions (Solomon 1991b) because they may not be representative of what actually happens in the field.

Although many questions regarding female reproductive suppression still remain, even less is known about social influences on reproduction of philopatric males. Male–male competition may affect young males in a way similar to that postulated to occur with reproductive competition between females. Fathers or other adult males may prevent younger males from breeding with any reproductively active female through dominance interactions or, possibly, through use of chemosignals. Alternatively, inbreeding avoidance also may occur in males. Philopatric males may delay reproduction while their mother or sisters are living at the natal nest. Again, after the potential for inbreeding is removed, philopatric males should breed with any female that immigrates into the social group.

Existing data from four species of cooperative breeders (naked mole-rats, Damaraland mole-rats, prairie voles, pine voles) show that males attain physiological maturity while at the natal nest (Faulkes, Abbott, & Jarvis 1991b; Bennett 1994; Mateo et al. 1994; Kerr 1994; see also Carter and Roberts this volume). For example, male prairie voles are capable of producing urine that activates females whether or not the males are in contact with family members (Mateo et al. 1994). These data suggest that males in contact with family members are physiologically mature (have high testosterone levels). Although it has been assumed that male offspring do not breed in their natal group, with the exception of naked mole-rats (Lacey and Sherman this volume), DNA fingerprinting will determine if this assumption is true.

In both naked mole-rats and pine voles, histological examination of testes showed that nonbreeding males undergo spermatogenesis. Seminiferous tubules containing spermatozoa were found in both breeders and nonbreeders (Faulkes & Abbott 1991; Kerr 1994). Furthermore, both nonbreeding and breeding male Damaraland mole-rats produce similar numbers of motile spermatozoa (Faulkes et al. 1994). Although nonbreeding male naked mole-rats are physiologically mature, they have lower levels of urinary testosterone, plasma luteinizing hormone, and fewer motile spermatozoa than do breeding males (Faulkes & Abbott 1991; Faulkes et al. 1991b). These data suggest that in male naked mole-rats, there may be physiological suppression of reproduction that is not seen in other male cooperatively breeding rodents. In naked mole-rats, the breeding male dominates the nonbreeding males (Lacey & Sherman this volume) suggesting that male–male competition may affect reproduction by subordinate males in this species. The functional explanation for male–male aggression is still not known.

Additionally, Faulkes and Abbott (this volume) have found that after removal of the breeding female, testosterone levels in nonbreeding male naked mole-rats increase significantly. This cannot be viewed as a simple case of inbreeding avoidance, because it has been shown that a high level of inbreeding occurs in naked mole-rat colonies (Reeve et al. 1990). With the exception of naked mole-rats, physiological differences between breeding and nonbreeding males appear to be more uncommon than those seen in females and suggest that the functional significance of the lack of or delay in reproduction may differ between males and females (see also Lacey and Sherman this volume).

8.5 Evolution and Maintenance of Alloparental Behavior

We will now consider hypotheses proposed to explain selection for alloparental behavior. The hypothesis seminal in initiating this line of research is Hamilton's (1964) inclusive fitness hypothesis. Inclusive fitness contains two components, and each can be divided into a current or future benefit: current direct fitness, future direct fitness, current indirect fitness, and future indirect fitness (Brown 1987). The direct fitness component is generally assumed to be equal to the number of offspring the breeder produces. The latter component of inclusive fitness is called "indirect fitness" because these benefits accrue indirectly to the individual because it shares genes by common descent with its nondescendent relatives. Grafen (1982) has shown that indirect fitness (in general, of the alloparent) should really be measured as the effect of one individual on the reproductive success of the other. Simply counting offspring produces an incorrect value for direct fitness. If all offspring are scored as direct fitness (for the parent), this results in double accounting, because the offspring are counted both as direct fitness for the parent and as indirect fitness for the offspring (also see Lucas et al. this volume). More recently, Creel (1990) has shown how to resolve the second problem: by subtracting a constant factor $e°$, Hamilton's social factor, from everyone's fitness. This factor is equal to the amount of help (i.e., the average number of additional offspring that accrue through alloparental behavior) received per capita in the population. Creel's solution eliminates the double accounting and was implicit in Hamilton's (1964) equations. There has been some discussion in the avian literature regarding the relative importance of these two components of fitness and the need to collect data on all potential direct and indirect fitness benefits (Brown 1987; Emlen & Wrege 1989; also see Lucas et al. this volume).

One potential direct fitness benefit of alloparental care would be that alloparents learn to be good parents by practicing skills associated with parental behavior (Lancaster 1971). This hypothesis predicts that experience as an alloparent results in an increased quantity or quality of future offspring. The strongest support for this prediction comes from a laboratory study by Salo and French (1989), who found that alloparental experience influenced reproductive performance, pup growth, and pup development in Mongolian gerbils. Age-matched pairs composed of at least one experienced alloparent produced their first litters significantly sooner than did pairs in which both breeders were inexperienced. Growth of pups in the first litter also was affected by previous alloparental experience. When both parents had previous alloparental experience, mean weights of pups on day 20 were 17.8 percent heavier than those of pups born to pairs in which only the mother had prior alloparental experience, suggesting that male experience as an alloparent was an important factor in pup growth. Development, as assessed by age of eye opening, showed the same pattern. Pups reared in pairs where fathers were experienced alloparents opened their eyes slightly, but significantly, sooner than did pups reared with inexperienced males.

Salo and French (1989) suggested that the presence of a male Mongolian gerbil with alloparental experience is probably beneficial for pup growth and development because he could compensate for the poor nests built by females with no previous alloparental experience. In support of this hypothesis, Elwood and Broom (1978) had observed that female Mongolian gerbils rearing pups in the absence of a male engaged in more nest building activity than did the female of a pair; single females built more-compact, high-sided nests than did pairs. This difference appeared to be compensation for loss of the male's contribution to keeping pups warm.

Slight beneficial effects of postweaning experience with younger siblings have been found in prairie voles. Wang (1991) did not find any behavioral differences between parents that had prior alloparental experience and those that did not. As with Mongolian gerbils, however, Wang (1991) found that prairie vole pups reared by two experienced alloparents developed slightly faster than those reared by inexperienced individuals. Pups opened their eyes sooner and moved in and out of the nest sooner than did pups reared by two inexperienced individuals. There was no difference in the day on which solid food was first eaten or when fur first appeared. Additionally, pup weights did not differ between the two treatments. There was only a tendency, which did not reach statistical significance, for larger litter size at birth and lower preweaning mortality when both parents had prior alloparental experience.

In pine voles, there were behavioral differences between parents that had previous alloparental experience and those that did not (Salek & Vandenbergh submitted), but these behavioral differences did not result in a strong effect on pup growth or survival. Additionally, alloparental experience did not have any long-term effect on parental responsiveness in California mice (Gubernick & Laskin 1994). Therefore, it appears that in at least three species postweaning experience with younger siblings did not result in increased parental responsiveness or effects on pup growth or survival. There is no a priori reason to expect that the most efficient way for a rodent to acquire parental skills is by remaining philopatric and behaving alloparentally instead of dispersing and breeding independently. The results from these studies suggest that future direct benefits may not be as important in rodents as in primates (Lancaster 1971; Nishida 1983; Hoage 1977, but see French 1994) or that the importance of direct benefits may differ greatly between rodent species.

Helping at the nest may also be selected as a result of indirect fitness benefits that accrue to the alloparents. Because alloparents share genes by common descent with breeders and offspring they help to raise, any direct benefits to the breeders or offspring accrue indirectly to alloparents in one or more ways. Alloparents may decrease the workload of breeders, which could result either in increased survival of breeders or in allowing breeders to initiate more reproductive attempts per breeding season. Alloparents may also increase the success of individual breeding attempts by increasing the quantity of quality of pups reared to weaning. These two ways of accruing indirect benefits may occur by means of distinctive behavioral pathways. French (pers. comm. 1993) has proposed that the former be regarded as compensatory parental care (freeing parental energy to enable breeders to produce the next litter sooner) and the latter as supplemental parental care (alloparents adding to the effort produced by breeders). Finally, indirect effects may change markedly with age as the degree of relatedness between breeder and alloparent decreases (Lucas et al. this volume). Therefore, potential indirect benefits should be greater in younger individuals, but this has not been examined in any rodent species to date.

Although the effects of alloparental behavior have not yet been quantified in naked mole-rats, Lacey and Sherman (this volume) have argued that the reason the breeding female can have four to five very large litters per year is that the alloparents do almost everything (foraging, tunnel building, colony defense etc.). Jarvis (1991) has shown that alloparents are not essential for successful rearing of offspring, which suggests that one or both of the breeders may benefit in terms of time or energy saved from the assistance provided by nonbreeding colony members.

In other cooperatively breeding rodents, the effects of juvenile alloparents on time budgets of parents have been examined in laboratory studies. Fried (1987) found that alloparents had no statistically significant effect on the time mothers or fathers spent with pine vole pups; however, there was a tendency for parents, especially fathers, to spend less time in the nest with pups when more alloparents were present. This was particularly noticeable during the latter part of the preweaning period.

In Mongolian gerbils and prairie voles, fathers also appeared to benefit most from the presence of alloparents. In the gerbil, both parents reduced nest-building activity in the presence of alloparents. Especially noticeable was the reduction in time that fathers spent nest building in the presence of two alloparents as compared to no alloparents (French pers. comm. 1993). In the presence of juvenile alloparents, fathers but not mothers spent significantly more time in nonparental behavior, that is, feeding, drinking, and foraging outside the nest (Solomon 1991b).

In all three studies, fathers appeared to benefit most in terms of decreased workload or increased time outside the nest. We would predict that this time may be used by fathers to maintain or improve their physical condition, patrol territorial boundaries, and repel male conspecifics in attempts to ensure fidelity of their mates, or to engage in extra-pair copulations to increase their own reproductive success. In prairie voles, it appears that males rarely use this opportunity to engage in extra-pair copulations. Only 4 percent of the nonresident male prairie voles that visit the home ranges and nests of social groups are from neighboring groups (Getz unpubl.), whereas the remainder are unpaired males. Also, preliminary DNA fingerprinting data (Bouzat pers. comm. 1993) shows that, although litters may be fathered by more than one male, such occurrences are infrequent in prairie voles.

A decreased workload may provide the mother with future reproductive benefits, one of which may be increased survival. Long-term, intensive monitoring is needed to see if this holds true for cooperatively breeding rodents. Mothers may also benefit from the presence of alloparents because the mothers could initiate more reproductive attempts per breeding season. One indication of this benefit would be a reduction in the length of time between successive litters. The data on interlitter interval from experimental laboratory manipulations provide mixed support for this hypothesis. Family groups of Mongolian gerbils containing alloparents did not produce litters any sooner than did pairs without alloparents (French 1994). For pine voles, there was no difference in interlitter interval in groups with no alloparents as compared to one or two alloparents. If three alloparents were present, the interlitter interval was reduced significantly (Powell & Fried 1992).

One of the potential drawbacks of most laboratory investigations is that families are housed under optimal laboratory conditions, that is, at ambient temperature, with ad lib access to food and water, and no threat of predation. Under the optimal conditions provided in many laboratory experiments, it may be difficult to detect benefits from the presence of alloparents. Benefits from alloparents may occur mainly when conditions are more natural, that is, when animals have to forage for food or when temperatures are more stressful. Solomon (1991b) therefore tested whether prairie vole alloparents gain indirect fitness benefits by housing family groups at 15° to 16°C, a temperature chosen to simulate burrow temperatures in April and November (when breeding does occur in the field). Under these conditions, the presence of alloparents resulted in more rapid production of the subsequent litter, especially when current litter size was large (Solomon 1991b). It is unclear whether more rapid production of litters in the presence of alloparents differs among species (i.e., is more important in some cooperatively breeding rodent species than in others) or was undetectable in previous experiments owing to the lack of an environmental challenge.

In addition to the compensatory parental care seen in many of these studies, there is also evidence of supplemental parental care. Nonbreeding naked mole-rats increase the time spent in huddling in the nest when pups are present (Lacey and Sherman this volume), thereby providing a stable thermal environment for pups (Lacey et al. 1991). Also, prairie and pine vole pups were alone in the nest for less time if alloparents were present (Solomon 1991b; Powell & Fried 1992; Thomas 1993).

It is possible that alloparents benefit indirectly by reducing preweaning mortality, thus increasing the number of offspring that survive to weaning. Neither presence of alloparents nor increased number of alloparents affected litter size at weaning in prairie voles, pine voles, or Mongolian gerbils (Fried 1987; Solomon 1991b; French 1994). In contrast, there was a positive relationship between group size and the number of offspring produced per female in prairie vole social groups in the field (McGuire pers. comm. 1994), although it is not clear whether this may have been the result of multiple-breeding females within groups.

Alternatively, alloparents may increase the success of individual breeding attempts by increasing the quality of offspring produced. One measure of offspring quality is body size at weaning. The only time that any effect of alloparents on preweaning pup growth has been found was when prairie voles were housed at environmentally challenging temperatures (Solomon 1991b). Under these conditions, pups reared in the presence of alloparents grew and developed faster than did those reared without alloparents (Solomon 1991b). In comparison, Wang (1991) found no differences in pup growth when prairie voles were housed at room temperature.

Results of studies on prairie voles (Solomon 1991a) and inspection of the rodent literature suggest that large body size at weaning may result in numerous potential benefits. Hypothesized benefits from large body size include benefits in growth, survival, mating preference, and reproduction. Prairie voles that were lighter at weaning tended to be lighter as adults (Solomon 1994). Therefore, although some compensatory growth occurs, small young do not completely catch up in size to larger individuals by the time they reach sexual maturity. In addition, large size at weaning increases immediate post-weaning survival in prairie voles (Solomon 1991b). Pups that survived were significantly heavier at weaning than those that died within 10 days of weaning. Large size is beneficial to both male and female prairie voles because they are preferred as social, and presumably, sexual partners (Solomon 1993). Finally, weaning weight of the female also affects the postweaning growth of her pups and the interval between her first and second litters (Solomon 1994). In a short-lived mammal such as the prairie vole, weaning weight may have a significant impact on lifetime reproductive success because larger females could produce more litters that contain larger pups at weaning.

Although Solomon (1991b) has shown that alloparents benefit indirectly through effects on their parents and younger siblings, no study has compared direct and indirect benefits under less than optimal conditions. Future studies are needed to compare the relative importance of direct and indirect fitness components of inclusive fitness in numerous cooperatively breeding rodents.

8.6 Conclusions

Although some of the hypotheses proposed to explain the evolution and maintenance of philopatry, reproductive suppression, and alloparental behavior have been addressed in some species, the study of rodent cooperative breeding is still in its infancy. There are at least eight areas in which the study of cooperative breeding should receive more attention.

1. Only four studies (for three species) have examined group structure throughout the year in a cooperatively breeding rodent (Braude 1991; Brett 1991b; Getz et al. 1993; Jarvis & Bennett 1993). Madison & McShea (1987) have shown that group structure in the noncooperatively breeding meadow vole, *M. pennsylvanicus,* changes seasonally. In the winter, large groups of nonbreeding individuals form. By spring, McShea and Madison (1984) found plural breeding in some groups. After these groups broke up, females became territorial and only one female occupied each nest. In contrast, naked mole-rats live in large colonies throughout the year. Cooperative breeding has been documented in prairie voles during both spring–summer and winter breeding seasons (Getz et al. 1993), but examination of seasonal patterns in other cooperatively breeding rodent species may provide additional insight.

2. For most species, we do not know the relatedness among individuals within groups, whether females tend to mate with males from their group, the frequency of extra-pair copulations, and other questions about the genetic structure of populations. Reeve et al. (1990) and Faulkes and Abbott (1990) have used DNA fingerprinting to address these questions in naked mole-rats.

In other species, DNA fingerprinting and related techniques will provide answers to many of these questions and probably some surprises, as have been found in avian studies (Rabenold et al. 1990).

3. The effect of habitat variables on cooperative breeding should be pursued. Jarvis et al. (1994) have proposed that patchy, unpredictably distributed food resources have favored natal philopatry in naked mole-rats. Social mole-rats are found in arid regions characterized by patchy distributions of geophytes, a primary food resource. Sociality may increase the probability of locating such dispersed food resources by an animal that must burrow through hard soil, an energetically expensive proposition, to find food. Jarvis et al. (1994) suggest that the erratic rainfall may impose severe constraints on dispersal and solitary living because during the short time when a mole-rat can easily move through the soil, it may not find enough food to sustain it until the next rainfall. In contrast, the more solitary species of mole-rats live in mesic environments where food resources are more evenly dispersed. Lacey and Sherman (this volume) suggest that other subterranean species may provide useful tests of the aridity–food-distribution hypothesis. Additionally, study of other rodents that live in underground nests or burrows but are not fully subterranean may prove informative.

4. We propose that the effects of habitat variables should also be studied intraspecifically; we need to study populations in less than optimal habitats and in habitats with varying degrees of patchiness (also see Koenig et al. 1992). Populations from different habitats, such as prairie voles from the more arid grasslands in Kansas as compared to the prairie voles studied by Getz and colleagues in alfalfa fields in Illinois, would provide information on the potential impact of ecological variables on dispersal and social structure. Using data collected during fall–winter, Getz et al. (1992) concluded that there were no habitat differences in the social organization of the prairie vole. However, habitat differences may affect prairie vole social structure at other times of the year. Additionally, habitat differences in other species may have had a profound effect on social structure but this is purely speculative at this time.

5. McGuire et al. (1993) and Getz et al. (1994) have described the degree and age of dispersal in prairie voles. In naked mole-rats, Braude's (1991) long-term mark recapture studies and Reeve et al.'s (1990) genetic analyses have documented the degree of philopatry in this species. These data are lacking in other cooperatively breeding rodents. Basic descriptive data are needed before experimental examination of hypotheses posed to explain the evolution and maintenance of philopatry can be fully tested.

6. As Stacey and Ligon (1991) have suggested, an investigation of the factors that influence philopatry should involve comparisons of related cooperatively

and noncooperatively breeding species. We need to identify ecological conditions that make the benefits of philopatry outweigh the costs in cooperatively breeding species. In noncooperative breeders, specific ecological factors should cause dispersal to be more beneficial than philopatry. There are at least two families of rodents in which cooperative and noncooperative members could be compared: Arvicolidae (voles and lemmings) and Bathyergidae (mole-rats).

7. The continuum in sociality described by Sherman et al. (1995) also is seen in avian cooperative breeders (Brown 1987) and social insects (Keller & Reeve 1994). Investigators have described the distribution of reproduction among potential reproductives as the reproductive skew in a group (Vehrencamp 1983; Keller & Reeve 1994; Sherman et al. 1995). The monopolization of reproduction by a single female within a group may be affected by multiple factors including kinship and resource availability. In turn, the control of reproduction may influence interactions among group members, the degree of philopatry, and the degree of alloparental care (see also Keller & Reeve 1994). Further attention to the interactions among these various factors is warranted.

8. Finally, laboratory investigations should consider manipulating variables to provide a challenging environment based on the natural history of the species. Comparison of studies conducted under optimal and environmentally challenging conditions may reveal benefits from the presence of alloparents that may otherwise remain undetected.

Acknowledgments

We thank Betty McGuire, Paul Sherman, and Peter Waser for constructive criticism on a previous version of this chapter. Nancy Solomon was supported by grants from the Committee on Faculty Research at Miami University and NIH MH 52471–01 and Lowell Getz by grants from NIH HD 09328 and NSF DEB 78–25864 during the writing of this chapter.

References

Agren, G., Zhou, Q., & Zhong, W. (1989). Ecology and social behaviour of Mongolian gerbils, *Meriones unguiculatus*, at Xilinhot, Inner Mongolia, China. *Anim. Behav.* 37:11–27.

Alexander, R. D., Noonan, K. M., & Crespi, B. J. (1991). The evolution of eusociality. In *The biology of the naked mole-rat*, ed. P. W. Sherman, J. U. M. Jarvis, & R. D. Alexander, pp. 3–44. Princeton: Princeton University Press.

Anthony, R. G., Simpson, D. A., Kelly, G. M., & Storm, G. L. (1986). Dynamics of pine vole populations in two Pennsylvania orchards. *Am. Midl. Nat.* 116:108–117.

Bennett, N. C. (1994). Reproductive suppression in social *Cryptomys damarensis* colonies – a lifetime of socially-induced sterility in males and females (Rodentia: Bathyergidae). *J. Zool., Lond.* 234:25–39.

Bennett, N. C., & Jarvis, J. U. M. (1988). The social structure and reproductive biology of colonies of the mole-rat *Cryptomys damarensis* (Rodentia, Bathyergidae). *J. Mamm.* 69:293–302.

Boyce, C. C. K., & Boyce, J. L. III (1988). Population biology of *Microtus arvalis*. I. Lifetime reproductive success of solitary and grouped breeding females. *J. Anim. Ecol.* 57:711–722.

Boyette, J. G. (1965). *A behavioral study of the pine mouse,* Pitymys pinetorum pinetorum *(Le Conte)*. PhD dissertation, North Carolina State University, Raleigh.

Brady, C. A., & Svendsen, G. E. (1981). Social behaviour in a family of beaver, *Castor canadensis*. *Biol. Behav.* 6:99–114.

Braude, S. (1991). *The behavior and demographics of the naked mole-rat,* Heterocephalus glaber. PhD dissertation, University of Michigan, Ann Arbor.

Brett, R. A. (1991a). The ecology of naked mole-rat colonies: Burrowing, food, and limiting factors. In *The biology of the naked mole-rat,* ed. P. W. Sherman, J. U. M. Jarvis & R. D. Alexander, pp. 137–184. Princeton: Princeton University Press.

Brett, R. A. (1991b). The population structure of naked mole-rat colonies. In *The biology of the naked mole-rat,* ed. P. W. Sherman, J. U. M. Jarvis, & R. D. Alexander, pp. 97–136. Princeton: Princeton University Press.

Brown, J. L. (1987). *Helping and communal breeding in birds.* Princeton: Princeton University Press.

Burda, H. (1990). Constraints of pregnancy and evolution of sociality in mole-rats. *Z. zool. Syst. Evolut.-forsch.* 28:26–39.

Burda, H. (1993). Evolution of eusociality in the Bathyergidae: The case of the giant mole-rats (*Cryptomys mechowi*). *Naturwissen.* 80:235–237.

Calhoun, J. B. (1963). *The ecology and sociology of the Norway rat* (PHS Publication No. 1008). Bethesda, Md.: U. S. Department of Health, Education and Welfare.

Carter, C. S., & Getz, L. L. (1985). Social and hormonal determinants of reproductive patterns in the prairie vole. In *Neurobiology,* ed. R. Gilles & J. Balthazart, pp. 18–36. Berlin: Springer-Verlag.

Carter, C. S., Getz, L. L., & Cohen-Parsons, M. (1986). Relationships between social organization and behavioral endocrinology in a monogamous mammal. *Adv. Study Behav.* 16:109–145.

Choate, T. S. (1972). Behavioural studies on some Rhodesian rodents. *Zool. Africana* 7:103–118.

Cornblower, T. R., & Kirkland, G. L. Jr. (1983). Comparisons of pine vole (*Pitymys pinetorum*) populations from orchards and natural habitats in southcentral Pennsylvania. *Proc. Penn. Acad. Sci.* 57:147–154.

Creel, S. (1990). How to measure inclusive fitness. *Proc. Royal Soc. Lond.* B 241:229–231.

Elwood, R. W., & Broom, D. M. (1978). The influence of litter size and parental behaviour on the development of Mongolian gerbil pups. *Anim. Behav.* 26:438–454.

Emlen, S. T. (1982). The evolution of helping I. An ecological constraints model. *Am. Nat.* 119:29–39.

Emlen, S. T., & Wrege, P. H. (1989). A test of alternate hypotheses for helping behavior in white-fronted bee-eaters of Kenya. *Behav. Ecol. Sociobiol.* 25:303–319.

Faulkes, C. G., & Abbott, D. H. (1990). Investigation of genetic diversity in wild colonies of naked mole-rats (*Heterocephalus glaber*) by DNA fingerprinting. *J. Zool., Lond.* 221:87–97.

Faulkes, C. G., Abbott, D. H., & Mellor, A. L. (1991). Social control of reproduction in breeding and non-breeding male naked mole-rats (*Heterocephalus glaber*). *J. Reprod. Fert.* 93:427–435.

Faulkes, C. G., Abbott, D. H., Liddell, C. E., George, L. M., & Jarvis, J. U. M. (1991a). Hormonal and behavioral aspects of reproductive suppression in female naked mole-rats. In *The biology of the naked mole-rat,* ed. P. W. Sherman, J. U. M. Jarvis, & R. D. Alexander, pp. 426–445. Princeton: Princeton University Press.

Faulkes, C. G., Abbott, D. H., & Jarvis, J. U. M. (1991b). Social suppression of reproduction in male naked mole-rats, *Heterocephalus glaber. J. Reprod. Fert.* 91:593–604.

Faulkes, C. G., Trowell, S. N., Jarvis, J. U. M., & Bennett, N. C. (1994). Investigation of numbers and motility of spermatozoa in reproductively active and socially suppressed males of two eusocial African mole-rats, the naked mole-rat (*Heterocephalus glaber*). and the Damaraland mole-rat (*Cryptomys damarensis*). *J. Reprod. Fert.* 100:411–416.

FitzGerald, R. W. (1984). *Population ecology and social biology of a free-ranging population of pine voles,* Microtus pinetorum. PhD dissertation, State University of New York at Binghamton.

FitzGerald, R. W., & Madison, D. M. (1983). Social organization of a free-ranging population of pine voles, *Microtus pinetorum. Behav. Ecol. Sociobiol.* 13:183–187.

French, J. A. (1994). Alloparents in the Mongolian gerbil: Impact on long-term reproductive performance of breeders and opportunities for independent reproduction. *Behav. Ecol.* 5:273–279.

Fried, J. J. (1987). *The role of juvenile pine voles* (Microtus pinetorum) *in the caretaking of their younger siblings.* MS thesis, North Carolina State University, Raleigh.

Fullerton, C., Berryman, J. C., & Porter, R. H. (1974). On the nature of mother–infant interactions in the guinea pig (*Cavia porcellus*). *Behaviour* 48:145–156.

Genelly, R. E. (1965). Ecology of the common mole-rat (*Cryptomys hottentotus*) in Rhodesia. *J. Mamm.* 46:647–665.

Gentry, J. B. (1968). Dynamics of an enclosed population of pine mice, *Microtus pinetorum. Res. Popul. Ecol.* 10:21–30.

Getz, L. L., Dluzen, D., & McDermott, J. L. (1983). Suppression of reproductive maturation in male-stimulated virgin female *Microtus* by a female urinary chemosignal. *Behav. Proc.* 8:59–64.

Getz, L. L., Hofmann, J. E., Klatt, B. J., Verner, L., Cole, F. R., & Lindroth, R. L. (1987). Fourteen years of population fluctuations of *Microtus ochrogaster* and *M. pennsylvanicus* in east central Illinois. *Can. J. Zool.* 65:1317–1325.

Getz, L. L., Gudermuth, D. F., & Benson, S. M. (1992). Pattern of nest occupancy of the prairie vole *Microtus ochrogaster* in different habitats. *Am. Midl. Nat.* 128:197–202.

Getz, L. L., McGuire, B., Pizzuto, T., Hofmann, J. E., & Frase, B. (1993). Social organization of the prairie vole (*Microtus ochrogaster*). *J. Mamm.* 74:44–58.

Getz, L. L., McGuire, B., Hofmann, J. E., Pizzuto, T., & Frase, B. (1994). Natal dispersal and philopatry in prairie voles (*Microtus ochrogaster*): Settlement, survival, and potential reproductive success. *Ethol. Ecol. Evol.* 6:267–284.

Goertz, J. W. (1971). An ecological study of *Microtus pinetorum* in Oklahoma. *Am. Midl. Nat.* 86:1–12.

Gourley, R. S. (1983). *Demography of* Microtus pinetorum. PhD dissertation, Cornell University, Ithaca, N. Y.

Grafen, A. (1982). How not to measure inclusive fitness. *Nature, Lond.* 298:425–426.

Gubernick, D. J., & Laskin, B. (1994). Mechanisms influencing sibling care in the monogamous biparental California mouse, *Peromyscus californicus. Anim. Behav.* 48:1235–1237.

Haigh, G. R. (1987). Reproductive inhibition of female *Peromyscus leucopus:* Female competition and behavioral regulation. *Am. Zool.* 27:867–878.

Hamilton, W. D. (1964). The evolution of social behaviour. *J. Theor. Biol.* 7:1–52.

Hansen, R. M. (1957). Communal litters of *Peromyscus maniculatus. J. Mamm.* 38:523.

Happold, M. (1976). Social behavior of the conilurine rodents (Muridae) of Australia. *Z. Tierpsychol.* 40:113–182.

Hellkamp, A. S. (1991). *The effect of age at pairing on survival, weight, and reproductive output in female pine voles* (Microtus pinetorum). MS thesis, North Carolina State University, Raleigh.

Hoage, R. J. (1977). Parental care in *Leontopithecus rosalia rosalia:* Sex and age differences in carrying behavior and the role of prior experience. In *The biology and conservation of the Callitrichidae,* ed. D. Kleiman, pp. 293–305. Washington, D.C.: Smithsonian Institution Press.

Hoogland, J. L. (1983). Black-tailed prairie dog coteries are cooperatively breeding units. *Am. Nat.* 121:275–280.

Hoogland, J. L., Tamarin, R. H., & Levy, C. K. (1989). Communal nursing in prairie dogs. *Behav. Ecol. Sociobiol.* 24:91–95.

Horsfall, F. Jr. (1964). Pine mouse invasion and reinfestation of orchards subsequent to removal of adjacent woody plant cover or the use of ground sprays. *Am. Soc. Hort. Sci.* 85:161–171.

Jacquot, J. J., & Vessey, S. H. (1994). Non-offspring nursing in the white-footed mouse, *Peromyscus leucopus. Anim. Behav.* 48:1238–1240.

Jannett, F. J. Jr. (1978). The density-dependent formation of extended maternal families of the montane vole, *Microtus montanus nanus. Behav. Ecol. Sociobiol.* 3:245–263.

Jarvis, J. U. M. (1981). Eusociality in a mammal: Cooperative breeding in naked mole-rat colonies. *Science* 212:571–573.

Jarvis, J. U. M. (1991). Reproduction of naked mole-rats. In *The biology of the naked mole-rat,* ed. P. W. Sherman, J. U. M. Jarvis & R. D. Alexander, pp. 384–425. Princeton: Princeton University Press.

Jarvis, J. U. M., & Bennett, N. C. (1993). Eusociality has evolved independently in two genera of bathyergid mole-rats – but occurs in no other subterranean mammal. *Behav. Ecol. Sociobiol.* 33:253–260.

Jarvis, J. U. M., O'Riain, J., Bennett, N. C., & Sherman, P. W. (1994). Mammalian eusociality: A family affair. *Trends Ecol. Evol.* 9:47–51.

Keller, L., & Reeve, H. K. (1994). Partitioning of reproduction in animal societies. *Trends Ecol. Evol.* 9:98–102.

Kerr, L. (1994). *Family influences on sexual maturation in male pine voles.* Honors thesis, Miami University, Oxford, Ohio.

Kingdon, J. (1974). *East African mammals. An atlas of evolution in Africa. II(B). Hares and rodents.* London: Academic Press.

Koenig, W. D., & Pitelka, F. A. (1981). Ecological factors and kin selection in the evolution of cooperative breeding in birds. In *Natural selection and social behavior,* ed. R. D. Alexander & D. W. Tinkle, pp. 261–280. New York: Chiron Press.

Koenig, W. D., Pitelka, F. A., Carmen, W. J., Mumme, R. L., & Stanback, M. T. (1992). The evolution of delayed dispersal in cooperative breeders. *Q. Rev. Biol.* 67:111–150.

König, B. (1994). Components of lifetime reproductive success in communally and solitarily nursing house mice – a laboratory study. *Behav. Ecol. Sociobiol.* 34:275–283.

Lacey, E. A., & Sherman, P. W. (1991). Social organization of naked mole-rat colonies: Evidence for divisions of labor. In *The biology of the naked mole-rat,* ed. P. W. Sherman, J. U. M. Jarvis, & R. D. Alexander, pp. 275–336. Princeton: Princeton University Press.

Lacey, E. A., Alexander, R. D., Braude, S. H., Sherman, P. W., & Jarvis, J. U. M. (1991). An ethogram for the naked mole-rat: Nonvocal behaviors, in *The biology of the naked mole-rat*, ed. P. W. Sherman, J. U. M. Jarvis, & R. D. Alexander, pp. 209–242. Princeton: Princeton University Press.

Lambin, X., & Krebs, C. J. (1991). Spatial organization and mating system of *Microtus townsendii*. *Behav. Ecol. Sociobiol.* 28:353–363.

Lancaster, J. B. (1971). Play-mothering: The relations between juvenile females and young infants among free-ranging vervet monkeys (*Cercopithecus aethiops*). *Folia primat.* 15:161–182.

Lemen, C. A., & Freeman, P. W. (1985). Tracking mammals with fluorescent pigments: A new technique. *J. Mamm.* 66:134–136.

Lepri, J. J., & Vandenbergh, J. G. (1986). Puberty in pine voles, *Microtus pinetorum*, and the influence of chemosignals on female reproduction. *Biol. Reprod.* 34:370–377.

Lovegrove, B. G. (1991). The evolution of eusociality in molerats (Bathyergidae): A question of risks, numbers, and costs. *Behav. Ecol. Sociobiol.* 28:37–45.

Macdonald, D. W. (1981). Dwindling resources and the social behaviour of capybaras (*Hydrochoerus hydrochaeris*) (Mammalia). *J. Zool., Lond.* 194:371–391.

Madison, D. M., & McShea, W. J. (1987). Seasonal changes in reproductive tolerance, spacing, and social organization in meadow voles: A microtine model. *Am. Zool.* 27:899–908.

Manning, C. J., Wakeland, E. K., & Potts, W. K. (1992). Communal nesting patterns in mice implicate MHC genes in kin recognition. *Nature, Lond.* 360:581–583.

Mateo, J. M., Holmes, W. G., Bell, A. M., & Turner, M. (1994). Sexual maturation in male prairie voles: Effects of the social environment. *Physiol. Behav.* 56:299–304.

McGuire, B., & Getz, L. L. (1991). Response of young female prairie voles (*Microtus ochrogaster*) to nonresident males: Implications for population regulation. *Can. J. Zool.* 69:1348–1355.

McGuire, B., & Getz, L. L. (1995). Communal nesting in prairie voles (*Microtus ochrogaster*): An evaluation of costs and benefits based on patterns of dispersal and settlement. *Can. J. Zool.* 73:383–391.

McGuire, B., Getz, L. L., Hofmann, J. E., Pizzuto, T., & Frase, B. (1993). Natal dispersal and philopatry in prairie voles (*Microtus ochrogaster*) in relation to population density, season, and natal social environment. *Behav. Ecol. Sociobiol.* 32:293–302.

McShea, W. J., & Madison, D. M. (1984). Communal nesting by reproductively active females in a spring population of *Microtus pennsylvanicus*. *Can. J. Zool.* 62:344–346.

Mennella, J. A., Blumberg, M. S., McClintock, M. K., & Moltz, H. (1990). Intra-litter competition and communal nursing among Norway rats: Advantages of birth synchrony. *Behav. Ecol. Sociobiol.* 27:183–190.

Meylan, A. (1977). Fossorial forms of the water vole, *Arvicola terrestris* (L.), in Europe. *EPPO Bulletin* 7:209–221.

Michener, G. R., & Murie, J. O. (1983). Black-tailed prairie dog coteries: Are they cooperatively breeding units? *Am. Nat.* 121:266–274.

Millar, J. S., & Derrickson, E. M. (1992). Group nesting in *Peromyscus maniculatus*. *J. Mamm.* 73:403–407.

Miller, D. H., & Getz, L. L. (1969). Life-history notes on *Microtus pinetorum* in central Connecticut. *J. Mamm.* 50:777–784.

Nevo, E. (1979). Adaptive convergence and divergence of subterranean mammals. *Ann. Rev. Ecol. Syst.* 10:269–308.

Nishida, T. (1983). Alloparental behavior in wild chimpanzees of the Mahale Mountains, Tanzania. *Folia primat.* 41:1–33.

Novak, M. (1977). Determining the average size and composition of beaver families. *J. Wildl. Mgt.* 41:751–754.

Nováková, V., & Babicky, A. (1977). Role of early experience in social behaviour of laboratory-bred female rats. *Behav. Proc.* 2:243–253.

Nowak, R. M. (1991). *Walker's mammals of the world.* Baltimore: Johns Hopkins University Press.

Ognev, S. I. (1964). *Mammals of the U.S.S.R. and adjacent countries.* Vol. VII Rodents. Jerusalem: Israel Program for Scientific Translations.

Ostermeyer, M. C., & Elwood, R. W. (1984). Helpers (?) at the nest in the Mongolian gerbil, *Meriones unguiculatus. Behaviour* 91:61–77.

Packer, C., Lewis, S., & Pusey, A. (1992). A comparative analysis of non-offspring nursing. *Anim. Behav.* 43:265–281.

Patenaude, F. (1983). Care of the young in a family of wild beavers, *Castor canadensis. Acta Zool. Fennica* 174:121–122.

Paul, J. R. (1970). Observations on the ecology, populations and reproductive biology of the pine vole, *Microtus pinetorum,* in North Carolina. *Reports of Investigations,* Illinois State Museum, No. 20:1–28.

Payman, B. C., & Swanson, H. H. (1980). Social influence on sexual maturation and breeding in the female Mongolian gerbil (*Meriones unguiculatus*). *Anim. Behav.* 28:528–535.

Pitts, R. M., & Garner, H. W. (1988). An observation of allomaternal nursing among captive *Baiomys taylori. Southwest. Nat.* 33:496–497.

Porter, R. H., & Doane, H. M. (1978). Studies of maternal behavior in spiny mice (*Acomys cahirinus*). *Z. Tierpsychol.* 47:225–235.

Powell, R. A., & Fried, J. J. (1992). Helping by juvenile pine voles (*Microtus pinetorum*), growth and survival of younger siblings, and the evolution of pine vole sociality. *Behav. Ecol.* 3:325–333.

Rabenold, P. P., Rabenold, K. N., Piper, W. H., Haydock, J., & Zack, S. W. (1990). Shared paternity revealed by genetic analysis in cooperatively breeding tropical wrens. *Nature, Lond.* 348:538–540.

Reeve, H. K., Westneat, D. F., Noon, W. A., Sherman, P. W., & Aquadro, C. F. (1990). DNA "fingerprinting" reveals high levels of inbreeding in colonies of the eusocial naked mole-rat. *Proc. Natl. Acad. Sci.* 87:2496–2500.

Ribble, D. O. (1991). The monogamous mating system of *Peromyscus californicus* as revealed by DNA fingerprinting. *Behav. Ecol. Sociobiol.* 29:161–166.

Rood, J. P. (1972). Ecological and behavioural comparisons of three genera of Argentine cavies. *Anim. Behav. Monog.* 5:1–83.

Saitoh, T. (1989). Communal nesting and spatial structure in an early spring population of the grey red-backed vole, *Clethrionomys rufocanus bedfordiae. J. Mamm. Soc. Japan* 14:27–41.

Salo, A. L., & French, J. A. (1989). Early experience, reproductive success, and development of parental behaviour in Mongolian gerbils. *Anim. Behav.* 38:693–702.

Saylor, A., & Salmon, M. (1971). An ethological analysis of communal nursing by the house mouse (*Mus musculus*). *Behaviour* 40:60–85.

Schadler, M. H. (1983). Male siblings inhibit reproductive activity in female pine voles, *Microtus pinetorum. Biol. Reprod.* 28:1137–1139.

Schadler, M. H. (1990). Social organization and population control in the pine vole, *Microtus pinetorum.* In *Social systems and population cycles in voles,* ed. R. H. Tamarin, R. S. Ostfeld, S. R. Pugh, & G. Bujalski, pp. 121–130. Basel: Birkhauser Verlag.

Seger, J. (1991). Cooperation and conflict in social insects. In *Behavioural ecology: An evolutionary approach* 3rd ed., ed. J. R. Krebs & N. B. Davies, pp. 338–373. Oxford, U.K.: Blackwell.

Sherman, P. W., Lacey, E. A., Reeve, H. K., & Keller, L. (1995). The eusociality continuum. *Behav. Ecol.* 6:102–108.

Solomon, N. G. (1984). *Allomaternal behavior in the albino rat: The influence of olfactory stimuli on recognition of pups.* MS thesis. Case Western Reserve University, Cleveland, Ohio.

Solomon, N. G. (1991a). *Indirect fitness benefits to philopatric juvenile prairie voles* Microtus ochrogaster. PhD dissertation, University of Illinois at Urbana–Champaign.

Solomon, N. G. (1991b). Current indirect fitness benefits associated with philopatry in juvenile prairie voles. *Behav. Ecol. Sociobiol.* 29:277–282.

Solomon, N. G. (1993). Body size and social preferences of male and female prairie voles, *Microtus ochrogaster. Anim. Behav.* 45:1031–1033.

Solomon, N. G. (1994). Effect of the pre-weaning environment on subsequent reproduction in prairie voles, *Microtus ochrogaster. Anim. Behav.* 48:331–341.

Solomon, N. G., Vandenbergh, J. G., & Sullivan, W. T. Jr. (in prep.). Social organization and reproduction in pine voles (*Microtus pinetorum*) in an orchard habitat. *J. Mamm.*

Solomon, N. G., & Vandenbergh, J. G. (1994). Management, breeding, and reproductive performance of pine voles. *Lab. Anim. Sci.* 44:612–616.

Stacey, P. B., & Ligon, J. D. (1987). Territory quality and dispersal options in the acorn woodpecker, and a challenge to the habitat-saturation model of cooperative breeding. *Am. Nat.* 130:654–676.

Stacey, P. B., & Ligon, J. D. (1991). The benefits of philopatry hypothesis for the evolution of cooperative breeding: Variation in territory quality and group size effects. *Am. Nat.* 137:831–846.

Taber, A. B., & Macdonald, D. W. (1992). Communal breeding in the mara, *Dolichotis patagonum. J. Zool., Lond.* 227:439–452.

Thomas, S. L. (1993). *Effect of juvenile helpers on parental care of younger siblings in pine vole pairs* (Microtus pinetorum). MS thesis, University of Massachusetts, Amherst.

Vehrencamp, S. L. (1983). A model for the evolution of despotic versus egalitarian societies. *Anim. Behav.* 31:667–682.

Wang, Z. (1991). *Effects of social environment and experience on parental care, behavioral development, and reproductive success of prairie voles* (Microtus ochrogaster). PhD dissertation, University of Massachusetts, Amherst.

Wang, Z., & Novak, M. A. (1992). Influence of the social environment on parental behavior and pup development of meadow voles (*Microtus pennsylvanicus*) and prairie voles (*M. ochrogaster*). *J. Comp. Psych.* 106:163–171.

Waser, P. M., & Jones, W. T. (1983). Natal philopatry among solitary mammals. *Q. Rev. Biol.* 58:355–390.

Wasser, S. K., & Barash, D. P. (1983). Reproductive suppression among female mammals: Implications for biomedicine and sexual selection theory. *Q. Rev. Biol.* 58:513–538.

Wilkinson, G. S., & Baker, A. E. M. (1988). Communal nesting among genetically similar house mice. *Ethology* 77:103–114.

Wolff, J. O. (1994). Reproductive success of solitarily and communally nesting white-footed mice and deer mice. *Behav. Ecol.* 5:206–209.

9

The Psychobiological Basis of Cooperative Breeding in Rodents

C. SUE CARTER and R. LUCILLE ROBERTS

9.1 Introduction

Cooperative mating systems require coordinated reproductive efforts among the members of a breeding unit. To achieve coordinated reproduction, cooperative mammalian species commonly possess mechanisms for reproductive and social suppression, limiting the number of reproductive individuals present and establishing a social hierarchy that provides structure for the breeding unit. Cooperative breeding systems are probably formed as a result of two primary factors: (1) ecological conditions that do not favor independent reproduction of subordinate members of the breeding unit, and/or (2) conditions under which the successful production of offspring by a breeding unit requires the effort of an extended family. The proximate factors resulting in philopatry, reproductive suppression, and alloparental behavior by subordinate members of the breeding unit are the topics of this chapter.

The importance of sociality and cooperative behavior to successful reproduction is particularly apparent in "monogamous" mammals, including animals as diverse as prairie voles (Carter, Getz, & Cohen-Parsons 1986; Carter & Getz 1993), Mongolian gerbils (Swanson 1985), elephant shrews (Rathbun 1979), dwarf mongoose (Rood 1980), aardwolves (Richardson 1987), and some New World primates, including marmosets and tamarins (Abbott et al. 1989; Snowdon 1990; Abbott 1993). Monogamy has been characterized by a complex of features including high levels of social behavior and pair bonds, exclusion of strangers from the family, biparental care, reproductive suppression of subordinate members of a family group, incest avoidance, and reduced sexual dimorphism (Kleiman 1977; Dewsbury 1981, 1988; Mendoza & Mason 1986). Although monogamy is not an essential component of a cooperative breeding system, the unique characteristics of monogamous mammals, such as the production of altricial young and reproductive suppression of offspring in the natal group, also favor the development of extended families and cooperative breeding. Therefore, some discussion of the psychobiology of monogamy is beneficial to our understanding of the proximate mechanisms for cooperative breeding in rodents.

9.2 Monogamy and Cooperative Breeding

Throughout this chapter we will use the monogamous prairie vole (*Microtus ochrogaster*) as a model system for the analysis of proximate mechanisms underlying the expression of reproductive suppression, monogamy, philopatry, and alloparental behavior. Prairie voles provide a particularly striking example of a species in which social factors regulate reproduction. The physiology of monogamy, reproductive suppression, and, to a lesser extent, philopatry and alloparental behavior have been researched in this species, and a relatively large literature exists on the proximate mechanisms for social behavior in prairie voles (Roberts 1994; Wang, Ferris, & DeVries 1994; Williams et al. 1994; Carter, DeVries, & Getz 1995). Where examples exist, research from other rodent species also will be discussed.

We use the term "monogamy" here in reference to a social system rather than a mating system. Both social organization and mating systems may be flexible, and "monogamous" mammals sometimes exhibit polyandrous or communal breeding systems. In prairie voles sexual exclusivity is not an absolute characteristic of "monogamy." However, most female prairie voles show a strong social preference for their established male partner (Carter et al. 1992; Williams, Catania, & Carter 1992). Female prairie voles are more selective in their social preferences, at least as measured by lateral physical (side-by-side) contact than they are in mating preferences. In nature, prairie vole pairs live together as long as both partners survive, but this may average only a few months. If one member of a pair dies, fewer than 20 percent of the survivors form new pairs (McGuire et al. 1993). Thus, the success of and the mechanisms responsible for an individual's first reproductive experience are of particular importance to reproductive fitness in this species. The first litter, produced by mechanisms that will be described, also may form the core of the extended family and the cooperative breeding unit. In prairie voles, reproductive activation of the female, including the induction of estrus and ovulation, depends almost totally on social cues from an unfamiliar male. Males show a parallel but less complete dependence on social cues to become reproductively active (Carter & Getz 1985). Monogamous pairs and their offspring form the core of a communal breeding group. Within these groups, reproductive suppression of the offspring, incest avoidance, social preferences for the familiar sexual partner, and active defense of the territory and mate promote the continued breeding of the original pair and the concurrent inhibition of reproduction in other members of the group. In this chapter we will describe a variety of physiological and behavioral mechanisms responsible for these adaptations, which in turn regulate monogamy and cooperative breeding.

9.3 Social Stimuli and Female Reproduction

The hypothalamic–pituitary–gonadal axis receives input from the vomero-nasal organ and olfactory system, providing a mechanism through which olfactory signals can mediate reproductive activation and suppression. The specific pathways and neurotransmitters involved have been examined in a variety of mammalian species, and there are outstanding reviews of this topic elsewhere (cf. Doty 1976; Vandenbergh 1988). Pheromonal input is known to regulate various aspects of reproduction, including the timing of puberty (Vandenbergh 1988), reproductive synchrony (McClintock 1987), reproductive suppression, reproductive activation, spontaneous abortion (Bruce 1959; Kenney, Evans, & Dewsbury 1977), and parental behavior (Gubernick 1990). Many aspects of reproduction are regulated or modulated by pheromonal factors in microtine rodents, as will be discussed.

9.3.1 Behavioral Requirements for Reproductive Activation in Female Prairie Voles

Prior to exposure to an unfamiliar male and irrespective of age, female prairie voles are functionally prepubescent and do not show the ovarian cycles typical of most laboratory rodents and primates (Richmond & Conaway 1969; Richmond & Stehn 1976). In montane voles (*Microtus montanus*) (Jannett 1980) and meadow voles (*Microtus pennsylvanicus*), which are not monogamous, there are indications of endogenous ovarian activation but not true "cycles" in the absence of males (Sawrey & Dewsbury 1985; Shapiro & Dewsbury 1990).

The young female prairie vole relies on chemical signals and other cues from a male to initiate the endocrine events that lead to behavioral estrus and mating (Carter et al. 1978b; Carter et al. 1980). In addition, a reproductively inexperienced male also may require stimuli from a novel female to induce his own reproductive activation (Carter et al. 1986).

When a reproductively naive female is placed with an unfamiliar male, they immediately engage in reciprocal sniffing (Getz & Carter 1980). During this time, chemical signals (pheromones) that stimulate the release of sex steroids are exchanged. Brief stimulation usually results in an increase in uterine weight (indicative of increased estrogen secretion) (Carter et al. 1980) but is not adequate to induce behavioral estrus in most females. However, if females that are briefly exposed to a novel male or his odors are left in a cage with male-soiled bedding, most will begin to show sexual receptivity within 24 to 48 hours. The continued presence of male pheromones is necessary to reinforce the processes responsible for behavioral estrus (Carter et al. 1987b).

An unfamiliar male is required to initiate the sequence of events leading to behavioral estrus in the female prairie vole. Familiar males, such as a father or brother, are not capable of inducing behavioral estrus (Carter & Getz 1985). Females do not direct their sniffing toward familiar males; however, if urine collected from a familiar male is applied to the nose of a young female, estrus can be induced (McGuire & Getz 1981). These findings suggest that the novel male is particularly effective in eliciting sniffing, which results in a transfer of pheromones and subsequent reproductive activation.

The absence of a spontaneous estrous cycle and the failure of familiar males to activate female estrus provides a means for controlling reproduction in young females within a family group. In the absence of ovarian activity, females may remain functionally prepubescent, subjugating their own reproduction while providing support for the communal family.

9.3.2 Neuroendocrine Correlates of Reproductive Activation in Female Prairie Voles

A single drop of male urine on the nose of a female prairie vole produces rapid changes in the neurotransmitter, norepinephrine, and in luteinizing hormone-releasing hormone (LHRH), in the posterior olfactory bulb (Dluzen et al. 1981; Carter et al. 1991). Both norepinephrine and LHRH have been implicated in the control of reproduction in other species. However, these substances are traditionally studied more deeply in the brain, within the hypothalamus, rather than in the olfactory bulb. The finding that urine could alter these hormones at a more peripheral level, in the olfactory bulb, suggests a direct role for the chemical senses in reproduction in prairie voles.

Removal of the olfactory bulb reduces the probability that females will come into estrus (Richmond & Stehn 1976; Williams et al. 1992). Selective removal of peripheral sensory receptors in the vomeronasal organ also inhibits estrus induction (Lepri & Wysocki 1987). Nerve cells from the vomeronasal organ project to the posterior part of the olfactory bulb, implicating the vomeronasal organ and its connections in the activation of ovarian function and estrus.

Brief contact with a male or male urine, presumably mediated by the olfactory system and the release of LHRH, results in the subsequent release of a surge of luteinizing hormone (LH) from the pituitary gland into the blood stream. In conjunction with other endocrine changes, LH begins a cascade of chemical and neural events that stimulate the ovary to secrete gonadal steroids, including estradiol (Cohen-Parsons & Carter 1987; Carter et al. 1989).

Estradiol is the most potent of the estrogens and usually is essential for estrus induction in rodents (Pfaff & Schwartz-Giblin 1988). Estrogen, of either ovarian or exogenous origin, acts on the nervous system to increase binding of estrogen to its nuclear receptors (Cohen-Parsons & Carter 1987) and may cause a subsequent down-regulation of estrogen receptors (ER), as identified by ER immunoreactivity (Hnatczuk et al. 1994). If the ovary is removed, female prairie voles will not come into behavioral estrus. Within about 24 hours, injections of estradiol restore the willingness of the female vole to mate, and ovariectomized females that receive repeated estrogen injections will remain in behavioral estrus for several days or longer.

Levels of estradiol in the ovary and progesterone levels in the ovary and in the blood have been measured in female prairie voles (Carter et al. 1989). Estrogen levels were elevated in females in heat and declined following mating. Progesterone levels were similar in estrous and nonestrous females, became elevated within a few hours after the onset of mating in the ovary, but did not rise in serum until at least 24 hours later. In such species as rats (Allen & Adler 1985) and nonmonogamous montane voles (Gray et al. 1976), which have a briefer period of behavioral estrus, progesterone levels in serum rise within a few hours following exposure to coital stimuli. Postcopulatory progesterone may facilitate sperm transport and concurrently shorten the female's period of behavioral receptivity. The finding in prairie voles that the secretion of progesterone from the ovary into serum is delayed (Carter et al. 1989), and thus presumably not available to other target tissues, could explain in part the exceptional capacity of female prairie voles to remain in behavioral estrus for more than a day.

In addition, prairie voles have high levels of the adrenal hormone, corticosterone. Corticosterone levels in prairie voles are 3 to 10 times those measured in rats and nonmonogamous montane voles (DeVries et al. 1995). The behavioral actions of corticosterone are incompletely known, and most behavioral studies of this hormone focus on changes in corticosterone levels as a function of exposure to putative stressors. However, corticosterone may also be behaviorally active (File, Vellucci, & Wendlandt 1979) and recently has been shown to be capable of stimulating behavioral receptivity in estrogen-primed female rats (McGinnes, Rutenberg, & Lumia 1993). Corticosterone and pro-gesterone share structural and functional similarities and may influence each other's receptors.

In many laboratory rodents, such as rats and hamsters, progesterone secreted just prior to ovulation plays a major role in timing the onset of behavioral estrus (Dluzen & Carter 1979; Carter et al. 1987a). Progesterone typically has a biphasic action, initially facilitating sexual behavior and subsequently inhibiting receptivity. The production of progestin receptors is correlated with the behavioral effects of progesterone and may be necessary for the behavioral actions attributed to progesterone. In prairie voles, as in other rodents, estrogen can enhance the production of progestin receptors (Cohen-Parsons & Carter 1988). However, in prairie voles progesterone does not reliably facilitate sexual behavior (Dluzen & Carter 1979). Progestin receptors induced by estrogen could play a role in the inhibition of female behavioral estrus. In addition, progestin receptors might be occupied by another hormone, such as the highly abundant steroid, corticosterone. On the basis of circumstantial evidence, we hypothesize that corticosterone and related neuroendocrine changes may support the behavioral processes necessary for initial estrus induction, possibly reducing fear of strangers (File et al. 1979) and making female prairie voles less reactive to copulatory stimuli.

9.3.3 Possible Functions of Extended Copulatory Interactions

In rats, hamsters, guinea pigs, and various other species, copulatory stimuli play an important role in regulating the duration of female sexual receptivity. For example, in guinea pigs mating usually is limited to a single ejaculation (Goldfoot & Goy 1970). In hamsters, unmated females normally stay in heat for about 15 to 20 hours, and as little as 45 minutes of mating terminates behavioral estrus (Carter & Schein 1971). In light of this work and similar findings in other rodents, we were surprised to find that coital stimulation had relatively little effect on subsequent sexual receptivity in naive female prairie voles. When females are in natural estrus, male and female prairie voles continued to engage in bouts of mating for approximately 30 hours (Witt et al. 1988). When females are given exogenous estrogen, they may remain in heat and mate intermittently for days (Carter et al. 1989).

Patterns of sexual behavior can influence the capacity of sperm to enter the female's reproductive tract and fertilize an egg (Adler 1969; Allen & Adler 1985). However, naive prairie voles continue to copulate for hours after they have met the requirements for pregnancy (Gray et al. 1974; Pierce et al. 1988). Extended copulatory interactions also may facilitate the formation of a social bond between the male and female. Relatively nonsocial voles, such as meadow voles (Roberts & Latchis unpublished data) and montane voles (Williams and Carter unpublished data), show lower levels of sexual behavior, and the male and female do not remain in physical contact following mating. Like the asocial hamster, meadow voles restrict their interactions to periods when the female is in behavioral estrus, and relatively brief mating is adequate to assure pregnancy.

Thus, work with prairie voles, particularly when viewed in the context of less social mammals, focuses attention on the importance of sexual interactions as pivotal events in the mammalian life cycle. If the male leaves or is expelled from the female's nest after mating, this may preclude the development of an extended family. Thus, behavioral interactions associated with an initial sexual experience are critical determinants of family structure and may be one of the proximate mechanisms responsible for monogamy and subsequent cooperative breeding. Reproductive hormones in turn mediate the expression of sexual behavior and postcopulatory social behaviors.

We postulate that prairie voles use a slightly different hormonal "cocktail" for estrus induction than do most rodents. This cocktail requires estrogen but apparently does not rely on progesterone. However, other hormones, such as corticosterone, may be behaviorally active. In combination these hormones could allow female prairie voles in their first estrus to remain in behavioral estrus and to mate for an unusually long period. This prolonged period of sexual receptivity, which is not abbreviated by mating or the inhibitory action of progesterone, might encourage sexual partners to remain together and thus enhance the formation of pair bonds.

9.3.4 Postpartum Estrus

Female prairie voles also come into behavioral estrus at approximately the time that they give birth. This postpartum estrus is presumably induced by hormonal events associated with pregnancy and may differ hormonally from male-induced estrus (Carter et al. 1989). In contrast to the prolonged sexual interactions that characterize the estrus of naive females, copulatory interactions during postpartum estrus are usually brief and sometimes limited to a single ejaculation (Witt et al. 1989). The demands of a new litter probably inhibit the expression of estrus behaviors. In addition, the short duration of postpartum estrus may be caused in part by a relatively transient secretion of ovarian estrogen.

In addition, during the postpartum period the production of another hormone, oxytocin, which is essential for birth and lactation, is probably elevated. Oxytocin facilitates female sexual behavior in rats (Caldwell 1992) but not in prairie voles (Witt, Carter, & Walton 1990). Female prairie voles respond to centrally administered exogenous oxytocin with declines in sexual receptivity. In addition, when female prairie voles remain with their infants, presumably nursing and maintaining the endogenous secretion of oxytocin, they also show an abbreviated estrus. Females in postpartum estrus reject social contact with strangers (Getz, Carter, & Gavish 1981; Witt et al. 1989), are usually paired with an established partner (Carter, Williams, & Witt 1990; McGuire et al. 1993), and thus may not "need" extended copulatory stimulation to form a new pair bond.

9.3.5 Reproductive Inhibition by Social Factors

9.3.5.1 Inhibition of estrus induction

Most female prairie voles that are exposed to male stimuli respond with increased uterine weight, indicative of reproductive activation. However, pheromones in the urine from other females are capable of inhibiting this activational process (Getz, Dluzen, & McDermott 1983). It is likely that at high population densities females living within communal groups are exposed to stimulation from male intruders (McGuire et al. 1993). Male stimulation, which might induce estrus in females living alone, is probably ineffective in females living in the presence of their mother and sisters. Young pine vole females (*Microtus pinetorum*) that were housed with their mothers or an unrelated, reproductively mature female since weaning experienced reduced reproductive activation upon the introduction of a novel male, although reproductive suppression did not occur among sexually naive females in the absence of a reproductively dominant female (Schadler 1990).

Pine voles are another monogamous, cooperatively breeding microtine rodent (Solomon & Getz this volume). Schadler (1983) demonstrated that female pine voles experienced significant reductions in estrus activation when housed with a novel male in the presence of their brothers. Active suppression of reproduction by male siblings has not been observed in prairie voles. Rather, reproductive activation of females by their brothers is believed not to occur in prairie voles because familiar animals do not spend time engaged in nasogenital grooming (McGuire & Getz 1981).

9.3.5.2 Incest avoidance

Breeding within a family group also is suppressed by incest avoidance (Gavish, Hofmann, and Getz 1984). In laboratory studies, when females are removed from the family long enough to allow estrus induction and are placed with either their father or brother, they will rarely mate, although such females will engage in sexual behavior with unfamiliar males. Mothers in postpartum estrus also will not mate with their sons (Carter & Getz 1985).

9.3.5.3 Female–female interactions and social stress

Social interactions in females also are affected by the sexual history of the female. In the laboratory, male–female prairie vole pairs generally produce litters, and aggression is rare (Gavish, Carter, & Getz 1981). When trios are composed of two sisters and an unfamiliar male, both females also usually produce offspring and cohabitation is peaceful. However, if trios are created of two unfamiliar females and a male, only one female reproduces and the second female typically dies before the other delivers her young. Under these conditions, both nonsibling females usually mate with the male. However, the reproductively successful female excludes from social contact with the male the female that eventually dies. The females may fight, but the cause of death is probably not physical aggression. In at least some cases, the second female seems to die of social "stress" (Firestone, Thompson, & Carter 1991).

We have assayed corticosterone levels in this species in groups composed of (1) two sibling females and a male, (2) two nonsibling females and a male, (3) male–female pairs, (4) singly housed females in male-induced estrus, and (5) singly housed females that failed to respond to male exposure with estrus induction. Baseline corticosterone levels are exceptionally high in prairie voles (approximately 1,000 ng/ml or more in an assay in which rats range from 60 to 100 ng/ml), and prairie voles do not show strong circadian variation in corticosterone production. Measurements taken three days after group assignment varied among the groups. Corticosterone levels remained high in females that did not come into estrus (nonreceptive females). Females that became behaviorally receptive (following exposure to male stimuli) usually showed reductions in corticosterone, but were not statistically different from females that were not in estrus. Females in nonsibling trios showed much larger reductions in corticosterone levels following trio formation, although corticosterone levels declined in both the female that eventually reproduced and the one that died. Corticosterone levels on the third day of pairing did not predict which female would survive, but the very low levels in both females probably reflect intense stress experienced by female prairie voles that were forced to interact with an unrelated female (Firestone, Grady, and Carter unpublished data).

9.4 Social Stimuli and Male Reproductive Behavior

9.4.1 Male reproductive activation

The effects of female exposure on male behavior and reproductive activation under a variety of social conditions have been described in male prairie voles (Richmond & Stehn 1976). Placing a male with a naive female, without copulatory interactions, is adequate to elicit in the male an increase in LH and a decrease in corticosterone (measured within 10 to 15 minutes) and increases in testosterone levels (measured within 24 hours) (Gaines et al. 1985; Carter et al. 1991; Bamshad, Novak, & DeVries 1994). In our own work, significant increases in testosterone were not observed within 10 minutes or after one or two ejaculations (about 1 hour). Testosterone levels in males allowed to engage in *ad libitum* sexual activity for 24 hours or even longer cohabitation (21 days) with a female did not differ from males that were simply exposed to a female and housed alone. However, in general, testosterone levels are highly variable in prairie voles (see Section 9.4.2). Declines in corticosterone in males following female exposure are more reliable, suggesting the possibility that the very high levels of corticosterone in this species may inhibit reproduction, while reductions in corticosterone may be part of the process required for reproductive activation.

9.4.2 Social Suppression of Male Reproduction

In social mammals the younger members of a family typically are nonreproductive while they remain in the natal group. For example, in prairie voles about 70 percent of the young males that escape juvenile mortality never emigrate from the family nest (McGuire et al. 1993). In addition, unrelated adults can join social groups, and at least some of these also remain nonreproductive. The suppression of reproduction may be important to the stability of a family group.

The probability of mating in male prairie voles is also highly variable, and (as in females) some males never show sexual behavior. We have found that living alone from weaning is associated with significantly higher levels of sexual performance on first exposure to a female at 55 days of age (Carter et al. 1991). Caging with a brother, sister, father, or mother produced comparable inhibitions of male sexual behavior, although all groups showed increases in the probability of mating after repeated (four daily) exposures to estrous females. Repeated exposure to receptive females usually increases the number of males that eventually show masculine sexual behavior, although a percentage (which varies from study to study) consistently fail to show sexual behavior even after repeated exposure to estrous females.

In many species, and especially in males, it appears that reproductive inhibition often occurs at the behavioral level rather than as a failure to produce gametes (Nelson et al. 1989). Male mammals require many weeks for sperm development and must "plan ahead" if they are to have viable sperm available when opportunities to mate occur. In contrast, access to a sexual partner might occur on short notice. Changes in social factors, such as the death of a parent, dominance relationships, or the availability of a novel suitable partner, have rapid effects on the expression of male sexual behavior or other indexes of masculine behavior (Swanson 1985).

Males caged with a brother from weaning show an increase in testosterone at around 45 days of age and then (if left only in the presence of males) show a decline by 55 and 75 days of age to very low levels of testosterone. In a preliminary study, we found evidence that the presence of female pheromones in the male's colony room (airspace) apparently prevented the age-related decline in testosterone (Witt et al. 1985). Luteinizing hormone levels also undergo an age-dependent decline under conditions of unisexual sibling housing (Roberts 1994).

Testosterone levels in adult males living in family groups remain very low (Carter et al. unpublished data), although exposure to a female can increase testosterone levels (Gaines et al. 1985; Bamshad et al. 1994). However, in our experience, testosterone levels in male prairie voles are highly variable and may be elevated in nonreproductive males, possibly because androgens are necessary to allow spermatogenesis to proceed. More predictably related to sexual behavior is a decline in corticosterone levels, which occurs immediately when naive males are exposed to a novel female and tends to last for several days (Carter et al. unpublished data).

In addition to familial influences, dominance relationship can also influence the male prairie vole's reproductive success. Females prefer dominant males (Shapiro & Dewsbury 1986). However, female preference for males is random prior to male–male confrontation and does not predict dominance, suggesting that behavioral–physiological events following male–male interactions alter the characteristics of the male in a manner that can be detected by females (Hastings et al. unpublished data).

In nonmonogamous mammals, such as domestic rats, males produce several times more testosterone than is necessary to assure either sperm production or male sexual behavior (Davidson & Levine 1972). However, in highly social mammals such as prairie voles, in which reproductive suppression is important, males may not engage in "excess" hormone production. Instead, a somewhat precarious balance probably exists between endocrine factors promoting reproductive activation and sexual motivation (such as androgens) and those encouraging reproductive suppression (such as adrenal corticoids). Of possible relevance to this hypothesis is the finding that basal glucocorticoid levels are also elevated in monogamous marmosets and tamarins (Chrousos et al. 1982). It has long been assumed that adrenal activity could contribute to the regulation of vertebrate reproduction and population dynamics (Christian

& Davis 1964), although interactions between the gonadal and adrenal axes are complex and many of these interactions remain to be described. In general, endocrinological data on social mammals are scarce, and our understanding of the factors responsible for reproductive suppression in either males or females is inadequate.

9.4.3 Factors Regulating Patterns of Male Sexual Behavior

One unusual feature of male sexual behavior in prairie voles is our finding that sexual activity with intermittent mating bouts can continue over a period of several days, with little evidence of behavioral "satiety" on the part of the male. The duration of mating bouts is strongly determined by the hormonal condition of the female. Mating may last only a few hours or less when males are tested with females in postpartum estrus (Witt et al. 1989), usually extends about 24 to 30 hours when females are in male-induced estrus (Witt et al. 1988), and can continue intermittently for 3 days or longer when males are tested with ovariectomized females in estrogen-induced heat (Witt et al. 1988). The mechanisms that account for the exceptional "potency" of male prairie voles have not been studied. However, these prolonged sexual interactions may serve to strengthen a pair bond and also could be a form of mate guarding.

9.5 Pair Bonding

The first indications of monogamy in prairie voles came from live-trapping studies (Getz & Carter 1980; Getz, Hofmann, & Carter 1987; Getz et al. 1993; McGuire et al. 1993). Males and females of this species share a territory and cohabit a common nest so long as they both remain alive. Male and female prairie vole pairs not only live in the same nest but are repeatedly captured together in the live-traps, sometimes bringing along their newly emerged young. Field studies of prairie vole social groups indicate that pairs may remain together throughout the breeding season and even during periods of reproductive quiescence. In contrast, under comparable conditions, meadow voles were rarely retrapped with the same partner and showed no indication of stable families. Female meadow voles occupied individual territories that they defended against other females. Male meadow voles, on the other hand, occupied large home ranges, overlapping the home ranges of other males, as well as the territories of more than one female.

In the laboratory, pair bonding is indexed by two easily defined behaviors: (1) partner preferences in a choice apparatus (Ferguson, et al. 1986; Shapiro et al. 1986; Newman & Halpin 1988), and (2) selective aggression directed toward intruders but not family members (Getz et al. 1981). Prior to having sexual experience, reproductively naive prairie voles are not aggressive toward either familiar animals or strangers of either sex. Following their first sexual experience, male and female pairs that are allowed to remain together become aggressive toward strangers (Getz et al. 1981; Gavish, Carter, & Getz 1983; Winslow et al. 1993). Established partners do not fight with each other but will threaten and attack intruders. In nature, this behavior presumably translates to defense of a territory, a mate, or the offspring. In the laboratory, we have used this model to examine physiological processes responsible for pair bonding.

In our first attempts to describe partner preferences in prairie voles, we used brief (10 minutes) tests in which females were permitted to spend time and/or mate with either their familiar partner or a stranger. In these trials, the female investigated and in some cases mated with a stranger as well as her partner. Recent results using DNA fingerprinting suggest that in nature female prairie voles do not show absolute sexual monogamy (Carter et al. 1990; Carter et al. 1992). However, a more careful examination of our laboratory results revealed that, although females were willing to mate with an unfamiliar male, they rarely remained in physical contact with the stranger after mating. To obtain a more reliable measure of preferences, females were given longer tests (three hours) in a relatively large T-maze containing either a familiar animal (partner) or an unfamiliar animal (stranger) (Williams et al. 1992a). Using this test procedure, we found that stable preferences for a male partner could be established if animals were allowed to interact sexually for 6 hours before testing. In pairs that did not mate, 6 hours of cohabitation did not result in a preference, and females were equally likely to spend time with either the familiar or an unfamiliar male. These results suggested the possibility that hormones or neurochemicals released during mating could influence partner preferences (Table 9.1, Figure 9.1). In contrast, meadow voles (Wilson 1982a) and montane voles show much lower levels of social contact than do prairie voles and do not show the development of clear partner preferences (Ferguson et al. 1986; Shapiro et al. 1986).

In our laboratory, DeVries and her colleagues have examined same-sex pair bonds in females. When pairs of unrelated adult females are housed together for 24 hours prior to preference testing, they show a preference for their partner versus a novel female. When a novel male is used as a stimulus animal, females divide their time approximately evenly between the male and the female partner with which they have cohabited. By contrast, when the female and male partner are both novel to the subject, they exhibit a prefer-

Table 9.1. *Factors influencing partner preference, intrasexual aggression, and parental care in prairie voles*

Test or Treatment	MALE			FEMALE		
	Partner preference	Same-sex aggression	Parental behavior	Partner preference	Same-sex aggression	Parental behavior
Cohabitation (6 hr)	No preference, or preference for unfamiliar female	Low	High	Familiar male preferred	Low	High
Sexual behavior	Familiar female preferred	High	High	Familiar male preferred	Low[a]	High
Gonadectomy & 24 hr cohab	Familiar female preferred	Low	Reduced	Familiar male preferred	Low	
Adrenalectomy & 24 hr cohab	No preference	Low-moderate		Familiar male preferred after as little as 1 hr cohabitation	Low	
Estrogen (2 days)				Familiar male preferred	Low, but slightly increased	
Corticosterone	Familiar female preferred (after 6 hr cohab–no mating)		High	Familiar male not preferred	Low	
Oxytocin (CNS)	No preference (low dose OT)[b]	Low		Familiar male preferred	Low	
Vasopressin (CNS)	Familiar female preferred	Moderate-high (without mating)	Very high			

[a]Females that are in late pregnancy (Getz et al. 1981) or postpartum estrus (Witt et al. 1989) are highly aggressive.
[b]Doses of OT used in males were matched to doses of vasopressin that were effective in increasing aggression in females (Williams et al. 1994)

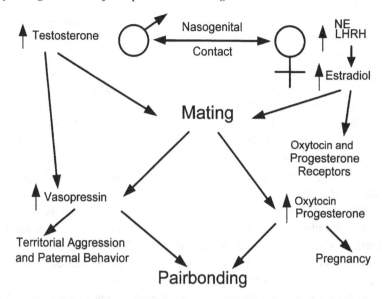

Figure 9.1. A flowchart of the course of putative events leading to the formation of a mating pair and pair bonding in male and female prairie voles and including factors associated with the activation of paternal behavior and territorial aggression in males.

ence for the male. This suggests that in prairie voles, the phenomenon we call pair bonding is not unique to heterosexual pairs, and the physiological mechanisms discussed in the next section might mediate interactions between members of communal groups other than the breeding pair.

9.5.1 Physiology of Partner Preference Formation in Female Prairie Voles

It has been known for many years that oxytocin is released as a result of genital stimulation such as that which occurs during birth or sexual behavior (reviewed by Carter 1992]. When female prairie voles received central injections of oxytocin, they engaged in more social behavior and less fighting with their male partner than did females receiving the saline control (Witt et al. 1990). (Oxytocin treatment also increased social contact in rats [Witt, Winslow, & Insel 1992] and squirrel monkeys [Winslow & Insel 1991], indicating that this action of oxytocin is not limited to monogamous mammals.) In sheep either vaginal stimulation or oxytocin administration can facilitate the formation of mother–infant bonds (Keverne & Kendrick 1992). These findings indicate that oxytocin facilitates affiliative behaviors and suggest the possibility that stimulation experienced during mating might release oxytocin, which in turn could hasten the formation of male–female social bonds (see Table 9.1 & Figure 9.1).

Using the partner preference paradigm described, we have found that infusion of oxytocin into the nervous system is associated with an increased preference in female prairie voles for males that are present when the oxytocin is given. When a drug that blocked oxytocin receptors was administered with oxytocin, the oxytocin treatment no longer produced a significant preference for the familiar partner. Thus, oxytocin released during mating may act directly to facilitate the formation of partner preferences, at least in females (Williams et al. 1994).

Cohabitation also facilitates the development of partner preferences, which may develop after a period of a day or less of living together (Williams, Catania, & Carter 1992). The mechanisms for cohabitation-induced partner preferences are not known. However, oxytocin can also be released by touch in rats (Stock & Uvnas-Moberg 1988), leaving open the possibility that cohabitation- and mating-induced partner preferences might have comparable underlying mechanisms.

Monogamous mammals exhibit patterns of oxytocin receptors different from those of nonmonogamous species (Insel, Gelhard, and Shapiro 1991; Witt, Carter, & Insel, 1991; Insel & Shapiro 1992; Insel et al. 1993). Species differences in patterns of oxytocin receptors support the hypothesis that oxytocin might also be involved in species differences in social organization.

9.5.2 Physiology of Partner Preferences in Male Prairie Voles

Male prairie voles also show partner preferences, although the parameters of these preferences differ from those observed in females. However, in males that were simultaneously housed with a female and treated centrally with vasopressin but not oxytocin, a significant preference was observed for the female present during vasopressin treatment (Winslow et al. 1993). In male prairie voles that are allowed to mate, it is possible to detect changes in vasopressin concentrations in selected brain areas, such as the lateral septum. In general, fiber density drops following mating, possibly due to the release of vasopressin. There is a much higher concentration of vasopressin in males than in females, and female prairie voles do not show mating-induced changes in central vasopressin (Bamshad, Novak, & DeVries 1993). Little is known regarding the behavioral role of vasopressin in females.

When reproductively naive males that have cohabitated and/or mated with one female for approximately six hours are tested for partner preferences, they may show no preference and may even exhibit a preference for the unfamiliar female. However, corticosterone treatment produced a dose-dependent reversal of this preference. Males spending as little as six hours with a female during corticosterone treatment showed a strong preference for the familiar partner (DeVries et al., under review). It is possible that the preference-enhancing effects of vasopressin (Winslow et al. 1993) are based on the release of corticosterone or vice versa. Alternatively, the response observed following corticosterone injections might be a nonspecific response, such as the formation of social bonds that is reported following intense stress, and may depend on as yet unidentified processes.

9.5.3 Physiology of Selective Aggression in Prairie Voles

Male aggression, especially toward other males, emerges within about 24 hours of the onset of *ad libitum* sexual activity and lasts for weeks or months (Winslow et al. 1993). Males with sexual experience but without well-established partners are also more aggressive than naive males, but they do not show the intensity of aggression that is typical of paired males (Gavish et al. 1983).

The onset of male aggression correlates with increases in central vasopressin (Bamshad et al. 1993) (see Figure 9.1). Treatment with vasopressin facilitated the onset of male–male aggression in unmated males, and the development of mating-induced aggression was blocked by treatment with a selective antagonist for vasopressin. Interestingly, once aggression was established, blocking the action of vasopressin did not reduce aggressivity, suggesting that vasopressin is necessary for the induction but not the maintenance of mating-induced aggression. Oxytocin and oxytocin antagonists were not effective in this model (Winslow et al. 1993).

Female aggression toward unfamiliar conspecifics of both sexes is seen in established breeding females during pregnancy and usually is seen in postpartum estrus (Getz et al. 1981; Witt et al. 1989). However, previously naive females are not very aggressive in the first few days of pregnancy. Oxytocin treatment tends to decrease aggression in females (Witt et al. 1990) and thus probably is not responsible for inducing the increased aggression seen following mating. The effects of vasopressin on female aggression have not been examined. Female aggression, particularly toward other females, is a characteristic of monogamous mammals (French & Inglett 1989). However, the physiology of this behavior in general is poorly understood and remains unknown in female prairie voles.

Nonmonogamous voles, such as montane voles, do not show mating-induced changes in aggression (Winslow et al. 1993). In addition, mating does not alter central vasopressin content in nonmonogamous meadow voles (Bamshad et al. 1993). Species differences in the distribution of vasopressin and oxytocin receptors correlate with social organization in voles (Insel & Shapiro 1992; Insel et al. 1993) and deer mice (Insel et al. 1991). These neurochemical patterns may reflect components of the physiological mechanisms responsible for species-specific patterns of social behavior.

9.5.4 Summary of Factors Regulating Pair Bonding

In summary, cohabitation and sexual interactions facilitate the formation of male–female pair bonds. We hypothesize that coordinated but sexually dimorphic physiological changes underlie two major components of pair-bond formation: partner preference development and selective intrasexual aggression. Oxytocin can facilitate the development of partner preferences in females but inhibits rather than enhances aggressivity. In male prairie voles, mating-induced changes in vasopressin may stimulate the development of partner preferences, the onset of postcopulatory aggression, and parental care (see Table 9.1 and Figure 9.1; see Section 9.6.1). Corticosterone treatment also facilitates partner preferences in males. It is likely that a number of other as yet unidentified neurochemicals, such as the endogenous opioids (Shapiro, Meyer, & Dewsbury 1990), the catecholamines, and possibly other peptides, play a role in pair-bond formation and maintenance.

9.6 Proximate Factors Associated with Alloparenting and Philopatry

Alloparenting and philopatry are two of the major characteristics of a cooperative breeding system. Philopatry literally means "love of the father" and is used in studies of cooperatively breeding species to refer to the fact that many young animals remain with the family rather than dispersing. Alloparenting is defined as "providing parental care to others." Thus, philopatric animals may remain with their parents past the age of weaning and puberty and direct caregiving or parental behaviors toward younger animals within the family group. This phenomenon is particularly apparent in prairie voles in which more than 70 percent of males and 75 percent of females remain in the natal family from birth until death (McGuire et al. 1993). Alloparenting may indirectly increase fitness in young prairie voles by reducing the workload of their parents (Wang & Novak 1994) or by increasing the survival of the related infants, which may in turn reproduce (Solomon 1991), or alloparenting may have direct benefits by increasing subsequent reproductive success in animals experienced with infants.

In comparison to our understanding of reproductive activation and suppression, relatively little is known regarding proximate mechanisms for alloparenting. However, the proximate causes for other types of parental behavior have been studied in a variety of rodent species (Rosenblatt, Siegel, & Mayer 1979; Numan 1988; Bridges 1990; Brown 1993; Gubernick, Schneider, & Jeannotte 1994). It has been suggested that the biological mechanisms for parental and alloparental behavior might be similar (Jamieson 1986, 1989). Although this topic is controversial, it is possible that the various mechanisms that have been implicated in maternal or paternal behavior might be involved in alloparenting as well.

9.6.1 Proximate Factors Associated with Parental Behavior

Sexually naive female and male rats do not show spontaneous parental responsiveness but can be induced to show parental behavior with a short latency by continuous exposure to infants over a period of days, a process called "concaveation" (Wiesner & Sheard 1933; Rosenblatt 1967). LeBlond (1938) demonstrated a similar phenomenon in mice. The period required for concaveation is significantly shortened by administering a hormonal regimen of estrogen and progesterone that mimics the hormones released during parturition, and it is likely that the hormonal events that occur at parturition alter the biological substrates that mediate maternal responsiveness.

Fleming and Luebke (1981; Fleming 1990) proposed that sexually naive rats had to overcome "timidity" in response to the novel infant stimulus in order to exhibit parental responsiveness. They suggested that this timidity is reduced by the hormones associated with pregnancy and parturition. They demonstrated this point by administering estradiol and progesterone to virgin female rats and examining their responses to novelty in emergence and open-field tests; these responses were compared to those of rats administered behaviorally inert cholesterol. The hormone-treated females emerged more quickly and were more active in the open field than were the cholesterol-treated females. This finding is consistent with studies that show that virgin females without hormonal priming respond more quickly to infants after removal of the olfactory bulb, whereas primiparous females respond more slowly after bulbectomy (Fleming & Rosenblatt 1974; Jirik-Babb et al. 1984). It has been suggested that female rats must overcome an aversion to infant odors before they can respond parentally and that the hormones associated with pregnancy somehow change the way the infants are perceived, allowing the females' timidity, or fear of the infants, to be overcome.

Sexually naive prairie voles typically do not require a concaveation period to exhibit parental care, possibly because they lack the timidity factor that prevents the immediate expression of parental behavior in sexually naive rats and mice. When infants are placed with a naive prairie vole, the majority of animals respond within one minute by licking and crouching over the infants. However, some individuals fail to respond to infants during their initial exposure, and approximately 20 percent of females and 11 percent of males attack infants (Roberts 1994). When tests are conducted in a two-chamber choice apparatus in which one cage contains the infant vole and the other cage is left empty, the incidence of attacks is almost completely eliminated (Roberts 1994). The infant stimulus may be perceived as being less threatening in the two-chamber choice apparatus because animals are able to leave the cage containing the infant and retreat to the empty cage if they choose. (It is under these conditions that all parental behavior tests are now conducted in our laboratory.) Prior experience with infants also eliminates the incidence of attacks in prairie voles (Roberts 1994). Young prairie voles that have been housed with the family unit and have participated in rearing a subsequent litter have never been observed to attack unfamiliar infants.

The foregoing studies suggest that novel stimuli from the infant might impede parental responsiveness and could even lead to attacks. Some species are more sensitive than others to novelty, and differential responses to novelty may be one major difference between species that show alloparental care and those that do not. In the absence of clear activational factors such as hormonal changes associated with pregnancy or birth (Numan 1988), other factors presumably cause some species, or even some individuals within a species, to show immediate short-latency parental responsiveness, whereas others do not.

Prolactin is present at parturition, plays a major role in lactation, and facilitates parental behavior in female rats (Bridges 1990). Although they do not lactate, male mammals also produce prolactin. Prolactin production is correlated with paternal behavior in the adult California mouse (*Peromyscus californicus*) (Gubernick & Nelson 1989) and the common marmoset (Dixson & George 1982). In these biparental species, serum prolactin levels are elevated in males that are parental and/or exposed to infants. Prolactin also has been implicated in the alloparental behavior of canids (Asa this volume), Harris's hawks (Mumme this volume), and callitrichid monkeys (Dixson & George 1982). If a common endocrine factor mediates maternal, paternal, and alloparental behavior, prolactin is a likely candidate; the possible role of prolactin in the mediation of alloparental behavior in cooperative rodents warrants further study.

The exceptional capacity of male prairie voles and other monogamous voles to show paternal care has been described in detail by several sources (Hartung & Dewsbury 1979; Thomas & Birney 1979; Wilson 1982b; Gruder-

Adams & Getz 1985; Oliveras & Novak 1986; Wang & Novak 1992, 1994). In prairie voles, the expression of parental behavior in adult males correlates with changes in vasopressin and androgens (see Table 9.1 and Figure 9.1). Castration in adulthood reduces parental behavior in sexually naive male prairie voles. Concurrent measurement of vasopressin immunoreactivity (AVP-ir) indicated that castration reduced the number of AVP-ir cells in the bed nucleus of the stria terminalis (BST) and medial amygdaloid nucleus (MA), as well as the density of AVP-ir fibers in the lateral septum (LS). Males that received testosterone implants at the time of castration exhibited normal parental care and vasopressin levels (Wang & DeVries 1994). Implants of vasopressin in the LS increased parental contact and the time spent crouching over the infants in comparison to saline-treated control males. Specific antagonists for vasopressin receptors (type V1a) reduced parental behaviors in normal males and also blocked the effects of exogenous vasopressin. These effects were site-specific to the LS, which receives input from the BST and MA; testosterone-dependent AVP-ir cells are located in the BST and MA (Wang et al. 1994). Wang and associates hypothesize that vasopressin might be released normally by events associated with the exposure to infants. In addition, the increased gonadal activity associated with mating might increase AVP release to levels that optimally stimulate neural circuits involved in paternal responsiveness. These studies offer the first indications of a neuroendocrine mechanism and circuitry that could support the expression of paternal behavior in prairie voles.

9.6.2 Prior Experience and Alloparental Behavior

Alloparental responding in prairie voles was examined as a function of various demographic factors such as age at the time of testing, sex of the animal, and social history. Each animal was reproductively naive, and subjects were tested between 21 and 60 days of age ($n = 370$). Serum levels of luteinizing hormone (LH) and corticosterone were also sampled to examine possible endocrine correlates of alloparental behavior (Roberts 1994). Although as described, the events associated with mating and pair bonding appear to facilitate male parental care in prairie voles, about 90 percent of reproductively naive males and 80 percent of sexually naive females from our colony were spontaneously parental. High levels of alloparenting were measured at 21 days of age and prior to reproductive activation, and the incidence of alloparenting was not related to age. The only factor that was significantly associated with an increased probability of alloparental behavior was prior experience with infants (Roberts 1994). Prairie voles that had a short (<4 days) period of experience in an extended family group exhibited significantly greater frequencies of alloparental responsiveness than did prairie voles that lacked experience with infants prior to testing. It also has been shown that cross-fostering of newborn meadow voles to prairie vole parents results in significantly greater parental contact by meadow voles of both sexes in adulthood (McGuire 1988).

These findings are consistent with research in juvenile California mice, demonstrating that residence in a family group with younger siblings significantly increased the incidence of alloparental responsiveness in comparison to residence with parents only or siblings only (Gubernick & Laskin 1994). Taken together, these results indicate that social learning, at least with infants, and reproductive maturity or experience are not essential for alloparental behavior. However, reproductive maturity and sexual experience (Bamshad et al. 1994) and experience with infants (Roberts 1994) may facilitate alloparenting, suggesting that multiple mechanisms may mediate parental behavior in prairie voles.

9.6.3 Pheromonal Factors and Parental Behavior

One factor contributing to parental responsiveness may be found in pheromonal secretions from a breeding female or the infants themselves. Chemosignals, especially in urine, are important to various aspects of social and reproductive behavior as well as to reproductive suppression (McClintock 1987; Vandenbergh 1988). It is possible that chemosignals affect parental responsiveness in rodents as well. Support for this hypothesis comes from the work of Gubernick and his colleagues, indicating that a chemosignal present in female urine induces paternal responsiveness in male California mice. Like prairie voles, California mice exhibit many features of monogamy (Gubernick & Alberts 1989; Gubernick 1990). However, in California mice, only 26 to 45 percent of males exhibit paternal behavior in the absence of their mate, whereas 80 to 95 percent of males exhibit parental behavior when exposed to excreta or urine from their mate. Furthermore, this response is specific to odor cues from the males' mates. The familiar female elicits paternal behavior at a rate higher than that seen following exposure to odors from either unfamiliar lactating females or familiar females with which the male has not mated (Gubernick 1990).

9.6.4 Organizational Factors and Parental Behavior

The presence of high levels of parental behavior in inexperienced, prepubertal prairie voles suggests that the substrates for these behaviors could be regulated by early development. Perinatal hormonal experiences during the period of sexual differentiation can have long-term consequences on many aspects of reproduction in other species. For example, Ward and her associates have found that prenatal stress demasculinizes and feminizes the behavior of male rats. Stress experienced by the mother during the last few days of gestation (Ward 1972; Ward & Ward 1985) or perinatal exposure to adrenal corticoids, including synthetic cortisone (Dahlof, Hard, & Larsson 1978), can reduce masculine sexual behavior and the anogenital distance in rats. Parental behavior in prenatally stressed male and female rats was studied by Kinsley and associates (1990). They found that response latencies to retrieve infants in prenatally stressed males were comparable to those of control females and that those of prenatally stressed females were comparable to those of control males. Thus, the behavioral mechanisms responsible for parental behavior can be altered by endogenous hormonal changes during development.

Further support for the hypothesis that prenatal hormonal events can contribute to increased parental responsiveness comes from recent research with prairie voles (Roberts 1994). Corticosterone was administered to pregnant prairie voles on days 12 to 20 of gestation, or to newborn infants on days 1 to 6 postpartum. Animals receiving these or various control treatments were subsequently tested for parental behavior at 24 or 42 days of age. Females, but not males, that had received postnatal corticosterone showed significantly less alloparental behavior. In constrast, males and females treated prenatally with corticosterone tended to spend prolonged periods in contact with infants, as did control subjects (Figure 9.2).

9.6.5 Organizational Factors and Philopatry

In addition to examining alloparenting, in the study described (Roberts 1994), we also measured the philopatric tendencies of juvenile prairie voles. At 50 days of age, animals were placed in a three-chamber preference-testing apparatus (similar to that used to study pair bonding), in which they could elect to spend time with a familiar same-sex sibling or a stranger of the same age and sex, or they could choose the empty chamber. The individual variability in these tests was large. However, untreated females tended to prefer the empty cage to the cage of either their siblings or the stranger (Figure 9.3). In contrast, females treated with corticosterone on postnatal days 1 to 6 spent significantly more time in the cage with the stranger, and females treated with corticosterone on prenatal days 12 to 20 spent significantly more time with their sibling in comparison to the untreated controls. Male prairie voles showed no significant alterations in their preference for sibling, stranger, or an empty

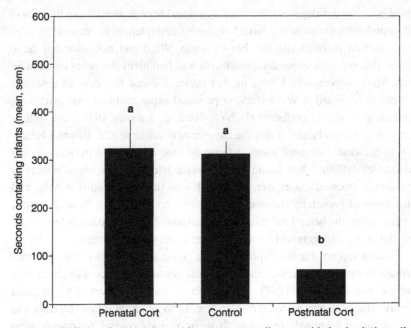

Figure 9.2. Effects of perinatal steroid treatments on alloparental behavior in juvenile female prairie voles. Postnatal corticosterone treatment (days 1–6) was associated with significant reductions in contact with infants, whereas animals exposed to exogenous (prenatal days 12–20) or endogenous (control) corticosterone tended to show extensive infant contact.

cage. These results suggest that in female prairie voles the tendency to associate with familiar versus unfamiliar animals can also be influenced by perinatal hormones. Collectively, results from this study offer support for the hypothesis that selected components of monogamy or philopatry, including alloparenting and social preferences, can be manipulated by hormonal experiences during perinatal development.

9.6.6 Population Differences in the Expression of Alloparenting

Population differences have been detected in paternal behavior in meadow voles (Storey & Snow 1987). Meadow voles originating in northern latitudes (Manitoba and Ontario, Canada) exhibit paternal behavior toward their own offspring, yet paternal behavior is rare in male meadow voles originating in the United States. Storey and her colleagues have demonstrated that female meadow voles from northerly populations are selective, allowing the sire of

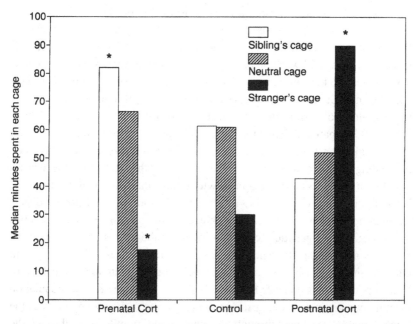

Figure 9.3. Effects of perinatal steroid treatments (see Figure 9.2) on social preferences in female prairie voles. A three-chamber preference apparatus was used to allow
females to spend time with either a sibling or stranger of the same sex, or alone. In
comparable tests there were no significant differences among treatment groups for
males. Sibling preferences may reflect behavioral tendencies associated with philopatry. Asterisks indicate significant differences relative to controls.

her litter to enter the nest containing infants, whereas strange males are
excluded (Storey, Bradbury, & Joyce 1994). Furthermore, females excluded
sires from the nest for a day after the birth of the litter if the sires were absent
during the pregnancy, whereas sires that were present throughout the pregnancy were permitted into the nest significantly more often. Clearly, this
research indicates that female meadow voles from northerly latitudes selec-
tively permit their mates to enter the nest only after the period of concaveation has passed, whereas other males are excluded. Further studies
showed that 24 hours of exposure to urine from the mate and pups reduced
infant-directed aggression but that physical contact with the infants was necessary to induce paternal behavior (Storey & Walsh 1994; Storey & Joyce
1995). It is possible that the population difference lies not in the behavior of
males but in the tolerance of females for males near the natal nest. Females
from northerly regions might permit males to enter the nest because of climatic demands, whereas meadow voles originating in warmer areas do not
tolerate males near their litters (Storey, Bradbury, & Joyce 1994).

Most of the research on prairie voles performed in our laboratory and elsewhere has employed animals that were originally captured in south-central
Illinois; strong evidence exists for monogamy and philopatry in these animals. However, field work on prairie voles from other regions and/or habitats
has suggested the hypothesis that the formation of communal groups
(Danielson & Gaines 1987) observed in animals from Illinois may not characterize prairie voles from different habitats. To test this hypothesis, we have
compared alloparental behavior in juvenile prairie voles from a breeding
stock originating near Lawrence, Kansas, to that of juveniles from a stock
originating near Urbana, Illinois. Juveniles of both sexes from the Kansas
population spent less time in contact with infants than did juveniles from the
Illinois population (Roberts 1994). The identification of intraspecific differences in behavior of this nature is a valuable tool that can be used to identify
the ecological and genetic factors that influence the expression of alternative
parenting and mating strategies. For example, prairie voles in Kansas exist in
a habitat that generally contains fewer food resources than does the habitat of
Illinois prairie voles (Cole & Batzli 1979; Getz 1985). The Kansas population
exhibits two distinct breeding seasons in the spring and autumn, whereas the
Illinois populations breed through the summer and, in mild years through the
winter. Consequently, population fluctuations in Kansas are of a shorter
amplitude, and population densities peak at lower levels in the Kansas habitat
than in the Illinois habitat (Danielson & Gaines 1987; Getz et al. 1993;
McGuire et al. 1993).

9.6.7 Summary of Proximate Mechanisms Responsible for Alloparenting

The behaviors associated with alloparenting are phenotypically similar to
maternal and paternal care. Among the factors that have been implicated in
parental behavior are adult levels of gonadal hormones, or hormonal changes
triggered by either sexual experience or parturition; juvenile animals engaged
in alloparenting would not typically have the benefit of any of these. Young
prairie voles that have continuous exposure to siblings through communal
rearing are less likely to attack infants than are animals that have been separated from infants at weaning, suggesting that fear of infants may interfere
with alloparenting. Juveniles remaining in the natal nest are exposed to stimuli such as pheromones from the infants and their parents and may indirectly
experience hormonal changes triggered by communal living or by reproductive processes in their parents. It is possible that pheromonal modulation of
the juvenile's physiology maximizes the expression of alloparenting.

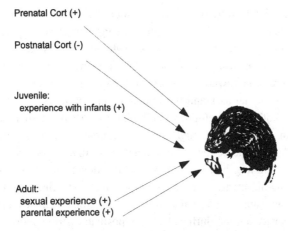

Prenatal Cort (+)

Postnatal Cort (-)

Juvenile:
experience with infants (+)

Adult:
sexual experience (+)
parental experience (+)

Figure 9.4. A summary of proximate mechanisms that may influence alloparental and parental behavior during the perinatal, juvenile, and adult (reproductively active) stages of life in prairie voles

However, high levels of alloparental behavior are seen in most juvenile prairie voles that are separated from the parents prior to the birth of a new litter. Thus, experience with infants and/or hormonal events that must be maintained by stimuli from the family are not essential to support alloparenting. Rapid changes in hormones or neurotransmitters, elicited by stimuli from an infant, may play a role in this behavior, but the nature of these physiological events remains to be described. Finally, as discussed, preliminary studies suggest that interactions among the gonadal and adrenal axes during perinatal development may regulate the expression of various characteristics of monogamy in prairie voles, including alloparenting. A summary schematic of the factors that may be important in the regulation of alloparental and parental care is presented in Figure 9.4.

9.7 Summary

Using prairie voles as a model system, it is possible to identify a variety of proximate processes that may contribute to cooperative breeding. Patterns of hormonal action in social or monogamous mammals such as prairie voles differ from those seen in polygynous species. In general, prairie voles and other monogamous mammals rely on social stimuli, mediated through relatively rapid physiological changes, for the induction and/or suppression of female and male sexual behavior. High levels of social behavior, including tolerance for postcopulatory interactions between a mating pair, may support the development of monogamy. Peptides such as oxytocin and vasopressin have been implicated in the neural control of mate preferences and the induction of selective aggression toward strangers. Interactions between the hypothalamic–pituitary–adrenal axis and the hypothalamic–pituitary–gonadal axis can also regulate both the development and adult expression of social and sexual behaviors. In other species, high levels of adrenal activity in the perinatal period can demasculinize males. We hypothesize that various components of monogamy, including biparental care and the development of a communal social system, might arise from developmental endocrine events, such as those that can cause demasculization in male rats. Females treated with exogenous corticosterone during the early postnatal period showed less alloparental behavior, whereas exposure to corticosterone during the prenatal period subsequently was associated with a tendency to exhibit high levels of alloparental behavior. Sibling contact was increased in animals exposed to prenatal corticosterone treatments, and contact with strangers increased in animals that had received postnatal corticosterone. Intraspecific variation in the expression of parental behavior has been observed in voles from different habitats. These studies suggest paradigms for the analysis of proximate mechanisms that may provide a substrate for philopatry and cooperative breeding.

Acknowledgments

We wish to express our gratitude to the many researchers cited here who have contributed to our understanding of vole biology and the biology of communal mammals. We wish to express particular thanks to Lowell Getz and Michael Gaines for access to prairie voles, and to Courtney DeVries and Susan Taymans for allowing us to describe unpublished results from their research. We are also grateful to Nancy Solomon and Jeff French for their editorial suggestions. This research was supported by NIH/NIMH (MH45836 and MH 01050) and NSF (IBN 9411216) to C. Sue Carter.

References

Abbott, D. H. (1993). Social conflict and reproductive suppression in marmoset and tamarin monkeys. In *Primate social conflict,* ed. W. A. Mason and S. P. Mendoza, pp. 331–372. Albany: State University of New York.
Abbott, D. H., Barrett, J., Faulkes, C. G., & George, L. M. (1989). Social contraception in naked mole-rats and marmoset monkeys. *J. Zool., Lond.* 219:703–707.

Adler, N. T. (1969). Effects of the male's copulatory behavior on successful pregnancy of the female rat. *J. Comp Physiol. Psych.* 69:613–622.

Allen, T. O., & Adler, N. T. (1985). Neuroendocrine consequences of sexual behavior in *Handbook of behavioral neurobiology,* Vol. 7, *Reproduction,* ed. N. Adler, D. Pfaff, & R. W. Goy, pp. 725–766. New York: Plenum Press.

Bamshad, M., Novak, M. A., & G. J. De Vries (1993). Species and sex differences in vasopressin innervation of sexually naive and parental prairie voles, *Microtus ochrogaster,* and meadow voles, *Microtus pennsylvanicus. Journal of Neuroendocrinol.* 5:247–255.

Bamshad, M., Novak, M. A., & G. J. De Vries (1994). Cohabitation alters vasopressin innervation and paternal behavior in prairie voles (*Microtus ochrogaster*). *Physiol. Behav.* 56:751–758.

Bridges, R. S. (1990). Endocrine regulation of parental behavior in rodents. In *Mammalian parenting: Biochemical, neurobiological and behavioral determinants,* ed. N. A. Krasnegor & R. S. Bridges, pp. 93–117. New York: Oxford University Press.

Bruce, H. M. (1959). An exteroreceptive block to pregnancy in the mouse. *Nature* 184:105.

Brown, R. E. (1993). Hormonal and experiential factors influencing parental behaviour in male rodents: An integrative approach. *Behavior. Proc.* 30:1–28.

Caldwell, J. D. (1992). Central oxytocin and female sexual behavior. In *Annals of the New York Academy of Sciences, Oxytocin in Maternal, Sexual and Social Behavior* 652:166–179.

Carter, C. S. (1992). Oxytocin and sexual behavior. *Neurosci. Biobehav. Rev.* 16:131–144.

Carter, C. S., DeVries, A. C., & Getz, L. L. (1995). Physiological substrates of monogamy: The prairie vole model. *Neurosci. Biobehav. Rev.* 19:303–314.

Carter, C. S., & Getz, L. L. (1985). Social and hormonal determinants of reproductive patterns in the prairie voles. In *Neurobiology,* ed. R. Giles & J. Balthazart, pp. 18–36. Berlin: Springer-Verlag.

Carter, C. S., & Getz, L. L. (1993). Monogamy in the prairie vole. *Sci. Amer.* 268:100–106.

Carter, C. S., Getz, L. L., & Cohen-Parsons, M. (1986). Relationships between social organization and behavioral endocrinology in a monogamous mammal. *Adv. Stud. Behav.* 16:109–145.

Carter, C. S., Getz, L. L., Gavish, L., McDermott, J. L., & Arnold, P. 1980. Male-related pheromones and the activation of female reproduction in the prairie vole (*Microtus ochrogaster*). *Biol. Reprod.* 23:1038–1045.

Carter, C. S., & Schein, M. W. (1971). Sexual receptivity and exhaustion in the female golden hamster. *Horm. Behav.* 2:191–200.

Carter, C. S., Williams, J. R., & Witt, D. M. (1990). The biology of social bonding in a monogamous mammal. In *Hormones, brain and behavior in vertebrates 2: Behavioral activation in males and females – social interactions and reproductive endocrinology,* ed. J. Balthazart, pp. 154–164. Basel: S. Karger.

Carter, C. S., Williams, J. R., Hasting, N., Paciotti, G. F., Tamarkin, L., & Insel, T. R. (1991). LH increases and corticosterone decreases within minutes following nonsexual exposure to the opposite sex in male and female prairie voles. *Conference on Reproductive Behavior Abstracts* 23:21.

Carter, C. S., Williams, J. R., Witt, D. M., & Insel, T. R. (1992). Oxytocin and social bonding. In *Annals of the New York Academy of Sciences, Oxytocin in Maternal, Sexual and Social Behavior* 652:204–211.

Carter, C. S., Williams, J. R., Witt, D. M., Matthews, C., & Dang, D. (1991). Rearing with family members suppresses reproductive development in male prairie voles. *Conference on Reproductive Behavior Abstracts* 23:22.

Carter, C. S., Witt, D. M., Auksi, T., & Casten, L. (1987a). Estrogen and the induction of lordosis in female and male prairie voles, *(Microtus ochrogaster). Horm. Behav.* 21:65–73.

Carter, C. S., Witt, D. M., Schneider, J., Harris, Z. L., & Volkening, D. (1987b). Male stimuli are necessary for female sexual behavior and uterine growth in prairie voles *(Microtus ochrogaster). Horm. Behav.* 21:74–82.

Carter, C. S., Witt, D. M., Manock, S. R., Adams, K. A., Bahr, J. M., & Carlstead, K. (1989). Hormonal correlates of sexual behavior and ovulation in prairie voles. *Physiol. Behav.* 46:941–948.

Christian, J. J., & Davis, D. E. (1964). Endocrines, behavior, and population. *Science* 145:1550–1560.

Chrousos, G. P., Renquist, D., Brandon, D., Eil, C., Pugeat, M., Vigersky, R., Cutler, G. B., Jr., Loriaux, D. L., & Lipsett, M. B. (1982). Glucocorticoid hormone resistance during primate evolution: Receptor-mediated mechanisms. *Proc. National Acad. Sci.* 79:2036–2040.

Cohen-Parsons, M., & Carter, C. S. (1987). Males increase serum estrogen and estrogen receptor binding in brain of female voles. *Physiol. Behav.* 39:309–314.

Cohen-Parsons, M., & Carter, C. S. (1988). Males increase progestin receptor binding in brain of female voles. *Physiol. Behav.* 42:191–197.

Cole, F. R., & Batzli, G. O. (1979). Nutrition and population dynamics of the prairie vole *(Microtus ochrogaster)* in central Illinois. *J. Anim. Ecol.* 48:455–470.

Dahlof, L.-G., Hard, E., & Larsson, K. (1978). Sexual differentiation of offspring of mothers treated with cortisone during pregnancy. *Physiol. Behav.* 21:673–674.

Danielson, B. J., & Gaines, M. S. (1987). Spatial patterns in two syntopic species of microtines, *Microtus ochrogaster* and *Synaptomys cooperi. J. Mammal.* 68:313–322.

Davidson, J. M., & Levine, S. (1972). Endocrine regulation of behavior. *Ann. Rev. Physiol.* 34:375–408.

DeVries, A. C., DeVries, M. B., Taymans, S. E., & Carter, S. C. (under review). Stress has sexually dimorphic effects on pair-bonding in prairie voles.

DeVries, A. C., DeVries, M. B., Taymans, S. E., & Carter, S. C. (1995). The modulation of pair-bonding by corticosterone in female prairie voles *(Microtus ochrogaster). Proc. Nat. Acad. Sci.* 92:7744–7748.

Dewsbury, D. A. (1981). An exercise in the prediction of monogamy in the field from laboratory data on 42 species of muroid rodents. *The Biologist* 63:138–162.

Dewsbury, D. A. (1988). The comparative psychology of monogamy. *Nebraska Symposium on Motivation.* 35:1–50.

Dixson, A. F., & George, L. (1982). Prolactin and parental behaviour in a male New World primate. *Nature* 299:551–553.

Dluzen, D. E., & Carter, C. S. (1979). Ovarian hormones regulating sexual and social behaviors in female prairie voles, *Microtus ochrogaster. Physiol. Behav.* 23:597–600.

Dluzen, D. E., Ramirez, V. C., Carter, C. S., & Getz, L. L. (1981). Male urine stimulates localized and opposite changes in luteinizing hormone-releasing hormones and norepinephrine within the olfactory bulb of female prairie voles. *Science* 212:573–575.

Doty, R. L. (1976). *Mammalian olfaction, reproductive processes and behavior.* New York: Academic Press.

Ferguson, B., Fuentes, S. M., Sawrey, D. K., & Dewsbury, D. A. (1986). Male preference for unmated versus mated females in two species of voles (*Microtus ochrogaster* and *M. montanus*). *J. Comp. Psychol.* 100:243–247.

File, S. E., Vellucci, S. V., & Wendlandt, S. (1979). Corticosterone – an anxiogenic or anxiolytic agent. *J. Pharmacol.* 31:300–305.

Firestone, K. B., Thompson, K. V., & Carter, C. S. (1991). Behavioral correlates of intra-female reproductive suppression in prairie voles, *Microtus ochrogaster*. *Behav. Neur. Biol.* 55:31–41.

Fleming, A. S. (1990). Hormonal and experiential correlates of maternal responsiveness in human mothers. In *Mammalian parenting: Biochemical, neurobiological and behavioral determinants*, ed. N. A. Krasnegor & R. S. Bridges, pp. 184–214. New York: Oxford University Press.

Fleming, A. S., & Luebke, C. (1981). Timidity prevents the virgin female rat from being a good mother: Emotionality differences between nulliparous and parturient females. *Physiol. Behav.* 27:863–868.

Fleming, A. S., & Rosenblatt, J. S. (1974). Olfactory regulation of maternal behavior in rats. II. Effects of peripherally induced anosmia and lesions of the lateral olfactory tract in pup-induced virgins. *J. Comp. Physiol. Psychol.* 86:233–246.

French, J. A., & Inglett, B. J. (1989). Female–female aggression and male indifference in response to unfamiliar intruders in lion tamarins. *Anim. Behav.* 37:487–497.

Gaines, M. S., Fugate, C. L., Johnson, M. L., Johnson, D. C., Hisey, J. R., & Quadagno, D. M. (1985). Manipulation of aggressive behaviour in male prairie voles (*Microtus ochrogaster*) implanted with testosterone in silastic tubing. *Can. J. Zool.* 63:2525–2528.

Gavish, L., Carter, C. S., & Getz, L. L. (1981). Further evidences of monogamy in the prairie vole. *Anim. Behav.* 29:955–957.

Gavish, L., Carter, C. S., & Getz, L. L. (1983). Male–female interactions in prairie voles. *Anim. Behav.* 31:511–517.

Gavish, L., Hofmann, J. E., & Getz, L. L. (1984). Sibling recognition in the prairie vole, *Microtus ochrogaster*. *Anim. Behav.* 32:362–366.

Getz, L. L. (1985). Habitats. In *Biology of New World* Microtus, ed. R. H. Tamarin, pp. 286–309. Special Publication Number 8: American Society of Mammalogists.

Getz, L. L., & Carter, C. S. (1980). Social organization in *Microtus ochrogaster*. *The Biologist* 62:56–69.

Getz, L. L., Carter, C. S., & Gavish, L. (1981). The mating system of the prairie vole *Microtus ochrogaster:* Field and laboratory evidence for pair-bonding. *Behav. Ecol. Sociobiol.* 8:189–194.

Getz, L. L., Dluzen, D. E., & McDermott, J. L. (1983). Suppression of reproductive maturation in male-stimulated virgin female *Microtus* by a female urinary chemosignal. *Behav. Proc.* 8:59–64.

Getz, L. L., Hofmann, J. E., & Carter, C. S. (1987). Mating system and population fluctuations of the prairie vole (*Microtus ochrogaster*). *Amer. Zool.* 27:909–920.

Getz, L. L., McGuire, B., Pizzuto, T., Hofmann, J. E., & Frase, B. (1993). Social organization of the prairie vole (*Microtus ochrogaster*). *J. Mammal.* 74:44–58.

Goldfoot, D. A., & Goy, R. W. (1970). Abbreviation of behavioral estrus in guinea pigs by coital and vaginocervical stimulation. *J. Comp. Physiol. Psychol.* 72:426–434.

Gray, G. D., Davis, H. N., Kenney, A. McM., & Dewsbury, D. A. (1976). Effect of mating on plasma levels of LH and progesterone in montane voles (*Microtus montanus*). *J. Reprod. Fert.* 47:89–91.

Gray, G. D., Zerylnick, M., Davis, H. N., & Dewsbury, D. A. (1974). Effects of variations in male copulatory behavior on ovulation and implantation in prairie voles, *Microtus ochrogaster. Horm. Behav.* 5:437–452.

Gruder-Adams, S., & Getz, L. L. (1985). Comparison of the mating system and paternal behavior in *Microtus ochrogaster* and *M. pennsylvanicus. J. Mammal.* 66:165–167.

Gubernick, D. J. (1990). A maternal chemosignal maintains paternal behaviour in the biparental California mouse, *Peromyscus californicus. Anim. Behav.* 39:916–923.

Gubernick, D. J., & Alberts, J. R. (1989). Postpartum maintenance of paternal behaviour in the biparental California mouse, *Peromyscus californicus. Anim. Behav.* 37:656–664.

Gubernick, D. J., & B. Laskin (1994). Mechanisms influencing sibling care in the monogamous biparental California mouse, *Peromyscus californicus. Anim. Behav.* 48:1235–1237.

Gubernick, D. J., & Nelson, R. J. (1989). Prolactin and paternal behavior in the biparental California mouse, *Peromyscus californicus. Horm. Behav.* 23:203–210.

Gubernick, D. J., Schneider, K. A., & Jeannotte, L. A. (1994). Individual differences in the mechanisms underlying the onset and maintenance of paternal behavior and the inhibition of infanticide in the monogamous biparental California mouse, *Peromyscus californicus. Behav. Ecol. Sociobiol.* 34:225–231.

Hartung, T. G., & Dewsbury, D. A. (1979). Paternal behaviour in six species of muroid rodents. *Behav. Neur. Biol.* 26:466–478.

Hnatczuk, O. C., Lisciotto, C. A., DonCarlos, L. L., Carter, C. S., & Morrell, J. I. (1994). Estrogen and progesterone receptor immunoreactivity (ER-IR & PR-IR) in specific brain areas of the prairie vole (*Microtus ochrogaster*) is altered by sexual receptivity and genetic sex. *J. Neuroendocrinol.* 6:89–100.

Insel, T. R., Gelhard, R., & Shapiro, L. E. (1991). The comparative distribution of forebrain receptors for neurohypophyseal peptides in monogamous and polygamous mice. *Neuroscience* 43:623–630.

Insel, T. R., & Shapiro, L. E. (1992). Oxytocin receptor distribution reflects social organization in monogamous and polygamous voles. *Proc. National Acad. Sci.* 89:5981–5985.

Insel, T. R., Winslow, J. T., Williams, J. R., Hastings, N., Shapiro, L. E., & Carter, C. S. (1993). The neurohypophyseal peptides in the central mediation of complex social processes – evidence from comparative studies. *Regulatory Peptides* 45:127–131.

Jamieson, I. G. (1986). The functional approach to behavior: Is it useful? *Amer. Naturalist* 127:195–208.

Jamieson, I. G. (1989). Behavioral heterochrony and the evolution of birds' helping at the nest: An unselected consequence of communal breeding? *Amer. Naturalist* 133:394–406.

Jannett, F. J. (1980). Social dynamics of the montane vole, *Microtus montanus*, as a paradigm. *The Biologist* 62:3–19.

Jikik-Babb, P., Manaker, S., Tucker, A. M., & Hofer, M. A. (1984). The role of the accessory and main olfactory systems in maternal behavior of the primiparous rat. *Behav. Neur. Biol.* 40:170–178.

Kenney, A., Evans, R. L., & Dewsbury, D. A. (1977). Postimplantation pregnancy disruption in *Microtus ochrogaster, M. pennsylvanicus* and *Peromyscus maniculatus. J. Reprod. Fert.* 49:365–367.

Keverne, E. B., & Kendrick, K. M. (1992). Oxytocin facilitation of maternal behavior in sheep. In *Annals of the New York Academy of Sciences, Oxytocin in Maternal, Sexual and Social Behavior.* 652:83–101.

Kinsley, C. H. (1990). Prenatal and postnatal influences in rodents. In *Mammalian parenting: biochemical, neurobiological and behavioral determinants,* ed. N. A. Krasnegor & R. S. Bridges. pp. 347–371. New York: Oxford University Press.

Kleiman, D. (1977). Monogamy in mammals. *Quarterly Review of Biology* 52:39–69.

LeBlond, C. P. (1938). Extra-hormonal factors in maternal behavior. *Proc. Soc. Exper. Biol. Med.* 38:66–70.

Lepri, J. J., & Wysocki, C. J. (1987). Removal of the vomeronasal organ disrupts the activation of reproduction in female voles. *Physiol. Behav.* 40:349–355.

McClintock, M. K. (1987). A functional approach to the behavioral endocrinology of rodents. In *Psychobiology of reproductive behavior,* ed. D. Crews, pp. 176–203. Prentice Hall: Englewood Cliffs, N.J.

McGinnis, M. Y., Rutenberg, E., & Lumia, A. R. (1993). Corticosterone facilitates lordosis behavior and proceptivity in estrogen-primed female rats. *Society for Neuroscience Abstracts* 23:586.

McGuire, B. (1988). Effects of cross-fostering on parental behavior of meadow voles (*Microtus pennsylvanicus*). *J. Mammal.* 69:332–341.

McGuire, B., Getz, L. L., Hofmann, J. E., Pizzuto, T., & Frase, B. (1993). Natal dispersal and philopatry in prairie voles (*Microtus ochrogaster*) in relation to population density, season, and natal social environment. *Behav. Ecol. Sociobiol.* 32:293–302.

McGuire, M. R., & Getz, L. L. (1981). Incest taboo between sibling *Microtus ochrogaster. J. Mammal.* 62:213–215.

Mendoza, S. P., & Mason, W. A. (1986). Contrasting responses to intruders and to involuntary separation by monogamous and polygynous New World monkeys. *Physiol. Behav.* 38:795–801.

Nelson, R. J., Frank, D., Bennett, S. A., & Carter, C. S. (1989). Simulated drought influences reproduction in male prairie voles. *Physiol. Behav.* 44:691–697.

Newman, K. S., & Halpin, Z. T. (1988). Individual odours and mate recognition in the prairie vole, *Microtus ochrogaster. Anim. Behav.* 36:1779–1787.

Numan, M. (1988). Maternal behavior. In *The physiology of reproduction,* ed. E. Knobil & J. Neill, pp. 1569–1645. New York: Raven Press.

Oliveras, D., & Novak, M. (1986). A comparison of paternal behaviour in the meadow vole *Microtus pennsylvanicus,* the pine vole, *M. pinetorum,* and the prairie vole, *M. ochrogaster. Anim. Behav.* 34:519–526.

Pfaff, D. W., & Schwartz-Giblin, S. (1988). Cellular mechanisms of female reproductive behaviors. In *The physiology of reproduction,* ed. E. Knobil & J. Neill, pp. 1487–1569. New York: Raven Press.

Pierce, J. D., Jr., Bryan, J. C., Deter, D., Ferguson, B., Sawrey, D. K., Taylor, S. A., & Dewsbury, D. A. (1988). Sexual activity and satiety over an extended observation period in prairie voles (*Microtus ochrogaster*). *J. Comp. Psychol.* 102:306–311.

Rathbun, G. B. (1979). The social structure and ecology of elephant-shrews. *Advances in Ethology* 20:1–77.

Richardson, P. R. K. (1987). Aardwolf mating system: Overt cuckoldry in an apparently monogamous mammal. *Suid-Afrikaanse Tydskrif vir Wetenskap* 83:405–410.

Richmond, M., & Conaway, C. H. (1969). Induced ovulation and oestrus in *Microtus ochrogaster. J. Reprod. Fert.,* Supplement 6:357–376.

Richmond, M., & Stehn, R. (1976). Olfaction and reproductive behavior in microtine rodents. In *Mammalian olfaction, reproductive processes and behavior,* ed. R. L. Doty, pp. 197–217. New York: Academic Press.

Roberts, R. L. (1994). *Sexual dimorphism and the cooperative breeding system of the prairie vole* (Microtus ochrogaster). Ph.D. dissertation, University of Maryland, College Park.

Rood, J. P. (1980). Mating relations and reproductive suppression in the dwarf mongoose. *Anim. Behav.* 28:143–150.

Rosenblatt, J. S. (1967). Nonhormonal basis of maternal behavior in the rat. *Science* 156:1512–1514.

Rosenblatt, J. S., Siegel, H. I., & Mayer, A. D. (1979). Progress in the study of maternal behavior in the rat: Hormonal, nonhormonal, sensory and developmental aspects. *Adv. Stud. Behav.* 10:225–311.

Sawrey, D. K., & Dewsbury, D. A. (1985). Control of ovulation, vaginal estrus, and behavioral receptivity in voles (*Microtus*). *Neurosci. Biobehav. Rev.* 9:563–571.

Schadler, M. H. (1983). Male siblings inhibit reproductive activity in female pine voles, *Microtus pinetorum. Biol. Reprod.* 28:1137–1139.

Schadler, M. H. (1990). Social organization and population control in the pine vole (*Microtus pinetorum*). In *Social systems and population cycles in voles,* ed. R. H. Tamarin, R. S. Ostfeld, S. R. Pugh, & G. Bujalska, pp. 121–130. Basel: Birkhauser, Verlag.

Shapiro, L. R., Austin, D., Ward, S. E., & Dewsbury, D. A. (1986). Familiarity and female mate choice in two species of voles (*Microtus ochrogaster* and *Microtus montanus*). *Anim. Behav.* 34:90–97.

Shapiro, L. R., & Dewsbury, D. A. (1986). Male dominance, female choice, and male copulatory behavior in two species of voles (*Microtus ochrogaster* and *Microtus montanus*). *Behav. Ecol. Sociobiol.* 18:267–274.

Shapiro, L. R., & Dewsbury, D. A. (1990). Differences in affiliative behavior, pair bonding, and vaginal cytology in two species of vole (*Microtus ochrogaster*) and (*M. montanus*). *J. Comp. Psychol.,* 104:268–274.

Shapiro, L. R., Meyer, M. E., & Dewsbury, D. A. (1990). Affiliative behavior in voles: Effects of morphine, naloxone, and cross-fostering. *Physiol. Behav.* 46:719–723.

Solomon, N. G. (1991). Current indirect fitness benefits associated with philopatry in juvenile prairie voles. *Behav. Ecol. Sociobiol.* 29:277–282.

Snowdon, C. T. (1990). Mechanisms maintaining monogamy in monkeys. In *Contemporary issues in comparative psychology,* ed. D. A. Dewsbury, pp. 225–251. Sunderland, Mass.: Sinauer.

Stock, S., & Uvnäs-Moberg, K. (1988). Increased plasma levels of oxytocin in response to afferent electrical stimulation of the sciatic and vagal nerves and in response to touch and pinch in anaesthetized rats. *Acta Physiologica Scandinavica* 132:29–34.

Storey, A. E., Bradbury, C. G., & Joyce, T. L. (1994). Nest attendance in male meadow voles: The role of the female in regulating male interactions with pups. *Anim. Behav.* 47:1037–1046.

Storey, A. E., & Joyce, T. L. (1995). Pup contact promotes paternal responsiveness in male meadow voles. *Anim. Behav.* 49:1–10.

Storey, A. E., & Snow, D. T. (1987). Male identity and enclosure size affect paternal attendance of meadow voles, *Microtus pennsylvanicus. Anim. Behav.* 35:411–419.

Storey, A. E., & Walsh, C. J. (1994). The role of physical contact and chemical cues from males and pups in the development of paternal responsiveness in meadow voles. *Behaviour* 131:139–151.

Swanson, H. H. (1985). Neuroendocrine control of population size in rodents with special emphasis on the Mongolian gerbil. In *Neurobiology,* ed. R. Giles & J. Balthazart, pp. 2–17. Berlin: Springer-Verlag.

Thomas, J. A., & Birney, E. C. (1979). Parental care and mating system of the prairie vole, *Microtus ochrogaster. Behav. Ecol. Sociobiol.* 5:171–186.

Vandenbergh, J. G. (1988). Pheromones and mammalian reproduction. In *The physiology of reproduction,* ed. E. Knobil & J. Neill, pp. 1679–1696. New York: Raven Press.

Wang, Z. X., & De Vries, G. J. (1994). Testosterone effects on paternal behavior and vasopressin immunoreactive projections in prairie voles (*Microtus ochrogaster*). *Brain Res.* 631:156–160.

Wang, Z. X., Ferris, C. F., & De Vries, G. J. (1994). The role of septal vasopressin innervation in paternal behavior in prairie voles (*Microtus ochrogaster*). *Proc. N. Y. Acad. Sci.* 91:400–404.

Wang, Z. X., & Novak, M. A. (1992). The influence of the social environment on parental behavior and pup development in meadow voles (*Microtus pennsylvanicus*) and prairie voles (*Microtus ochrogaster*). *J. Comp. Psychol.* 106:163–171.

Wang, Z. X., & Novak, M. A. (1994). Alloparental care and the influence of father presence on juvenile prairie voles (*Microtus ochrogaster*). *Anim. Behav.* 47:281–288.

Ward, I. L. (1972). Prenatal stress feminizes and demasculinizes the behavior of males. *Science* 175:82–84.

Ward, I. L., & Ward, O. B. (1985). Sexual behavior differentiation: Effects of prenatal manipulations in rats. In *Handbook of behavioral neurobiology,* Vol. 7, *Reproduction,* ed. N. Adler, D. Pfaff, & R. W. Goy, pp. 77–98. New York: Plenum Press.

Wiesner, B. P., & Sheard, N. M. (1933). *Maternal behavior in the rat.* Edinburgh: Oliver and Boyd.

Williams, J. R., Catania, K. C., & Carter, C. S. (1992). Development of partner preferences in female prairie voles (*Microtus ochrogaster*): The role of social and sexual experience. *Horm. Behav.* 26:339–349.

Williams, J. R., Insel, T. R., Harbaugh, C. R., & Carter, C. S. (1994). Oxytocin centrally administered facilitates formation of a partner preference in female prairie voles (*Microtus ochrogaster*). *J. Neuroendocrinol.* 6:247–250.

Williams, J. R., Slotnick, B. M., Kirkpatrick, B. W., & Carter, C. S. (1992). Olfactory bulb removal affects partner preference development and estrus induction in female prairie voles. *Physiol. Behav.* 52:635–639.

Wilson, S. (1982a). The development of social behaviour between siblings and nonsiblings of the voles *Microtus ochrogaster* and *Microtus pennsylvanicus. Anim. Behav.* 30:426–437.

Wilson, S. (1982b). Parent–young contact in prairie and meadow voles. *J. Mammal.* 63:300–305.

Winslow, J. T., & Insel, T. R. (1991). Social status in pairs of male squirrel monkeys determines response to central oxytocin administration. *J. Neurosci.* 11:2032–2038.

Winslow, J. T., Hastings, N., Carter, C. S., Harbaugh, C. R., & T. R. Insel. (1993). A role for central vasopressin in pair bonding monogamous prairie voles. *Nature* 365:545–548.

Witt, D. M., Carter, C. S., Carlstead, K., & Read, L. (1988). Sexual and social interactions preceding and during male-induced oestrus in prairie voles (*Microtus ochrogaster*). *Anim. Behav.* 36:1465–1471.

Witt, D. M., Carter, C. S., Chayer, R., & Adams, K. (1989). Patterns of behavior during postpartum oestrus in prairie voles, *Microtus ochrogaster. Anim. Behav.* 39:528–534.

Witt, D. M., Carter, C. S., & T. R. Insel. (1991). Oxytocin receptor binding in female prairie voles: endogenous and exogenous oestradiol stimulation. *J. Neuroendocrinol.* 3:155–161.

Witt, D. M., Carter, C. S., & Walton, D. (1990). Central and peripheral effects of oxytocin administration in prairie voles (*Microtus ochrogaster*). *Pharmacol. Biochem. Behav.* 37:63–69.

Witt, D. M., Davis, S. L., Indovina, A., Kleinman, M. B., & Carter, C. S. (1985). Social influences on male reproductive development in the prairie vole. *Conference on Reproductive Behavior Abstracts* 17:89.

Witt, D. M., Winslow, J. T., & Insel, T. R. (1992). Enhanced social interactions in rats following chronic, centrally infused oxytocin. *Pharmacol. Biochem. Behav.* 43:855–856.

10

Cooperative Breeding in Naked Mole-Rats: Implications for Vertebrate and Invertebrate Sociality

EILEEN A. LACEY and PAUL W. SHERMAN

10.1 Introduction

The naked mole-rat (*Heterocephalus glaber*) occupies an unusual place among social vertebrates. Like the other species considered in this volume, naked mole-rats are cooperative breeders. These small, virtually hairless rodents live in subterranean colonies in which nonreproductive individuals routinely groom, feed, and protect the offspring of reproductive colony members (Jarvis 1981, 1991; Lacey & Sherman 1991). At the same time, naked mole-rats are eusocial, meaning "truly social" (Batra 1966). Although this term has generally been reserved for insect societies (e.g., ants, termites, and some bee and wasp species) in which there is "cooperation in caring for the young, reproductive division of labor with more or less sterile individuals working on behalf of individuals engaged in reproduction, and overlap of at least two generations of life stages capable of contributing to colony labor" (Hölldobler & Wilson 1990, p. 638), *H. glaber* also exhibits these diagnostic characteristics. Consequently, evolutionary explanations for both cooperative breeding and eusociality must apply to naked mole-rats.

Although characterizing *H. glaber* as cooperatively breeding and eusocial seems redundant, this dual description is currently necessary because of the apparent divergence between studies of social evolution in vertebrates and invertebrates. Despite repeated attempts to draw attention to behavioral similarities between eusocial insects and cooperatively breeding birds and mammals (e.g., Vehrencamp 1979; Andersson 1984; Alexander, Noonan, & Crespi 1991; Emlen et al. 1991; Lacey & Sherman 1991; Krebs & Davies 1993), studies of vertebrate and invertebrate sociality have proceeded more or less independently. Whereas evolutionary analyses of insect eusociality have focused primarily on what Evans (1977) termed "intrinsic" (genetic) selective factors (e.g., Hamilton 1964; Trivers 1985; Seger 1991), analyses of vertebrate cooperative breeding have more frequently emphasized the role of "extrinsic" (ecological) selective factors (e.g., Koenig & Pitelka 1981; Brown 1987; Emlen 1991; Koenig et al. 1992). The naked mole-rat, the most widely accepted example of vertebrate eusociality (e.g., Seger 1991; Gadagkar 1994), promises to play a pivotal role in efforts to unite occurrences of alloparental care under a single theoretical umbrella.

Here we review the social biology of the naked mole-rat and present new information regarding colony fissioning and the effects of age, reproductive status, and colony composition on individual growth and behavior. Data are drawn from both our own ongoing studies and those of an international contingent of colleagues. We join Jarvis, O'Riain, Bennett, and Sherman (1994) in suggesting that the social system of *H. glaber* evolved in response to (1) ecological constraints on dispersal and independent reproduction that promoted coloniality and (2) patterns of kinship and reproductive biology that favored cooperation within colonies. We argue that this evolutionary scenario should apply to all cooperative breeders, vertebrate and invertebrate, and we discuss the implications of this integrative approach for future analyses of social behavior. Finally, we review evidence of sociality in other subterranean rodents and identify several "target" taxa for future studies of the evolution of group living and alloparental behavior.

10.2 Natural History of Naked Mole-Rats

Naked mole-rats inhabit arid regions of Kenya, Ethiopia, and Somalia in northeastern Africa (Figure 10.1). The animals belong to the family Bathyergidae, which consists of five genera and at least 12 species that collectively occur throughout much of sub-Saharan Africa (Honeycutt et al. 1991; Jarvis & Bennett 1991). All bathyergids are fossorial and exhibit morphological traits associated with life underground, including short limbs, cylindrical bodies, minute eyes, and reduced ear pinnae. The habitats used by bathyergids differ substantially with respect to rainfall, soil type, and food distribution (Jarvis & Bennett 1991). Pronounced differences in social behavior are also evident: whereas species in the genera *Heliophobius, Georychus,* and *Bathyergus* are solitary, *Heterocephalus* and several species of *Cryptomys* are social (Bennett & Jarvis 1988; Bennett 1989, 1990; Jarvis & Bennett 1991, 1993; Jarvis et al. 1994). This marked behavioral variation among closely related taxa provides an ideal opportunity to examine ecological and life-history correlates of sociality.

Among cooperatively breeding mammals, African mole-rats are distinguished by their extreme specialization for subterranean life. Whereas most cooperatively breeding carnivores rear their young in subterranean dens, all

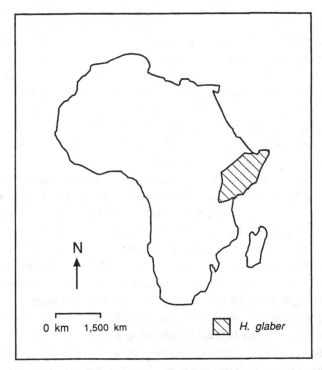

Figure 10.1. The geographic distribution of the naked mole-rat (*Heterocephalus glaber*). The species' range includes portions of Kenya, Ethiopia, and Somalia in eastern Africa.

live and forage primarily above ground (e.g., black-backed jackals, *Canis mesomelas:* Moehlman 1979; African wild dogs, *Lycaon pictus:* Frame et al. 1979; Creel & Creel 1995; dwarf mongooses, *Helogale parvula:* Rood 1978). Similarly, although cooperatively breeding pine voles (*Microtus pinetorum*) construct grass- or soil-covered tunnels, they often forage and travel above ground (FitzGerald & Madison 1983; Solomon 1994). In contrast, naked mole-rats, like other bathyergids, are completely subterranean. Although individual *H. glaber* have occasionally been sighted above ground (Braude 1991), such episodes appear to be rare and may be associated with severe environmental disturbances (e.g., heavy rains or the destruction of burrow systems). That *H. glaber* does not routinely occur on the surface is also suggested by the animals' poor visual and auditory acuity (Heffner & Heffner 1993), minimal thermal insulation (Jarvis 1978), and poor thermoregulatory ability (McNab 1966; Buffenstein & Yahav 1991).

Typically, above-ground evidence of *H. glaber* is restricted to volcano-shaped mounds 10 to 20 centimeters high that are created as the animals eject loose soil from their burrows. Because such soil is generated primarily during tunnel excavation, rates of mound production ("volcanoing") can be used to estimate rates of burrowing. On a daily basis, mound production is most common at dawn and dusk (Braude 1991; Brett 1991b). Annually, mound production is associated with rainy periods (October to December and February to May at Mtito Andei in southeastern Kenya) (Brett 1991b), presumably because rain softens the rock-hard soil in which the animals live, making burrowing energetically feasible (Jarvis et al. 1994).

Most tunnel excavation occurs as the animals forage for subterranean bulbs and tubers (Brett 1991b). To locate new food sources, naked mole-rats dig until they encounter an appropriate food item. The animals are apparently unable to detect tubers through the soil, often narrowly missing one or more large geophytes before burrowing directly into a tuber of the same species (Brett 1991b). As will be discussed, this mode of foraging, when combined with limited and unpredictable opportunities for burrow expansion, has important implications for the evolution of cooperative breeding.

Owing to their subterranean *modus vivendi*, the behavior of individual naked mole-rats cannot be easily observed in the field. If a tunnel system is opened (e.g., to insert a transparent observation window), the animals block the disturbed tunnel with soil and retreat to distant portions of the burrow system. Radiotelemetry has been used to monitor the movements of free-living mole-rats (Brett 1986, 1991b), but radio fixes provide only crude information regarding behavior. Thus, our knowledge of social behavior and alloparental care in bathyergid mole-rats is based primarily upon observations of captive colonies housed in artificial tunnel systems (summarized in Sherman, Jarvis, & Alexander 1991; Sherman, Jarvis, & Braude 1992). Although field observations of behavior are not possible, studies of free-living colonies by R. A. Brett (1986, 1991a,b) and S. H. Braude (1991, pers. comm.) provide essential ecological and demographic information that links laboratory observations of behavior to the appropriate environmental and evolutionary contexts.

10.3 Social Organization of Naked Mole-Rats

Naked mole-rats exhibit a number of intriguing behavioral parallels with cooperatively breeding vertebrates and eusocial insects. To underscore these similarities, our review of naked mole-rat behavior is divided into sections that correspond to the traditional criteria for insect eusociality (Wilson 1971; Hölldobler & Wilson 1990). In addition, comparative data from social verte-

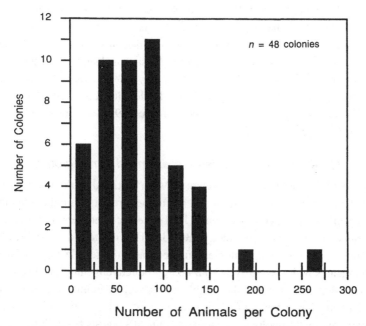

Figure 10.2. Sizes of 48 free-living naked mole-rat colonies. Data are from Braude (1991); only data from colonies thought to be complete (i.e., from which all animals were captured) are shown.

brates are included in each section. By emphasizing parallels between cooperatively breeding birds, mammals, and insects, we hope to facilitate a more integrated approach to studies of complex, cooperative societies (see also Andersson 1984; Sherman et al. 1995).

10.3.1 Coloniality

10.3.1.1 Naked mole-rats live in groups

Colonies captured in the field range in size from fewer than 10 to more than 295 individuals (Brett 1991a; Braude 1991), with a mean of 75 to 80 animals per colony (Figure 10.2). In contrast, group sizes for most other cooperatively breeding mammals are much smaller. Colonies of the eusocial Damaraland mole-rat (*Cryptomys damarensis*) contain fewer than 45 animals (Bennett & Jarvis 1988; Jarvis & Bennett 1993), and group sizes for other mammalian cooperative breeders rarely exceed two dozen individuals (e.g., wolves, *Canis lupus:* Harrington, Mech, & Fritts 1983; dwarf mongooses: Rood 1983; Creel

& Waser this volume; wild dogs: Creel & Creel 1995; see also Jennions & Macdonald 1994). Whereas colony sizes for naked mole-rats approximate those found in some paper wasps (e.g., *Polistes fuscatus:* Downing & Jeanne 1985; Reeve 1991; *Ropalidia marginata:* Gadagkar 1991) and ants (e.g., *Leptothorax longispinosus:* Herbers 1984), they do not reach the extremes found among other ants (e.g., *Atta sexdens:* Hölldobler & Wilson 1990), yellowjacket wasps (e.g., *Vespula vulgaris:* Greene 1991), or honey bees (e.g., *Apis mellifera:* Seeley 1985).

10.3.1.2 Naked mole-rat colonies are extended family groups

Mark and recapture studies conducted in Meru National Park (northcentral Kenya) from 1986 to 1991 suggest that recruitment results from the retention of young in their natal colony (Braude 1991). Replacement reproductives are also drawn from within the colony. Braude (1991) detected three breeder replacements in free-living colonies; in each case, the female that assumed breeding status had previously been a nonreproductive member of that same colony. In the lab, naked mole-rats of both sexes, including littermate siblings, have become breeders in their natal colony. This combination of natal philopatry and within-colony breeder replacement suggests that colony mates are typically close genetic relatives.

Molecular genetic analyses support this prediction. Using band-sharing frequencies calculated from multilocus DNA fingerprints, Reeve et al. (1990) estimated that the mean coefficient of relatedness among colony mates at Mtito Andei was 0.81. Low levels of intracolonial variation have also been reported for allozymes and mitochondrial DNA (Honeycutt et al. 1991), as well as minisatellite DNA and major histocompatibility genes (Faulkes, Abbott, & Mellor 1990). Although some intercolony exchange of genetic material probably occurs, the coefficient of inbreeding calculated by Reeve et al. (1990), $F = 0.45$, is the highest yet reported for free-living mammals. Indeed, Reeve et al. (1990) estimated that in nature ≥ 80 percent of matings occur between siblings or between parents and offspring.

10.3.1.3 New colony formation is poorly understood

Brett (1991a) proposed that colonies form when either (1) existing groups fission or (2) solitary dispersers meet, mate, and initiate a new social group. Clearly, these alternatives have important implications for patterns of inbreeding and a colony's genetic structure. At present, some evidence exists for both processes. Regarding (1), Brett (1991a) captured a small group of mole-rats (11 adults, 4 subadults, 10 pups) located approximately 60 meters from a larger colony of 93 individuals. Although tunnel connections between these groups were not evident, the presence of old volcanos in the intervening area suggested that the adults in the smaller group had originally been part of the larger colony.

Regarding (2), Braude (1991, pers. comm.) discovered five small colonies at Meru, each of which contained only a breeding pair and one or two litters of recently weaned young. One adult in each colony was marked and had originally been captured in a nearby larger colony. In all cases, the second adult was unmarked. Because Braude consistently marked mole-rats in this area from 1986 to 1995, he believed that the unmarked animals must have immigrated from colonies located outside of his study site. These intriguing data imply that colonies sometimes form when opposite-sex dispersers meet and establish residence together in a burrow system (see also O'Riain, Jarvis, & Faulkes 1996).

Additional insights into (1) are provided by an apparent colony fission that occurred at Cornell University. Colony A was captured near Mtito Andei in December 1979; ever since, the animals have been housed in an artificial burrow system constructed of Plexiglas (Lacey & Sherman 1991). For 11 years, colony mates shared the same nest box. In June 1990, however, the animals began using two different nest boxes. Daily scan sampling indicated that each nest box was occupied by a different subset of individuals (Figure 10.3) and that the two subcolonies concentrated their activities in different regions of the 32-meter-long burrow system. Members of the two subcolonies reacted aggressively toward one another; agonistic interactions between subcolonies occurred significantly more often than expected, given the number of individuals in each group ($\chi^2 = 15.2$, $n = 53$, $df = 1$, $p < 0.01$). Whereas a new dominant female eventually arose within the larger subcolony, no female ever achieved behavioral dominance in the smaller subcolony. After 6.5 months the subcolonies recombined, with the dominant female from the larger subgroup assuming the role of breeder. However, had we been able to provide the animals with sufficient tunnel space, it is possible that the two groups would have formed discrete colonies.

The timing of these events is intriguing. The two subgroups in Colony A first became evident two to three weeks after the death of the colony's breeding female (Figure 10.4). At the time of her demise, Colony A contained individuals from six lab-born litters, aged 1 to 8 years (see Figure 10.3). The three oldest litters were offspring of the colony's original breeding male and female, both of which died in 1986 (see Figure 10.4). The remaining litters were conceived during copulations between the colony's second breeding female and the second and third breeding males (see Figure 10.4). Although

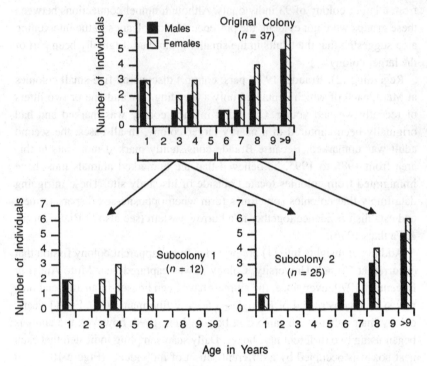

Figure 10.3. The age and sex compositions of the two subcolonies formed during an apparent colony fission event that occurred at Cornell University in 1990. Subcolony 1 consisted primarily of individuals from the three most recent litters of lab-born pups; subcolony 2 consisted primarily of individuals from three older lab-born litters, plus a number of adults captured when the colony was originally trapped in Kenya (1979). The mean age of individuals in subcolony 1 (≥2.9 yr) was significantly less than that of individuals in subcolony 2 (≥7.9 yr) (Mann-Whitney $U = 25$, $n = 12.26$, $p < 0.0001$).

exact genetic relationships among these litters are unknown, members of litters 1 to 3 would certainly have been more closely related to either their own offspring or those of their siblings (i.e., other members of litters 1 to 3) than to the offspring of individuals with entirely different parentage (litters 4 to 6; see Figure 10.4). This distinction appears to be important; the subgroups formed when Colony A "fissioned" (see Figure 10.3) paralleled this asymmetry in relatedness.

Kinship has been implicated in group fissioning in social insects (e.g., honey bees: Getz, Bruckner, & Parisian 1982), as well as in some societies of humans (e.g., Yanomamö: Chagnon 1979) and nonhuman primates (e.g., *Macaca sinica:* Dittus 1988). Our observations indicate that in Colony A, fis-

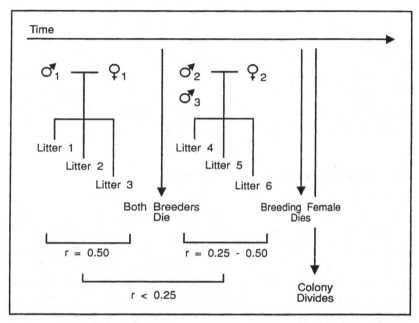

Figure 10.4. A summary of the changes in colony composition that preceded the apparent colony fission event in 1990 (see Figure 10.3). Following the death of the colony's second breeding female, colony members segregated to form two distinct subgroups. At the time of fissioning, the "parent" colony contained six litters of lab-born pups plus a number of wild-caught adults. Whereas litters 1 to 3 had been born to the colony's original breeding pair, litters 4 to 6 had been born to the colony's second breeding female (whose death initiated this series of events) and her two mates. Although exact genetic relationships between these litters were not known, pups in litters 1 to 3 were more closely related to one another than to pups in litters 4 to 6. Consequently, the emergence of a new breeder from either litters 1 to 3 or 4 to 6 would have generated marked asymmetries in relatedness between the new breeder's offspring and nonbreeders from the other matriline. These observations suggest that kinship may play a role in initiating formation of new colonies.

sioning was precipitated by intracolonial conflicts between kin groups. This finding is somewhat unexpected, given that colony members are so closely related and that there is so little genetic variation within colonies. Perhaps the genetic payoffs of becoming a breeder are so great and the costs of forfeiting direct reproduction are so high that even relatively small differences in relatedness substantially influence the behavior of individual naked mole-rats.

Our current knowledge of the demography of free-living *H. glaber* is based on fieldwork by Brett (1986, 1991a,b) and Braude (1991, pers. comm.). Braude reported that a substantial proportion of colony members disappear each year and that disappearance rates vary with reproductive status. The mean probability that a breeding female was present in consecutive years was 0.93 versus only 0.43 for nonbreeding animals. Indeed, nonbreeding males and females were unlikely to be trapped in their natal colony for >2 years. The proportion of disappearances caused by dispersal rather than mortality is unknown. Mortality is difficult to detect, and thus data on the rate and timing of colony formation may provide the only practical means of estimating the probability of successful dispersal by free-living mole-rats.

10.3.2 Reproductive Division of Labor

In most naked mole-rat colonies, only one female and one to three males reproduce (Jarvis 1981, 1991; Lacey & Sherman 1991). The breeding female is readily identified by her large size, elongate body, perforate vagina, and 10 to 14 enlarged teats (Jarvis 1991). Reproductive differences among females are associated with pronounced differences in behavior. The breeding female is aggressively dominant over other colony members (Reeve & Sherman 1991; Rymond 1991; Reeve 1992; Schieffelin & Sherman 1995), and she is the only female to exhibit behavioral estrus or to engage in sexual behavior.

Restricted reproduction in naked mole-rat colonies is not an artifact of captivity. Among six free-living colonies from Mtito Andei, Jarvis (1985) and Brett (1991a) found only one breeding female per colony. At Meru, Braude (1991) found that 18 of 20 free-living colonies (90%) each contained only a single reproductive female. The remaining two Meru colonies contained two breeding females each; this condition persisted for several years, indicating that it was not owing simply to a transition between breeders. Of 53 lab colonies housed at the University of Cape Town, Cornell University, and the University of Michigan, 47 (89%) contained only a single breeding female each (Sherman et al. 1992).

In the field (Brett 1991a) and in the lab (Jarvis 1991; Lacey & Sherman 1991), breeding females give birth as often as every 2 to 3 months. Field-born litters contain a mean of 9.7 ($n = 7$; Brett 1991a) to 14.0 pups ($n = 21$; Braude 1991), with a range of 4 to 22 pups per litter. Lab-born litters average between 7.7 ($n = 26$;) and 12.3 pups ($n = 84$; Jarvis 1991), with a range of 1 to 27 pups per litter. Thus, a reproductive female can produce at least 30 to 60 offspring per year. Comparable data on male reproductive success are not available because (1) breeding females sometimes mate with several males during a single estrous period and (2) attempts to determine paternity by DNA fingerprinting have been stymied by exceptionally low levels of intracolonial genetic variability.

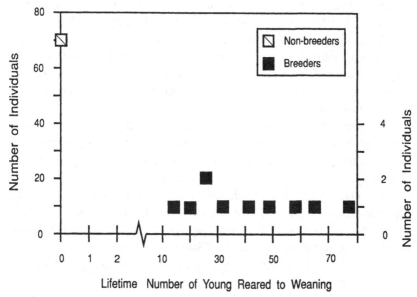

Figure 10.5. Lifetime reproductive success for females in seven captive colonies of naked mole-rats housed at Cornell University and the University of Michigan. Between 1979 and 1991, a total of 350 pups were born to the 10 breeding females in these colonies; no young were born to the remaining 70 females in these colonies.

Pronounced short-term differences in direct reproductive success should generate substantial intracolonial differences in individual lifetime reproductive success. Although lifetime reproductive success has not been quantified for free-living *H. glaber,* data on reproductive success are available for 80 females in seven captive colonies housed at Cornell and Michigan. Within each colony, individual lifetime reproductive success was bimodally distributed: a small number of females produced many young, whereas the majority produced none (Figure 10.5). Again, comparable data for males are not available. However, because only one to three males per colony mate, individual lifetime reproductive success for males should also be bimodally distributed.

Asymmetries in lifetime reproduction appear to be more extreme among captive *H. glaber* than among other mammalian cooperative breeders. Among free-living Mongolian gerbils (*Meriones unguiculatus*: Agren, Zhou, & Zhong 1989) and silver-backed jackals (Moehlman 1983), tenure as a nonreproductive alloparent is effectively an ontogenetic stage through which many individuals pass before attempting to breed independently. Individual dwarf mongooses (Rood 1983, 1990; Creel & Waser this volume) and wolves

(Mech 1983) may remain in their natal group for a considerably greater proportion of their lifetime, but alloparents often eventually reproduce directly by (1) mating while still subordinates, (2) becoming breeders within their natal group, or (3) dispersing to groups lacking breeders. In contrast, these options appear to be only rarely available to individual naked mole-rats, and most colony members probably achieve reproductive success only by helping to rear collateral kin.

Nonbreeding naked mole-rats are not physiologically sterile. If a breeder dies or is removed from a colony, it is replaced by a previously nonbreeding animal of the same sex. The endocrinological bases for this suppression are reviewed by Faulkes et al. (1991) and Faulkes and Abbott (this volume). Behaviorally, reproductive differences among females are maintained by aggressive interactions with the breeding female (Faulkes et al. 1991). Behavioral aspects of reproductive suppression among males are not as well understood. In addition to dominance interactions, physical size and mate choice by the breeding female may play a role in maintaining reproductive differences among males.

In the lab, changes in a female's reproductive status are preceded by rapid changes in her body weight. Following the loss of a breeding female, nonreproductive females may gain weight for several weeks to months before becoming sexually receptive (Jarvis 1991; Lacey & Sherman 1991). Field studies suggest a similar pattern of prereproductive weight gain. Braude (1991) reported that replacement breeders in two of three field colonies exhibited dramatic increases in body weight before becoming reproductive; the third replacement breeder was already among the largest animals in her colony and did not undergo a period of rapid growth prior to mating.

Males may also increase in body weight before becoming breeders (Lacey & Sherman 1991). However, whereas the breeding female typically remains among the largest (heaviest) animals in her colony throughout her tenure as breeder, breeding males frequently begin losing weight after achieving reproductive status. For example, Jarvis, O'Riain, and McDaid (1991) reported that breeding males ($n = 5$) in their study colonies lost 17 to 30 percent of their maximum body weight and that males that remained breeders for extended periods (several years or more) became visibly emaciated. Similarly, a reproductive male in one of the Cornell colonies lost 59 percent of his maximum body weight while continuing to behave as a breeder.

In addition to becoming heavier, the bodies of newly reproductive females become more elongate because of the addition of bone to the vertebrae (Jarvis 1991). This spinal elongation, which to our knowledge is unique among mammals, occurs in both field (Braude 1991) and lab colonies of naked mole-rats (Jarvis 1991). The resulting changes in body shape may allow females pregnant with large litters to move through their colonies' tunnel systems (Jarvis 1991). In contrast, no such skeletal changes are evident among newly reproductive males.

Little is known regarding the phenotypic traits that characterize future breeders. Field studies of breeder succession are virtually impossible, and experimental manipulations of lab colonies have been hampered by the extreme aggression and mortality that often follow the removal of a breeding female. The few field and lab data available indicate that replacement breeders can come from among either the lightest or heaviest nonbreeders in a colony (Braude 1991; Jarvis 1991; Sherman unpubl. data) and that they may or may not be the offspring of the individuals they replace. Not all colony members, including not all litter mates, respond to the death of a same-sex breeder by gaining weight and competing for dominance (Lacey & Sherman 1991), indicating that even within litters the probability of breeding varies. Jarvis (1991) has suggested that dominance hierarchies arise among litter mates early in ontogeny, perhaps during wrestling and tooth-fencing bouts among juveniles; these hierarchies may in turn influence the reproductive options available to adults.

Among males, breeder succession also appears to be influenced by the identity of the current reproductive female (or group of females vying to fill a breeding vacancy). Following the death of a breeding female, would-be successors often attack other females and, surprisingly, certain males (Jarvis 1991; Lacey & Sherman 1991). Potential breeding females appear to target specific males for attack, often shoving and biting the same individual repeatedly. Of seven such protracted intersexual conflicts observed at Cornell, five resulted in the death of the targeted male. All of these conflicts were initiated by females. The males attacked were either (1) mates of the previous breeding female ($n = 3$) or (2) males that were closely associated (i.e., anogenital nuzzled frequently) with rival females ($n = 4$).

We do not know why potential breeding females kill males that are affiliated with reproductive competitors. Mortality per se may be an artifact of captivity; in nature, animals that are attacked may escape by retreating to distant portions of the burrow system or dispersing. Nevertheless, the reasons for females' attacks are puzzling, given that their victims could presumably contribute to colony maintenance and defense. Perhaps female–male pairs cooperate much more extensively to achieve reproductive dominance than is currently recognized. If so, males that consort with rival females may themselves be reproductive competitors (Alexander 1991). Alternatively, attacks by breeding females may represent a form of mate choice that maintains high levels of inbreeding within colonies (Hamilton 1992), thereby ensuring that nonbreeding colony members receive significant inclusive fitness benefits for contributing to a female's reproductive efforts.

10.4 Cooperative Care of Young

Nonbreeding males and females contribute to the reproductive efforts of breeders both directly and indirectly. The most obvious manifestation of direct assistance is alloparental care. Shortly before a litter is born, nonbreeders of both sexes increase the amount of time spent huddling in the colony's communal nest. Increased huddling continues until shortly after the young are weaned (age: 3 to 4 weeks) and appears to provide pups with a stable thermal environment (Lacey et al. 1991). While in the nest, nonbreeders frequently nudge, handle, and groom young (Figure 10.6). Nonbreeders also retrieve pups that fall out of the nest, transport pups when the colony moves to a new nest, and evacuate pups from the nest during colony disturbances.

As pups approach weaning, they begin to eat caecotrophes (partially digested fecal pellets) produced by adults. Allocoprophagy provides pups with nutrients and the endosymbiotic gut flora required for cellulose digestion (Jarvis 1991). Pups routinely solicit and obtain caecotrophes from nonbreeders of both sexes but rarely solicit from breeders of either sex (see Figure 10.6; also Jarvis 1991; Lacey & Sherman 1991). Allocoprophagy continues until the young are fully weaned, after which pups consume food items carried to the nest by nonbreeders, as well as forage directly for themselves.

Nonbreeding colony members also contribute indirectly to pup production by foraging for breeders and by defending, maintaining, and extending the colony's burrow system. In the lab, nonbreeders of both sexes are the primary participants in all colony maintenance and defense activities, including transporting food. Although food-carriers sometimes eat what they transport, most food items are carried to the nest, where they are consumed by other colony members, including the breeding female and her mates (Lacey & Sherman 1991). Nonbreeders also collect nesting material, excavate foraging tunnels, clear the burrow system of debris (e.g., from cave-ins), and attack predators and unfamiliar mole-rats that attempt to enter the colony (Lacey & Sherman 1991). In contrast, breeders seldom participate in these activities (Jarvis 1981; Payne 1982; Lacey & Sherman 1991).

Many of the alloparental behaviors exhibited by naked mole-rats also occur in other cooperatively breeding mammals. For example, alloparental pine voles (Powell & Fried 1992), prairie voles (Solomon 1991; Solomon & Getz this volume), Damaraland mole-rats (Jarvis & Bennett 1993), dwarf mon-

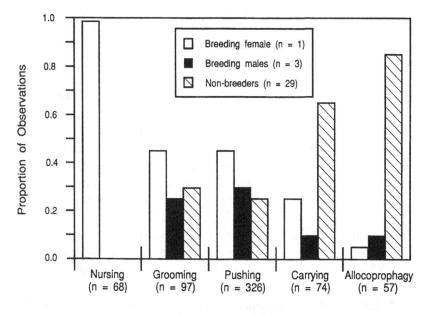

Pup Care Behaviors

Figure 10.6. Participation in five pup care behaviors by breeding and nonbreeding members of a captive naked mole-rat colony at Cornell University. The behaviors quantified included nursing, grooming, pushing, and carrying pups, as well as providing pups with caecotrophes (allocoprophagy). Although the function of pushing is unknown, it is most frequently directed toward pups that are successfully reared to weaning, and it appears to be a routine part of juvenile care (see Lacey et al. 1991).

gooses (Rood 1983; Creel & Waser this volume), and callitrichid primates (Tardif this volume) all groom, tend (babysit), and transport young. Alloparental provisioning and defense of young occur in many cooperatively breeding carnivores, including coyotes (*Canis latrans:* Bekoff & Wells 1982, 1986), African wild dogs (Malcolm & Marten 1982), red foxes (*Vulpes vulpes:* Macdonald 1980), and silver-backed jackals (Moehlman 1983). However, behavioral differentiation between breeders and nonbreeders in these species is not as extreme as in *H. glaber.* In this regard, naked mole-rats more closely resemble some small-colony social insects in which both behavior and body size differ with reproductive status (e.g., paper wasps, *Polistes* spp.: Reeve 1991, and sweat bees, *Lasioglossum* spp.: Michener & Brothers 1974).

The effects of alloparental behavior on pup production and survival have not been quantified for naked mole-rats. Presumably, alloparental care increases the number of young that can be reared by a colony. Alloparental care is not required for pup survival, however, as a breeding pair can rear young to weaning in both the lab (Jarvis 1991; pers. obs.) and the field (Braude 1991). Ideally, the effects of alloparental care on breeder fitness should be quantified for free-living colonies by (1) examining correlations between the number of alloparents per colony and juvenile recruitment and (2) determining how experimental reductions of alloparent numbers affect juvenile survival (e.g., Emlen 1991; Jennions & Macdonald 1994).

Determining how alloparental behavior affects nonbreeder fitness is more difficult, as there are several mechanisms by which alloparents may benefit. For example, alloparental behavior may increase (1) the survival of breeders and their offspring (i.e., inclusive fitness), (2) the probability of being allowed to remain in the safety of the natal burrow system, (3) the likelihood of establishing a new colony by "budding off" a portion of the parental burrow system, or (4) the probability that, if they get a chance to breed, alloparents will be more successful in rearing young as a result of increased parental experience (Woolfenden & Fitzpatrick 1990; Jennions & Macdonald 1994). Although none of these possibilities can be ruled out, the combination of limited direct reproduction and high intracolony relatedness suggests that inclusive fitness benefits are important in maintaining alloparental behavior in naked mole-rats.

10.5 Colony Regulation

One fascinating aspect of naked mole-rat behavior that does not fall under the preceding headings is the regulation of colony activity. In her initial description of naked mole-rat sociality, Jarvis (1981) suggested that nonbreeders can be divided into two or three discrete "castes" – frequent workers, infrequent workers, and nonworkers – that vary with respect to both body size and participation in colony maintenance. Subsequent studies at Cape Town, Cornell, Michigan, and London have confirmed that the behavior of nonbreeders varies significantly with body size (Payne 1982; Isil 1983; Faulkes et al. 1991; Jarvis et al. 1991; Lacey & Sherman 1991). Whereas small nonbreeders are the primary participants in colony maintenance activities (e.g., gathering food, building nests, cleaning tunnels), large nonbreeders are the primary participants in colony defense activities (e.g., guarding the nest, threatening snakes and unfamiliar conspecifics). However, contrary to Jarvis's three-caste hypothesis, the behavior of animals at Cornell, Michigan, and London varied continuously as a function of body weight (Payne 1982; Isil 1983; Faulkes et al. 1991; Lacey & Sherman 1991), providing no evidence of discrete morphological castes.

The proximate factors associated with behavioral variation among non-breeders have been the subject of some disagreement. Although Jarvis (1981) did not specify a mechanism of caste determination, her discussion of size differences among nonbreeders suggests that castes reflect individual growth rates: whereas fast-growing individuals may become nonworkers and eventually breeders, slow-growing individuals permanently remain frequent workers (Jarvis et al. 1991). In contrast, Lacey and Sherman (1991) suggested that size-related differences in behavior represent an age polyethism in which the behavior of all nonbreeders gradually shifts from maintenance to defense activities as individuals grow older and larger. Long-term behavioral data from known-age mole-rats reared at Cornell indicate that in established colonies nonbreeders usually switch from "small animal" (i.e., maintenance) to "large animal" (i.e., defense) activities at 30 to 40 months of age (Figure 10.7).

Jarvis (1981) and Lacey and Sherman (1991) presented what appear to be conflicting data regarding the relationship between body weight and age. We suggest that this disagreement arose because body weight is labile and is influenced by multiple factors, including age, past reproductive opportunities, and colony composition. As discussed previously, several colony members exhibit rapid increases in body weight following the death of a breeding animal. Only a few of these individuals eventually become breeders; body weights of the remaining animals stabilize at postgrowth levels, rather than declining to pregrowth values. As a result, colony mates, including litter mates, whose body weights were similar prior to the loss of a breeding animal, can differ greatly in size following the emergence of a replacement breeder. Thus, intracolonial differences in body weight reflect past reproductive opportunities as well as differences in age.

Among both juveniles and adults, growth rates (and hence body weights) may vary in response to colony composition. For example, Jarvis et al. (1991) noted that pups reared in small colonies (i.e., colonies containing few nonbreeders) grew faster and achieved larger adult body weights than pups reared in larger colonies. Sherman (unpubl. data) observed that, following the deaths of five large nonbreeders from one captive colony, a group of six small nonbreeders gained weight rapidly and began participating in defense activities. Conversely, following the experimental removal of a group of six small nonbreeders from a different colony, rates of weight gain by four of the remaining nonbreeders in the colony decreased; the body weights of these individuals stabilized and these animals continued to perform maintenance activities until they were much older than maintenance workers in unmanipulated colonies.

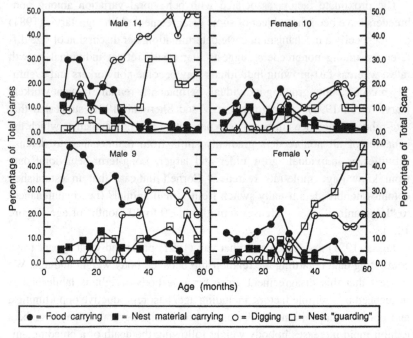

Figure 10.7. Evidence of age polyethism in naked mole-rats. Among four lab-born members of a single colony (two males, two females) housed at Cornell University, individual participation in two types of maintenance activities declined with age, whereas participation in two types of defensive activities increased with age. Participation in each behavior was quantified during 13 sampling periods conducted over 60 months; the same data collection procedures were used during each sampling period (for details, see Lacey & Sherman 1991). To facilitate comparisons of various behaviors, the number of food carries and nest material carries by an individual during a given sampling period were converted to a percentage of the total number of each type of carry observed. Similarly, the numbers of scans in which an individual was observed digging or facing out of the nest (i.e., "guarding") were converted to a percentage of the total number of scans completed for each behavior during a given sampling period. To minimize any interdependence of data points for different animals resulting from this procedure, copious quantities of food and nesting material and numerous digging and "guarding" sites (nest exit tunnels) were provided during each trial. However, even partial dependence of data points for different colony members should not have produced the consistent switch from maintenance to defensive behaviors evident here.

Variation in growth rates caused by variable demographic conditions may explain seemingly contradictory information regarding body weight and age. At Cornell, where colony composition remained relatively stable in two of the three colonies monitored by Lacey and Sherman (1991), body weight and behavior were significantly correlated with age. In contrast, the compositions of Jarvis, O'Riain, and McDaid's (1991, p. 413–416) study colonies were considerably more variable. For example, membership in Lerata 4 (in which body weight was not consistently correlated with age) varied because of mortality, frequent attempts at reproductive supersedure, and the experimental removal of animals. In two other colonies ("300" and "700") in which body weight was not consistently correlated with age (Jarvis et al. 1991), several cohorts of pups had been removed to simulate predation. Although Jarvis, O'Riain, and McDaid (1991) concluded that behavioral differences among nonbreeders were not the product of age polyethism, we believe that this issue warrants further investigation in light of the apparently complex relationships among age, body size, and colony structure. We suggest that the behavior of nonbreeding naked mole-rats does vary with age but that this variation is often obscured by behavioral and size changes associated with reproductive opportunities, colony composition, or both.

In addition to body size and colony composition, the behavior of nonbreeding naked mole-rats is strongly influenced by the actions of the breeding female. Colonies are active 24 hours per day and do not exhibit circadian activity cycles or entrain to artificial light cycles (Davis-Walton & Sherman 1994). Colony activity is stimulated by the breeding female, the most active and aggressive individual within a colony (Reeve & Sherman 1991; Rymond 1991; Reeve 1992). As the breeding female moves through the tunnel system, she uses her muzzle to push and shove nonbreeding animals. Reeve (1992) found that nonbreeders respond to these shoves by increasing their participation in colony maintenance activities and that even temporary removal of the breeding female results in a significant decrease in the rates at which maintenance activities are performed. Thus, although breeding females rarely participate directly in maintenance or defense activities, they regulate the maintenance and defense efforts of others.

Aggression by the breeding female is most frequently directed toward larger and more distantly related colony members (Reeve & Sherman 1991) – the individuals that are least likely to engage spontaneously in colony maintenance activities when the breeding female is absent and most likely to attempt supercedure even when she is present. Reeve and Sherman (1991) and Reeve (1992) suggested that aggression by the breeding female incites activity by these otherwise "lazy" animals. This need to incite nonbreeder activity may reflect reproductive conflicts of interest within colonies. Whereas the breeding female's reproductive output should be greatest when all colony members engage in maintenance and defense activities, larger or more distantly related nonbreeders may benefit more by refraining from risky or exhausting activities and instead resting, growing, and waiting for reproductive opportunities to arise.

Reeve's (1992) hypothesis assumes that size is positively correlated with reproductive potential. No field data are available to test this assumption, thus underscoring the need for detailed studies of the reproductive options available to different size (and age) classes of colony members. Clearly, the social structure of naked mole-rat colonies represents the outcome of complex interactions among numerous factors. Behavioral variation among colony members likely reflects both individual reproductive interests and the sometimes competing force of colony-level interests (Lacey & Sherman 1991). The result is a dynamic balance of cooperation and conflict that varies as conditions within the colony vary.

10.6 Evolution of Cooperative Breeding in Naked Mole-Rats

The evolution of cooperative breeding can be viewed as a two-step process (Brown 1987; Emlen 1991; Emlen et al. 1991; Koenig et al. 1992). The first step is the formation of social groups, generally by natal philopatry. When ecological or demographic conditions render dispersal sufficiently costly (or group living sufficiently beneficial), individuals may do better to remain in their natal group rather than to attempt to disperse and breed on their own. Environmental conditions may cause young to remain with their parents well beyond the age at which reproduction is possible. As a result, groups are typically composed of two or more generations of reproductively mature, related individuals. The proportion of group members that reproduce directly varies among species, providing an important axis for comparative studies of cooperatively breeding taxa (see below).

The second step in the evolution of cooperative breeding is the elaboration of alloparental care. Once philopatry and group living have become established, young that remain in their natal group may benefit by assisting the reproductive efforts of others (Emlen & Wrege 1989; Emlen et al. 1991; Jennions & Macdonald 1994). The most obvious reproductive benefit of alloparental behavior is the increased production and survival of nondescendant kin (frequently siblings). Other possible benefits to alloparents include continued access to the safety of the natal area and an increased probability of future direct reproductive success due to either experience at parenting or inheritance of an established breeding site. These benefits are not mutually exclusive; alloparental behavior may simultaneously increase both an individual's indirect and (future) direct reproductive success.

This two-step scenario is useful as a conceptual framework for separating the reproductive costs and benefits of philopatry from those of alloparental care. However, these steps are not independent. As discussed by Creel and Waser (this volume), the potential fitness consequences of alloparental care may influence whether an individual remains in its natal group; if associated benefits are large, individuals should be less likely to attempt dispersal and independent reproduction. Conversely, the probability of successful dispersal and independent reproduction should affect the occurrence of alloparental behavior; individuals with virtually no chance of reproducing on their own should assume greater risks while engaged in alloparental care than individuals with a higher probability of direct reproduction. Thus, in many cooperatively breeding species, tendencies to delay dispersal and assist the reproductive efforts of others should be plastic and should vary with the dispersal and reproductive options available to individuals.

Within the Bathyergidae, coloniality appears to have evolved in response to several factors, including (1) the distribution of critical food resources, (2) the energetic costs of burrowing, and (3) the danger of predation. Of these, (1) and (2) have received the greatest attention (e.g., Jarvis 1978; Lovegrove & Wissel 1988; Lovegrove 1991). Recent comparative studies of naked and Damaraland mole-rats underscore the connection between food distribution, habitat type, and cooperative breeding (Jarvis & Bennett 1993; Jarvis et al. 1994). Both species consume irregularly distributed bulbs and tubers, patches of which represent superabundant food sources that can be reached only by extensive burrowing. Although naked and Damaraland mole-rats occur in different soil types, the substrates used by both species can be efficiently excavated only when recently saturated by rain.

The costs of burrowing to locate new food resources, combined with the brief periods during which tunnel excavation is possible, led Jarvis et al. (1994) to suggest that individual dispersers would be unlikely to locate enough food to sustain themselves through long and unpredictable dry periods. Consequently, individuals may be forced to remain in their natal colony. In support of this hypothesis, Jarvis et al. (1994) noted that, whereas most species of bathyergids occur in mesic habitats and are solitary, the only species that are abundant in xeric habitats are colonial, cooperative breeders.

The aridity–food distribution hypothesis developed by Jarvis et al. (1994) represents the most comprehensive explanation for mole-rat coloniality proposed to date. Other factors that have been suggested as important in the evolution of bathyergid coloniality include body size, metabolic rate, reproductive biology (Burda 1990; Lovegrove 1991), and predation (Alexander 1991; Alexander, Noonan, et al. 1991). Because the first three factors are expected to influence social behavior primarily by altering the costs of burrow excavation and foraging, their effects are subsumed by the aridity–food distribution hypothesis. In contrast, predation represents a distinct cost of dispersal (or benefit of philopatry) that is not directly influenced by constraints on foraging or tunnel excavation. Jarvis et al. (1994) acknowledged the importance of predation in social evolution but noted that no obvious differences exist between the types of predators on solitary and social bathyergid species. However, quantitative data regarding both the rates of predation on solitary versus social mole-rats and the relative effectiveness of group defense are needed to assess the role of predation in bathyergid social evolution.

With regard to alloparental care, inclusive fitness benefits probably play an important role in shaping intracolonial interactions. The apparent rarity of successful colony formation by *H. glaber* suggests that an individual's best chance of reproducing directly may be to remain in its natal colony and wait for a breeder to die. This scenario may be especially applicable to males, given that colonies frequently contain multiple breeding males and that rates of breeder turnover appear to be greater for males than for females. However, the observed combination of extremely restricted breeding, large colony size, and relatively low breeder mortality implies that the vast majority of naked mole-rats never reproduce directly. These demographic patterns, in conjunction with the high genetic relatedness within colonies, suggest that nonbreeding *H. glaber* may reap substantial inclusive fitness benefits from assisting the reproductive efforts of breeders. Unfortunately, field studies of inclusive fitness in naked mole-rats lag behind studies of many of the other taxa described in this volume (see also Jennions & Macdonald 1994) because of the logistical challenges of monitoring demography and reproductive success in a subterranean species.

10.7 Implications for Vertebrate and Invertebrate Sociality

Studies of bathyergid mole-rats have important implications for analyses of social evolution in vertebrates and invertebrates. The realization that naked and Damaraland mole-rats meet the traditional criteria for eusociality has forced behavioral ecologists to reconsider both the definition and the taxonomic distribution of this type of social system. Whereas the term "eusocial" has generally been reserved for insects in the orders Hymenoptera (ants, bees, wasps) and Isoptera (termites), the last decade of research has revealed remarkable similarities between these societies and societies of cooperatively breeding vertebrates. As our understanding of the complexity of vertebrate and invertebrate sociality has matured, distinctions between eusociality and cooperative breeding have become blurred.

To accommodate variation in insect social systems, several authors have suggested using diagnostic modifiers to characterize eusociality (e.g., "primitive" vs. "advanced": Michener 1974; Gadagkar 1994; "morphological" vs. "behavioral": Kukuk 1994). Alternatively, there have been two recent attempts to redefine eusociality, one seeking to greatly restrict (Crespi & Yanega 1995) and the other to greatly expand usage of this term (Sherman et al. 1995). We favor the latter approach: eusociality is not a discrete phenomenon but instead represents a continuum along which societies of cooperatively breeding vertebrates and invertebrates can be arrayed according to the degree of skew in individual lifetime reproductive success within social groups (Figure 10.8; see also Vehrencamp 1983a,b; Keller & Perrin 1995). This scheme emphasizes behavioral similarities between cooperatively breeding vertebrates and invertebrates (see also Brown 1975, pp. 198–207) and unites all occurrences of alloparental care under a single terminological umbrella. Thus, the traditional distinction between eusociality and cooperative breeding disappears, leaving in its place the realization that vertebrate and invertebrate social systems are not qualitatively different but vary only quantitatively with respect to the same underlying evolutionary principles (e.g., Hamilton's rule; Grafen 1991).

In addition to underscoring similarities between vertebrate and invertebrate societies, studies of social bathyergids have forced biologists to reexamine the relative importance of extrinsic and intrinsic factors (sensu Evans 1977) in the evolution of cooperative breeding. Because eusociality was traditionally thought to occur only among hymenopteran and isopteran insects, evolutionary explanations for this type of social system have frequently focused on genetic attributes unique to these groups (e.g., Hamilton 1964, 1972; Charnov 1978; Bartz 1979; Lacy 1980, 1984; Trivers 1985). In contrast, evolutionary analyses of vertebrate cooperative breeding have tended to emphasize the effects of ecological or demographic conditions on dispersal and natal philopatry (e.g., Koenig & Pitelka 1981; Brown 1987; Emlen 1991; Koenig et al. 1992). Because naked and Damaraland mole-rats are both cooperatively breeding *and* eusocial, evolutionary analyses of their behavior must integrate these intrinsic and extrinsic explanations for sociality.

Our current understanding of bathyergid behavior suggests that extrinsic factors set the stage for social evolution by constraining dispersal and reproductive options, thereby leading to natal philopatry and coloniality. Intrinsic factors, in turn, shape the nature of social interactions among colony members by determining how and to what extent individuals benefit from specific competitive and cooperative behaviors, including attempts at direct reproduction (see Reeve & Keller 1995). This explanatory framework is not exclusive to bathyergid mole-rats but should apply to all vertebrate and invertebrate cooperative breeders.

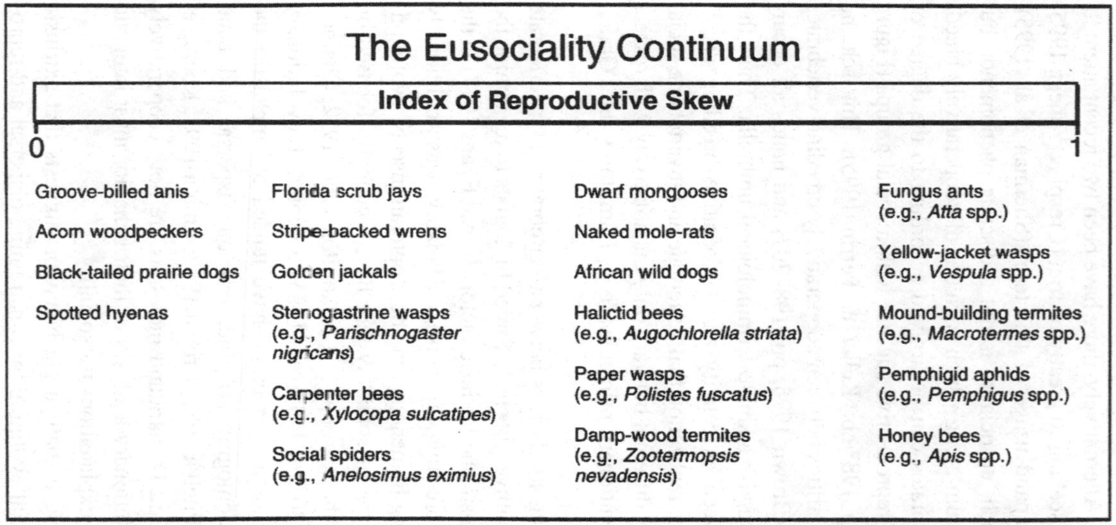

The Eusociality Continuum

Index of Reproductive Skew

Figure 10.8. Predicted locations of selected cooperatively breeding societies of vertebrates and invertebrates along a common scale (an index of the skew in lifetime reproductive success among members of a social group; see Sherman et al. 1995; Keller & Perrin 1995). The skew is 0 when lifetime reproductive success is equal among members of a social group; the skew is 1 when reproduction within a group is restricted to a single individual of each sex and helpers never breed. When skews vary considerably among conspecific groups or populations, a species is best represented as a segment of the continuum denoting the intraspecific range of skew values. Indexes of reproductive skew may be calculated for male group members only, females only, or both sexes, depending on which animals participate in alloparental care. In this figure, only females are considered, and societies that exhibit apparently similar reproductive skews are grouped together. This figure was modified slightly from Sherman et al. (1995).

10.8 Sociality in Other Subterranean Rodents

None of the ecological factors believed to favor sociality in African mole-rats is unique to the Bathyergidae. Arid environments, patchy food distributions, high costs of burrow excavation, and low levels of genetic variability also characterize other subterranean taxa. If hypotheses regarding the evolution of sociality in bathyergids (e.g., Alexander et al. 1991; Jarvis et al. 1994) are correct, then similar combinations of selective factors should favor sociality in other subterranean rodents. Already, comparative studies of naked and Damaraland mole-rats have revealed ecological similarities that may help to explain behavioral parallels between these species (Jarvis et al. 1994). Here we briefly review the occurrence of sociality (defined as multiple adults sharing a single burrow system) among subterranean rodents, with the intent of identifying promising taxa for comparative studies of social evolution.

Among the rodent taxa identified as subterranean by Nevo (1979), sociality occurs in at least 14 species in five genera in the families Muridae (subfamily Arvicolinae), Octodontidae, Ctenomyidae, and Bathyergidae (Table 10.1). Social behavior has been most extensively documented for African mole-rats, due primarily to the efforts of J. U. M. Jarvis and her students. Within the Bathyergidae, coloniality occurs in at least seven species of *Cryptomys* and in the monotypic genus *Heterocephalus* (Jarvis et al. 1994). Phylogenetic analyses of the Bathyergidae (Allard & Honeycutt 1992) suggest that sociality has arisen independently in *Heterocephalus* and *Cryptomys*. If so, studies of social *Cryptomys* can be used to evaluate hypotheses regarding sociality in *Heterocephalus*. However, because the aridity–food distribution hypothesis (Jarvis et al. 1994) was developed using information from both naked and Damaraland mole-rats, comparative analyses of these species do not provide a robust test of this argument.

In contrast, studies of two South American genera, *Spalacopus* (coruros) and *Ctenomys* (tuco-tucos), do allow independent evaluations of the factors favoring sociality in subterranean rodents. Behavioral data on coruros (*S. cyanus*) are scarce, but available information suggests a number of intriguing parallels with social mole-rats. First, coruros appear to be colonial. Reig (1970) trapped 15 animals, including multiple adults, in a single burrow system; two of six adult females trapped were pregnant at the time of capture. Second, the ecology of *Spalacopus* resembles that of the social bathyergids. The colony of coruros trapped by Reig (1970) occurred in semiarid habitat in western Chile, where the animals fed on the roots and stems of a patchily distributed species of lily. Because coruros and African mole-rats are taxonomically and geographically distinct, these behavioral and ecological similarities must reflect evolutionary convergence.

Table 10.1. *The occurrence of sociality in rodent taxa commonly recognized as subterranean*

Family (common name)	Genera	Geographic distribution	Social species reported	References on sociality
Bathyergidae (African mole-rats)	*Bathyergus*	Africa	0	Jarvis & Bennett 1991
	Cryptomys		8	Jarvis et al. 1994
	Georychus		0	Sherman et al. 1991
	Heliophobius		0	
	Heterocephalus		1	
Ctenomyidae (Tuco-tucos)	*Ctenomys*	South America	3	Pearson & Christie 1985 Lacey & Braude 1993 Reig et al. 1990
Octodontidae (Octodonts)	*Spalacopus*	South America	1	Reig 1970
Muridae				
Myospalacinae (Voles)	*Myospalax*	Asia	0	
Arvicolinae (Voles)	*Ellobius*	Asia	0	
	Prometheomys		1	Grzimek 1975
Spalacinae (Blind mole-rats)	*Spalax*	Northern Africa, SE Europe, Asia	0	
Rhizomyinae (Bamboo rats)	*Cannomys*	Africa, SE Asia	0	
	Rhizomys		0	
	Tachyoryctes		0	
Geomyidae (Pocket gophers)	*Geomys*	North and Central America	0	
	Orthogeomys		0	
	Pappogeomys		0	
	Thomomys		0	
	Zygogeomys		0	

Whereas *Spalacopus* is a monotypic genus that occurs only in western Chile, the genus *Ctenomys* contains more than 30 species and occurs throughout much of sub-Amazonian South America (Redford & Eisenberg 1992; Wilson & Reeder 1993). Tuco-tucos are generally thought to be solitary (Reig et al. 1990). Although anecdotal reports of sociality exist for several species, quantitative evidence of group living is available only for the colonial tuco-tuco (*Ctenomys sociabilis:* Pearson & Christie 1985). Recent field studies indicate that *C. sociabilis* burrow systems are inhabited by multiple adults and juveniles (Lacey & Braude 1993). All adult females in a colony lactate simultaneously, and there is no evidence of a reproductive division of labor among colony members. Thus, comparisons between *C. sociabilis* and social African mole-rats may yield new insights into the selective factors favoring plural versus singular breeding in social groups (e.g., Vehrencamp 1983a,b; Keller & Reeve 1994). At the same time, behavioral diversity within *Ctenomys* (as in *Cryptomys*) should allow us to examine the extrinsic and intrinsic factors associated with marked social differences among closely related species.

Subterranean rodents exhibit a number of ecological and life-history differences that may be affiliated with interspecific variation in sociality. For example, whereas social African mole-rats forage exclusively on subterranean food items, tuco-tucos emerge briefly from their burrows to crop above-ground vegetation. This difference may significantly alter foraging costs and benefits, perhaps making it easier for lone tuco-tucos to locate and obtain sufficient quantities of food. Other potentially important factors include the seasonality and predictability of the environment, the ability of individual females to produce multiple litters per year, and the nature and frequency of predation. For example, because naked mole-rats reproduce year-round, opportunities to augment inclusive fitness are nearly continuous, and individuals specializing as alloparents may benefit substantially relative to members of species that reproduce only once per year (e.g., some *Ctenomys* spp.)

Interspecific differences in demography and population genetic structure may also influence sociality. Whereas naked mole-rats are highly inbred, the same does not appear to be true of Damaraland mole-rats (Jarvis et al. 1994). Although the genetic structure of *C. damarensis* colonies has not been quantified, consistent outbreeding is suggested by observations that (1) in the lab individuals refuse to mate consanguineously and (2) in the field colonies frequently disband following the death of a breeding animal. Greater outbreeding (and the associated reduction in intracolonial relatedness) may help to explain why Damaraland mole-rats do not exhibit the same extremes of reproductive and behavioral specialization seen in naked mole-rats.

Comparative studies of subterranean rodents may also help to explain the apparent absence of sociality in some groups, notably pocket gophers (Geomyidae) and blind mole-rats (Spalacidae). Analyses of why sociality does *not* occur in these families represent a logical complement to analyses of the selective factors favoring coloniality and cooperative breeding in other taxa. If the aridity–food distribution hypothesis is generally applicable across taxa, we predict that sociality will not occur in subterranean species in which (1) food is evenly distributed throughout the habitat and (2) costs of burrowing are relatively low and consistent throughout the year. Furthermore, we predict that cooperative group attacks will have little effect on the predators of solitary species.

Clearly, much remains to be learned concerning social evolution in subterranean rodents. Studies of naked mole-rats have generated testable hypotheses regarding interactions between ecology, demography, genetics, and behavior. Given this theoretical framework, comparative studies will greatly enhance our understanding of the factors favoring coloniality and alloparental care. As the nature and extent of sociality among subterranean mammals become clearer, we suspect that many new and exciting insights into social evolution will be discovered in the soil beneath our feet.

10.9 Summary

By any axis used to characterize mammalian sociality, the behavior of naked mole-rats is extreme. Colony sizes are larger, behavioral specializations are greater, and reproductive differences are more pronounced in *H. glaber* than in any other known mammal. Coloniality in this species appears to be an evolutionary response to ecological constraints on dispersal and independent reproduction imposed by (1) patchy, unpredictably distributed locally abundant food resources, (2) high costs of burrow excavation, and (3) unpredictable opportunities for burrow expansion. Within colonies, high coefficients of genetic relatedness and extremely restricted opportunities for direct reproduction suggest that individuals benefit from alloparental care primarily through gains in inclusive fitness.

Among nonbreeding *H. glaber*, behavior varies with size. Whereas smaller individuals are the primary participants in colony maintenance tasks, larger individuals are the primary participants in colony defense activities. Long-term observations of unmanipulated laboratory colonies indicate that individual nonbreeders switch from maintenance to defense activities at the age of 30 to 40 months. However, both behavior and body weight are labile, and this apparent age polyethism is frequently obscured by behavioral and size changes associated with changes in colony composition. Factors influencing individual growth rates include colony size, the availability of reproductive opportunities, and the simultaneous loss of several nonbreeders of the same size class.

In nature, new colonies may form when existing colonies fission, when lone dispersers meet, or both. During 1993, we documented an apparent fission event among members of a captive colony. Fissioning occurred following the death of the colony's breeding female. Colony members segregated into two subgroups, each composed of similarly aged animals. The kin structures of these subgroups suggest that intracolonial conflicts arising from asymmetries in relatedness are important in colony formation, as has been reported for other social mammals.

Comparative analyses suggest that the social systems of cooperatively breeding vertebrates and eusocial invertebrates are fundamentally similar. Although these societies differ quantitatively with respect to the degree of reproductive skew within groups, they are not qualitatively different phenomena. Instead, cooperative breeding and eusociality form a continuum: cooperatively breeding birds and mammals are eusocial, just as eusocial insects are cooperative breeders. Consequently, hypotheses generated to explain the evolution of eusociality must account for alloparental behavior in both invertebrates and vertebrates.

Comparative studies of subterranean rodents provide important opportunities for testing evolutionary hypotheses generated for social African mole-rats. In particular, studies of two South American species, the coruro (*Spalacopus cyanus*) and the colonial tuco-tuco (*Ctenomys sociabilis*), should yield significant new insights into the ecological, demographic, and genetic factors favoring coloniality and alloparental care in subterranean rodents.

Acknowledgments

We thank Nancy Solomon and Jeffrey French for inviting us to participate in the 1992 Animal Behavior Society Symposium at Queens University on *Cooperative Breeding in Mammals*. For assistance with mole-rat care and data collection we thank Jennifer Davis-Walton, Timothy Judd, and John Schieffelin. Earlier versions of this chapter were greatly improved by comments from Stanton Braude, Jennifer Jarvis, Ronald Mumme, Hudson Reeve, and Peter Waser. During the preparation of this manuscript, Eileen Lacey was supported by a Postdoctoral Fellowship funded by the National Science Foundation Research Training Grant awarded to the Animal Behavior Group at the University of California, Davis; Paul Sherman's research was supported by grants from the National Science Foundation and Cornell University.

References

Agren, G., Zhou, Q., & Zhong, W. (1989). Ecology and social behaviour of Mongolian gerbils, *Meriones unguiculatus*, at Xilinhot, inner Mongolia, China. *Anim. Behav.* 37:11–27.

Alexander, R. D. (1991). Some unanswered questions about naked mole-rats. In *The biology of the naked mole-rat*, ed. P. W. Sherman, J. U. M. Jarvis & R. D. Alexander, pp. 446–465. Princeton: Princeton University Press.

Alexander, R. D., Noonan, K. M., & Crespi, B. J. (1991). The evolution of eusociality. In *The biology of the naked mole-rat*, ed. P. W. Sherman, J. U. M. Jarvis, & R. D. Alexander, pp. 3–44. Princeton: Princeton University Press.

Allard, M. W., & Honeycutt, R. L. (1992). Nucleotide sequence variation in the mitochondrial 12s rRNA gene and the phylogeny of African mole-rats (Rodentia: Bathyergidae). *Mol. Biol. Evol.* 9:27–40.

Andersson, M. (1984). The evolution of eusociality. *Ann. Rev. Ecol. Syst.* 15:165–189.

Bartz, S. H. (1979). Evolution of eusociality in termites. *Proc. Natl. Acad. Sci.* 76:5764–5768.

Batra, S. W. T. (1966). Nests and social behavior of halictine bees of India (Hymenoptera: Halictidae). *Indian J. Entomol.* 28:375–393.

Bekoff, M., & Wells, M. C. (1982). Behavioral ecology of coyotes: Social organization, rearing patterns, space use, and resource defense. *Z. Tierpsychol.* 60:281–305.

Bekoff, M., & Wells, M. C. (1986). Social ecology and behavior of coyotes. *Adv. Study Behav.* 16:251–338.

Bennett, N. C. (1989). The social structure and reproductive biology of the common mole-rat, *Cryptomys h. hottentotus* and remarks on the trends in reproduction and sociality in the family Bathyergidae. *J. Zool. Lond.* 219:45–59.

Bennett, N. C. (1990). Behaviour and social organization in a colony of the Damaraland mole-rat *Cryptomys damarensis*. *J. Zool., Lond.* 220:225–248.

Bennett, N. C., & Jarvis, J. U. M. (1988). The social structure and reproductive biology of colonies of the mole-rat, *Cryptomys damarensis* (Rodentia, Bathyergidae). *J. Mamm.,* 69:293–302.

Braude, S. H. (1991). *The behavior and demographics of the naked mole-rat*, Heterocephalus glaber. PhD dissertation, University of Michigan, Ann Arbor.

Brett, R. A. (1986). *The ecology and behaviour of the naked mole-rat* (Heterocephalus glaber Rüppell) (Rodentia: Bathyergidae). PhD dissertation, University of London.

Brett, R. A. (1991a). The population structure of naked mole-rat colonies. In *The biology of the naked mole-rat*, ed. P. W. Sherman, J. U. M. Jarvis, & R. D. Alexander, pp. 97–136. Princeton: Princeton University Press.

Brett, R. A. (1991b). The ecology of naked mole-rat colonies: Burrowing, food, and limiting factors. In *The biology of the naked mole-rat*, ed. P. W. Sherman, J. U. M. Jarvis, & R. D. Alexander, pp. 137–184. Princeton: Princeton University Press.

Brown, J. L. (1975). *The evolution of behavior.* New York: Norton.

Brown, J. L. (1987). *Helping and communal breeding in birds.* Princeton: Princeton University Press.

Buffenstein, R., & Yahav, S. (1991). Is the naked mole-rat *Heterocephalus glaber* an endothermic yet poikilothermic mammal? *J. therm. Biol.* 16:227–232.

Burda, H. (1990). Constraints of pregnancy and evolution of sociality in mole-rats. *Z. zool. Syst. Evolut.-forsch.* 28:26–39.

Chagnon, N. A. (1979). Mate competition, favoring close kin, and village fissioning among the Yanomamö indians. In *Evolutionary biology and human social behavior: An Anthropological perspective,* ed. N. A. Chagnon & W. Irons, pp. 86–132. North Scituate, Mass.: Duxbury Press.

Charnov, E. L. (1978). Evolution of eusocial behavior: Offspring choice or parental parasitism? *J. theor. Biol.* 75:451–465.

Creel, S., & Creel, N. M. (1995) Communal hunting and pack size in African wild dogs, *Lycaon pictus. Anim. Behav.* 50:1325–1339.

Crespi, B. J., & Yanega, D. (1995). The definition of eusociality. *Behav. Ecol.* 6:109–115.

Davis-Walton, J., & Sherman, P. W. (1994). Sleep arrhythmia in the eusocial naked mole-rat. *Naturwissen.* 80:272–275.

Dittus, W. P. J. (1988). Group fission among wild toque macaques as a consequence of female resource competition and environmental stress. *Anim. Behav.* 36:1626–1645.

Downing, H. A., & Jeanne, R. L. (1985). Communication of status in the social wasp *Polistes fuscatus* (Hymenoptera: Vespidae). *Z. Tierpsychol.* 67:78–96.

Emlen, S. T. (1991). Evolution of cooperative breeding in birds and mammals. In *Behavioural ecology: An evolutionary approach,* ed. J. R. Krebs & N. B. Davies, pp. 301–337. Oxford, U.K.: Blackwell.

Emlen, S. T., Reeve, H. K., Sherman, P. W., Wrege, P. H., Ratnieks, F. L. W., & Shellman-Reeve, J. (1991). Adaptive versus nonadaptive explanations of behavior: The case of alloparental helping. *Am. Nat.* 138:259–270.

Emlen, S. T., & Wrege, P. H. (1989). A test of alternative hypotheses for helping behavior in white-fronted bee-eaters of Kenya. *Behav. Ecol. Sociobiol.* 25:303–319.

Evans, H. E. (1977). Extrinsic versus intrinsic factors in the evolution of insect sociality. *BioScience* 27:613–617.

Faulkes, C. G., Abbott, D. H., & Mellor, A. L. (1990). Investigation of genetic diversity in wild colonies of naked mole-rats (*Heterocephalus glaber*) by DNA fingerprinting. *J. Zool., Lond.* 221:87–97.

Faulkes, C. G., Abbott, D. H., Liddell, C. E., George, L. M., & Jarvis, J. U. M. (1991). Hormonal and behavioral aspects of reproductive suppression in female naked mole-rats. In *The biology of the naked mole-rat,* ed. P.W. Sherman, J. U. M. Jarvis, & R. D. Alexander, pp. 426–445. Princeton: Princeton University Press.

FitzGerald, R. W., & Madison, D. M. (1983). Social organization of a free-ranging population of pine voles, *Microtus pinetorum. Behav. Ecol. Sociobiol.* 13:183–187.

Frame, L. H., Malcolm, J. R., Frame, G. W., & van Lawick, H. (1979). Social organization of African wild dogs (*Lycaon pictus*) on the Serengeti Plains, Tanzania, 1967–1978. *Z. Tierpsychol.* 50:225–249.

Gadagkar, R. (1991). *Belonogaster, Mischocyttarus, Parapolybia,* and independent-founding *Ropalidia.* In *The social biology of wasps,* ed. K. G. Ross & R.W. Matthews, pp. 149–190. Ithaca: Cornell University Press.

Gadagkar, R. (1994). Why the definition of eusociality is not helpful to understand its evolution and what should we do about it. *Oikos* 70:485–488.

Getz, W. M., Brückner, D., & Parisian, T. R. (1982). Kin structure and the swarming behavior of the honeybee *Apis mellifera. Behav. Ecol. Sociobiol.* 10:265–270.

Grafen, A. (1991). Modelling in behavioural ecology. In *Behavioural ecology: An evolutionary approch,* ed. J. R. Krebs & N. B. Davies, pp. 5–31. Oxford, U.K.: Blackwell.

Greene, A. (1991). *Dolichovespula* and *Vespula*. In *The social biology of wasps,* ed. K. G. Ross & R. W. Matthews, pp. 263–308. Ithaca: Cornell University Press.

Grzimek, B., ed. (1975). *Grzimek's animal life encyclopedia. Mammals,* I–IV, Vols. 10–13. New York: Van Nostrand.

Hamilton, W. D. (1964). The genetical evolution of social behaviour I, II. *J. theor. Biol.* 7:1–52.

Hamilton, W. D. (1972). Altruism and related phenomena, mainly in social insects. *Ann. Rev. Ecol. Syst.* 3:193–232.

Hamilton, W. D. (1992). Oedipal mating. Reply to Braude and Lacey. *The Sciences* 32:5.

Harrington, F. H., Mech, L. D., & Fritts, S. H. (1983). Pack size and wolf pup survival: Their relationship under varying ecological conditions. *Behav. Ecol. Sociobiol.* 13:19–26.

Heffner, R. S., & Heffner, H. E. (1993). Degenerate hearing and sound localization in naked mole-rats (*Heterocephalus glaber*), with an overview of central auditory structures. *J. Comp. Neurol.* 331:418–433.

Herbers, J. M. (1984). Queen–worker conflict and eusocial evolution in a polygynous ant species. *Evolution* 38:631–643.

Hölldobler, B., & Wilson, E. O. (1990). *The ants.* Cambridge, Mass.: Harvard University Press.

Honeycutt, R. L., Nelson, K., Schlitter, D. A., & Sherman, P. W. (1991). Genetic variation within and among populations of the naked mole-rat: Evidence from nuclear and mitochondrial genomes. In *The biology of the naked mole-rat,* ed. P. W. Sherman, J. U. M. Jarvis, & R. D. Alexander, pp. 195–208, Princeton: Princeton University Press.

Isil, S. (1983). *A study of social behavior in laboratory colonies of the naked mole-rat* (Heterocephalus glaber *Rüppell; Rodentia, Bathyergidae*). MS thesis, University of Michigan, Ann Arbor.

Jarvis, J. U. M. (1978). Energetics of survival in *Heterocephalus glaber* (Rüppell), the naked mole-rat (Rodentia: Bathyergidae). *Bull. Carnegie Mus. Nat. Hist.* 6:81–87.

Jarvis, J. U. M. (1981). Eusociality in a mammal: Cooperative breeding in naked mole-rat colonies. *Science* 212:571–573.

Jarvis, J. U. M. (1985). Ecological studies on *Heterocephalus glaber,* the naked mole-rat, in Kenya. *Natl. Geogr. Soc. Res. Rep.* 20:429–437.

Jarvis, J. U. M. (1991). Reproduction of naked mole-rats. In *The biology of the naked mole-rat,* ed. P. W. Sherman, J. U. M. Jarvis, & R. D. Alexander, pp. 384–425. Princeton: Princeton University Press.

Jarvis, J. U. M., & Bennett, N. C. (1991). Ecology and behavior of the family Bathyergidae. In *The biology of the naked mole-rat,* ed. P. W. Sherman, J. U. M. Jarvis, & R. D. Alexander, pp. 66–96. Princeton: Princeton University Press.

Jarvis, J. U. M., & Bennett, N. C. (1993). Eusociality has evolved independently in two genera of bathyergid mole-rats – but occurs in no other subterranean mammal. *Behav. Ecol. Sociobiol.* 33:253–260.

Jarvis, J. U. M., O'Riain, M. J., & McDaid, E. (1991). Growth and factors affecting body size in naked mole-rats. In *The biology of the naked mole-rat,* ed. P. W. Sherman, J. U. M. Jarvis, & R. D. Alexander, pp. 358–383. Princeton: Princeton University Press.

Jarvis, J. U. M., O'Riain, M. J., Bennett, N. C., & Sherman, P. W. (1994). Mammalian eusociality: A family affair. *Trends Ecol. Evol.* 9:47–51.

Jennions, M. D., & Macdonald, D. W. (1994). Cooperative breeding in mammals. *Trends Ecol. Evol.* 9:89–93.

Keller, L., & Perrin, N. (1995). Quantifying the level of eusociality. *Proc. Roy. Soc. Lond. B.* 260:311–315.

Keller, L., & Reeve, H. K. (1994). Partitioning of reproduction in animal societies. *Trends Ecol. Evol.* 9:98–102.

Koenig, W. D., & Pitelka, F. A. (1981). Ecological factors and kin selection in the evolution of cooperative breeding in birds. In *Natural selection and social behavior,* ed. R. D. Alexander & D. W. Tinkle, pp. 261–280. New York: Chiron Press.

Koenig, W. D., Pitelka, F. A., Carmen, W. J., Mumme, R. L., & Stanback, M. T. (1992). The evolution of delayed dispersal in cooperative breeders. *Q. Rev. Biol.* 67:111–150.

Krebs, J. R., & Davies, N. B. (1993). *An introduction to behavioural ecology,* 3rd ed. Oxford, U.K.: Blackwell.

Kukuk, P. F. (1994). Replacing the terms "primitive" and "advanced": New modifiers for the term "eusocial." *Anim. Behav.* 47:1475–1478.

Lacey, E. A., & Braude, S. H. (1993). Social behavior, biogeography, and ecology of the colonial tuco-tuco (*Ctenomys sociabilis*). Unpubl. res. rep. to Administración de Parques Nacionales, Delegación Técnica Regional Patagonica, Argentina.

Lacey, E. A., & Sherman, P. W. (1991). Social organization of naked mole-rat colonies: Evidence for divisions of labor. In *The biology of the naked mole-rat,* ed. P. W. Sherman, J. U. M. Jarvis, & R. D. Alexander, pp. 275–336. Princeton: Princeton University Press.

Lacey, E. A., Alexander, R. D., Braude, S. H., Sherman, P. W., & Jarvis, J. U. M. (1991). An ethogram for the naked mole-rat: Non-vocal behaviors. In *The biology of the naked mole-rat,* ed. P. W. Sherman, J. U. M. Jarvis, & R. D. Alexander, pp. 209–242. Princeton: Princeton University Press.

Lacy, R. C. (1980). The evolution of eusociality in termites: A haplodiploid analogy? *Am. Nat.* 116:449–451.

Lacy, R. C. (1984). The evolution of termite eusociality: Reply to Leinaas. *Am. Nat.* 123:876–878.

Lovegrove, B. G. (1991). The evolution of eusociality in molerats (Bathyergidae): A question of risks, numbers, and costs. *Behav. Ecol. Sociobiol.* 28:37–45.

Lovegrove, B. G., & Wissel, C. (1988). Sociality in molerats: Metabolic scaling and the role of risk sensitivity. *Oecologia* 74:600–606.

McNab, B. K. (1966). The metabolism of fossorial rodents: A study of convergence. *Ecology* 47:712–733.

Macdonald, D. W. (1980). Social factors affecting reproduction among red foxes. In *Biogeographica,* Vol. 18: *The red fox,* ed. E. Zimen, pp. 123–175. The Hague: Junk.

Malcolm, J. R., & Marten, K. (1982). Natural selection and the communal rearing of pups in African wild dogs (*Lycaon pictus*). *Behav. Ecol. Sociobiol.* 10:1–13.

Mech, L. D. (1983). Age, season, distance, direction, and social aspects of wolf dispersal from a Minnesota pack. In *Mammalian dispersal patterns,* ed. B. D. Chepko-Sade & Z. T. Halpin, pp. 55–74. Chicago: University of Chicago Press.

Michener, C. D. (1974). *The social behavior of bees.* Cambridge, Mass.: Harvard University Press.

Michener, C. D., & Brothers, D. J. (1974). Were workers of eusocial Hymenoptera initially altruistic or oppressed? *Proc. Natl. Acad. Sci.* 71:671–674.

Moehlman, P. D. (1979). Jackal helpers and pup survival. *Nature, Lond.*
277:382–383.

Moehlman, P. D. (1983). Socioecology of silverbacked and golden jackals (*Canis
mesomelas* and *C. aureus*). In *Advances in the study of mammalian behavior*, ed.
J. F. Eisenberg & D. G. Kleiman, pp. 423–452. Special Publication 7. Lawrence,
Kans.: American Society of Mammalogists.

Nevo, E. (1979). Adaptive convergence and divergence of subterranean mammals.
Ann. Rev. Ecol. Syst. 10:269–308.

O'Riain, M. J., Jarvis, J. U. M., & Faulkes, C. G. (1996). A dispersive morph in the
naked mole-rat. *Nature, Lond.* 380:619–621.

Payne, S. F. (1982). *Social organization of the naked mole-rat* (Heterocephalus
glaber): *Cooperation in colony labor and reproduction.* MS thesis, University of
California, Santa Cruz.

Pearson, O. P., & Christie, M. I. (1985). Los tuco-tucos (genero *Ctenomys*) de los
Parques Nacionales Lanin y Nahuel Huapi, Argentina. *Hist. Nat.* 5:337–343.

Powell, R. A., & Fried, J. J. (1992). Helping by juvenile pine voles (*Microtus pinetorum*), growth and survival of younger siblings, and the evolution of pine vole
sociality. *Behav. Ecol.* 3:325–333.

Redford, K. H., & Eisenberg, J. F. (1992). *Mammals of the Neotropics: The southern
cone.* Chicago: University of Chicago Press.

Reeve, H. K. (1991). *Polistes.* In *The social biology of wasps*, ed. K. G. Ross & R. W.
Matthews, pp. 99–148. Ithaca: Cornell University Press.

Reeve, H. K. (1992). Queen activation of lazy workers in colonies of the eusocial
naked mole-rat. *Nature, Lond.* 358:147–149.

Reeve, H. K., & Keller, L. (1995). Partitioning of reproduction in mother–daughter
versus sibling associations: A test of optimal skew theory. *Am. Nat.*
145:119–132.

Reeve, H. K., & Sherman, P. W. (1991). Intracolonial aggression and nepotism by the
breeding female naked mole-rat. In *The biology of the naked mole-rat*, ed. P. W.
Sherman, J. U. M. Jarvis, & R. D. Alexander pp. 337–357. Princeton: Princeton
University Press.

Reeve, H. K., Westneat, D. F., Noon, W. A., Sherman, P. W., & Aquadro, C. F.
(1990). DNA "fingerprinting" reveals high levels of inbreeding in colonies of the
eusocial naked mole-rat. *Proc. Nat. Acad. Sci.* 87:2496–2500.

Reig, O. A. (1970). Ecological notes on the fossorial octodont rodent *Spalacopus
cyanus* (Molina). *J. Mamm.* 51:592–601.

Reig, O. A., Busch, C., Ortells, M. O., & Contreras, J. R. (1990). An overview of the
evolution, systematics, population biology, cytogenetics, molecular biology, and
speciation in *Ctenomys.* In *Evolution of subterranean mammals at the organis-
mal and molecular levels*, ed. E. Nevo & O. A. Reig, pp. 71–96. New York: Liss.

Rood, J. P. (1978). Dwarf mongoose helpers at the den. *Z. Tierpsychol.* 48:277–287.

Rood, J. P. (1983). The social system of the dwarf mongoose. In *Advances in the
study of mammalian behavior*, ed. J. F. Eisenberg & D. G. Kleiman, pp.
454–488. Special Publication 7. Lawrence, Kans.: American Society of
Mammalogists.

Rood, J. P. (1990). Group size, survival, reproduction, and routes to breeding in dwarf
mongooses. *Anim. Behav.* 39:566–572.

Rymond, M. A. (1991). *Aggression and dominance in the naked mole-rat*
(Heterocephalus glaber). MS thesis, University of Michigan, Ann Arbor.

Schieffelin, J. S., & Sherman, P. W. (1995). Tugging contests reveal feeding hierarchies in naked mole-rat colonies. *Anim. Behav.* 49:537–541.

Seeley, T. D. (1985). *Honeybee ecology.* Princeton: Princeton University Press.
Seger, J. (1991). Cooperation and conflict in social insects. In *Behavioural ecology: An evolutionary approach,* ed. J. R. Krebs & N. B. Davies, pp. 338–373. Oxford, U.K.: Blackwell.
Sherman, P. W., Jarvis, J. U. M., & Alexander, R. D., eds. (1991). *The biology of the naked mole-rat.* Princeton: Princeton University Press.
Sherman, P. W., Jarvis, J. U. M., & Braude, S. H. (1992). Naked mole rats. *Sci. Amer.* 267:72–78.
Sherman, P. W., Lacey, E. A., Reeve, H. K., & Keller, L. (1995). The eusociality continuum. *Behav. Ecol.* 6:102–108.
Solomon, N. G. (1991). Current indirect fitness benefits associated with philopatry in juvenile prairie voles. *Behav. Ecol. Sociobiol.* 29:277–282.
Solomon, N. G. (1994). Eusociality in a microtine rodent. *Trends Ecol. Evol.* 9:264.
Trivers, R. (1985). *Social evolution.* Menlo Park, Calif.: Benjamin–Cummings.
Vehrencamp, S. L. (1979). The roles of individual, kin, and group selection in the evolution of sociality. In *Handbook of behavioral neurobiology,* Vol. 3, *Social behavior and communication,* ed. P. R. Marler & J. G. Vandenbergh, pp. 351–394. New York: Plenum Press.
Vehrencamp, S. L. (1983a). Optimal degree of skew in cooperative societies. *Am. Zool.* 23:327–335.
Vehrencamp, S. L. (1983b). A model for the evolution of despotic versus egalitarian societies. *Anim. Behav.* 31:667–682.
Wilson, D. E., & Reeder, D. M., eds. (1993). *Mammal species of the world: A taxonomic and geographic reference,* 2nd ed. Washington, D.C.: Smithsonian Institution Press.
Wilson, E. O. (1971). *The insect societies.* Cambridge, Mass.: Harvard University Press.
Woolfenden, G. E., & Fitzpatrick, J. W. (1990). Florida scrub jays: A synopsis after 18 years of study. In *Cooperative breeding in birds: Long-term studies of ecology and behavior,* ed. P. B. Stacey & W. D. Koenig, pp. 241–266. Cambridge: Cambridge University Press.

11

The Physiology of a Reproductive Dictatorship: Regulation of Male and Female Reproduction by a Single Breeding Female in Colonies of Naked Mole-Rats

CHRISTOPHER G. FAULKES and DAVID H. ABBOTT

11.1 Introduction

The eusocial naked mole-rat, *Heterocephalus glaber,* provides the most extreme mammalian example of social regulation of reproduction. One large, dominant female (the "queen": Jarvis 1981), controls the reproduction of all males and females in a colony that may contain hundreds of individuals. Even though these small (20–80 g), highly specialized fossorial rodents live up to 18 years in captivity, in the wild most stand no chance of reproducing in their lifetime, because there is little opportunity to attain queen or consort status or to disperse from their natal colony (Brett, 1991a; Sherman, Jarvis, & Braude, 1992; Jarvis et al., 1994). As Sherman and colleagues discuss in this volume, the naked mole-rat is the mammalian equivalent of a eusocial insect such as the termites or wasps.

The naked mole-rat exhibits the classical characteristics of eusociality (Michener 1969; Wilson 1975). Colonies contain overlapping generations, and there is a clear division of reproductive labor in which only a few individuals produce offspring, while infertile group members cooperate in rearing offspring and protecting and servicing the colony (Jarvis 1981, 1991; Faulkes et al. 1991; Lacey & Sherman 1991). Naked mole-rats live in a closed society, where their subterranean habitat, eusociality, and extreme specializations to a fossorial life style (e.g., degenerate vision and poikilothermy), and xenophobic aggression to conspecifics from surrounding colonies lead to an apparent lack of dispersal and consequently to high levels of inbreeding (Buffenstein & Yahav 1991; Jarvis & Bennett 1991; Sherman et al. 1992; Jarvis et al. 1994). Because of the extremely high genetic relatedness within a colony (Faulkes, Abbott, & Mellor 1990; Reeve et al. 1990; Faulkes et al. in prep) nonbreeding naked mole-rats

302

contribute to their own fitness by helping their parents or siblings rear offspring (inclusive fitness: Hamilton 1964).

Naked mole-rats have evolved highly specialized behavioral and physiological responses to their social surroundings, and although specialized reproductive responses to environmental cues are not uncommon in mammals, it is the extreme and lifelong social restriction of breeding in naked mole-rats that is unusual. Such social determination of reproduction is typical for singular, cooperatively breeding mammals, like the naked mole-rat, in which only one female usually breeds despite the presence of multiple adult females in the group (Emlen 1991; Jennison & MacDonald 1994).

11.2 Environmental Regulation of Mammalian Reproduction

Many mammals live in changeable environments where conditions are not always optimal or even adequate for successful reproduction. Food can become scarce, ambient temperature can fall to critical levels, populations can become overcrowded, and in socially living species, too few helpers may be available to aid in rearing offspring. The effects of such environmental factors on the timing and modulation of reproductive processes to maximize offspring survival are well documented (Bronson & Heideman 1994).

Annual changes in daylength or in the availability of food are striking examples of environmental cues that occur on a regular basis and provide reliable predictors for the timing of successful reproduction. The proximate cueing of these ultimate factors is provided, respectively, by changes in daylength resulting in visually mediated changes in the circadian pattern of melatonin secretion from the pineal gland (Lincoln 1981; Karsch et al. 1984) or by the initial ingestion of increased quantities of food or of particular secondary plant compounds in newly emergent vegetation (Bronson 1989). These factors then trigger changes in reproductive physiology, behavior, and metabolism, resulting in the appropriate seasonal timing of courtship, mating, parturition, and weaning of offspring.

Cues from an animal's social environment may also act independently or in conjunction with cues from the physical environment. For instance, estrus in sheep at the beginning of the breeding season can be advanced and synchronized by olfactory, visual, and behavioral cues from rams (Signoret & Lindsay 1982; Cohen-Tannoudji, Locatelli, & Signoret 1986), and the same effect is induced in red deer hinds, *Cervus elaphus,* by auditory cues from stags (McComb 1987). Increased release of luteinizing hormone (LH) from the anterior pituitary gland (presumably stimulated by the release of gonadotropin-releasing hormone [GnRH] from the neuroendocrine hypothalamus) can occur in ewes within 10 to 20 minutes of exposure to such male cues and results in ovarian stimulation and ovulation (Martin, Oldham, & Lindsay 1980). In males, similar enhancement of reproduction can be elicited by social cues, as illustrated by the effect of female presence on the acceleration of reproductive maturation in male deermice, *Peromyscus maniculatus* (Bronson 1985). Social cues can also regulate reproductive success in quite a different way by suppressing reproduction.

11.2.1 Social Status and Reproductive Success

Inhibition of reproduction in animals that would otherwise be fertile (being of sufficient chronological age and body weight for maturity and being at an appropriate season or time of year for breeding) is frequently found to be the product of low social rank (Bronson 1989). Competition within the social environment is therefore a strong determinant of reproductive success, independently of the action of social cues affecting appropriate timing of reproduction.

Socially mediated influences may intervene at any one of a number of stages in the reproductive process in either sex (Wasser & Barash 1983; Abbott, Barrett, & Faulkes 1990). Low-ranking females may exhibit the following characteristics:

1. Delayed puberty or menarche (e.g., captive saddleback tamarin, *Saguinus fuscicollis:* Epple & Katz 1980; French, this volume).
2. Infrequent or inhibited ovulation (e.g., wild and captive naked mole-rats: Faulkes, Abbott, & Jarvis 1990);
3. Inhibited pituitary LH and (probably) altered hypothalamic GnRH release (e.g., captive naked mole-rats: Faulkes, Abbott, Jarvis, & Sherriff 1990).
4. Implantation block (e.g., captive white-footed mouse, *Peromycus leucopus:* Haigh, Cushing, & Bronson 1988).
5. Delayed first birth (e.g., wild savannah baboons, *Papio cynocephalus:* Altmann, Hausfater, & Altmann 1988).
6. Increased miscarriage, premature delivery, and prolonged interbirth interval (e.g., wild savannah baboons: Wasser & Starling 1988);
7. Impaired lactation (e.g., wild red deer: Clutton-Brock, Guinness, & Albon 1982; Clutton-Brock, Albon, & Guinness 1986), sometimes in combination with postpartum harassment of both mother and infant (e.g., wild savannah baboons: Rhine, Wasser, & Norton 1988).

In males, low rank can similarly result in the following specific reproductive failures:

1. Delayed puberty or maturation of secondary sexual characteristics (e.g., semi–free-ranging mandrills, *Mandrillus sphinx:* Wickings & Dixson 1992).

2. Impaired spermatogenesis (e.g., captive naked mole-rats: Faulkes et al. 1994 and this chapter).
3. Reduced circulating testosterone concentrations (e.g., captive lesser mouse lemur, *Microcebus murinus:* Perret 1992).
4. Impaired testicular steroidogenic (and possibly paracrine) function (e.g., captive common marmoset, *Callithrix jacchus:* Abbott 1993; Sheffield, Abbott, & O'Shaungnessy 1989);
5. Impaired pituitary LH and (probably) hypothalamic GnRH release (e.g., captive naked mole-rat: Faulkes, Abbott, & Jarvis 1991 and this chapter).
6. Impaired ejaculatory ability (e.g., captive talapoin monkey, *Miopithecus talapoin:* Keverne et al. 1984).

Such diverse manifestation of socially induced reproductive impairments in both sexes probably reflects the different degrees of reproductive inhibition found among species and the particular adaptation each has made during the evolution of social regulatory mechanisms. All the impairments have been found to be reversible, and such changes occur rapidly with alterations in the social environment (see Abbott & George [1991], which illustrates the manipulation of ovulatory function by altering social rank). The precise cues responsible for triggering reproductive inhibition also vary. They can range from predominantly olfactory cues in subordinate male lesser mouse lemurs (Perret, 1992) to harassment in low-ranking female savannah baboons (Rhine et al. 1988) and can involve a combination of olfactory, visual, and behavioral cues and actions, as in subordinate female common marmosets (Barrett, Abbott, & George 1993).

11.3 Cooperatively Breeding Species

Cooperative breeding has evolved independently many times among mammalian and nonmammalian species (Emlen 1991; Jennison & MacDonald 1994). It occurs in many forms and patterns of social organization and includes social groups in which several males and females regularly breed (plural breeders) (e.g., multiple collections of monogamous pairs or polygynandrous mating systems in birds: Emlen 1991), and groups in which only one dominant female usually breeds (singular breeders) (e.g., marmoset and tamarin, naked mole-rat, dwarf mongoose, silver-backed jackal, pine vole, and prairie vole: chapters in this volume). The latter form of cooperative breeding is more common among both birds and mammals (Creel & Waser 1991).

The characteristics that most clearly distinguish singular cooperative breeders from competitive breeders (species in which breeding females raise their infants unaided, e.g., red deer and savannah baboons) are (1) the common exclusion of all but one dominant female and one to two dominant males from successfully producing or raising offspring, and (2) the prolonged retention of mature but nonbreeding offspring in natal groups (Wasser & Barash 1983; MacDonald & Carr 1989; Emlen 1991; Abbott, Barrett, & George 1993). In singular cooperative breeders, the failure of subordinate animals to raise offspring appears to occur because of competition for resources that limit their breeding opportunities (resource dispersal hypothesis: Emlen 1991; MacDonald & Carr 1989; reproductive suppression model: Wasser & Barash 1983) and because their retention within a group increases their chances of survival and those of the breeding female's offspring (philopatry hypothesis: Waser, Creel, & Lucas 1994; Sussman & Garber 1987; reproductive suppression model: Wasser & Barash 1983). The evolution of singular cooperative breeding systems in which social rank is an extreme determinant of reproductive success results in the elevation of the social environment to providing the predominant proximate cues for *timing* reproductive effort. This is exemplified by the rapid reproductive and sexual activation in subordinates following the death or absence of the same-sexed dominant (e.g., naked mole-rats: Faulkes, Abbott, & Jarvis 1990; Margulis, Saltzman, & Abbott 1995). Such precise timing of reproductive effort would enable subordinates to engage expeditiously in the intense, intrasexual competition for vacant dominant positions within groups. Cooperative breeders have specialized in the evolution of social contraception (Abbott 1984) to time attempts at reproduction and minimize investment in unsuccessful reproductive attempts in a fashion analogous to that of competitive seasonal breeders. Both have evolved precise physiological responses to reliable proximate cues.

11.4 African Mole-Rats and the Occurrence of Cooperative Breeding and Reproductive Suppression

The bathyergid rodents are unique among mammals in the array of social strategies adopted by the member species, ranging from solitary to eusocial (Table 11.1). The spectrum of sociality displayed by the Bathyergidae reaches an extreme in the eusocial naked mole-rat. The Damaraland mole-rat is now also known to be eusocial, although the maximum group sizes attained are less than in the naked mole-rat (Jarvis & Bennett 1993). In the social species listed in Table 11.1, group sizes are smaller and are more transient than in the eusocial species. Phylogenetic studies based on variation in the nucleic acid sequence in the mitochondrial 12S rRNA gene have revealed the naked mole-rat to be the most basal representative of the family, diverging early in evolu-

Table 11.1. *Group sizes and degree of sociality for the family*
Bathyergidae

Species	Group size/ sociality	Reference
Hellophobius argenteocinereus (silvery mole-rat)	solitary	Jarvis & Sale 1971
Georychus capensis (dune mole-rat)	solitary	Du Toit et al. 1985
Bathyergus suillus (Cape dune mole-rat)	solitary	Jarvis & Sale 1971; Davies & Jarvis 1986
Bathyergus janetta (Namaqua dune mole-rat)	solitary	Jarvis & Bennett 1991
Cryptomys bocagei	solitary?	Jarvis, Bennett, Aguillar, & Faulkes, unpubl. data
Cryptomys mechowi (plant mole-rat)	solitary?	Jarvis, Bennett, Aguillar, & Faulkes, unpubl. data
Cryptomys ochraceocinereus	unknown	–
Cryptomys foxi	unknown	–
Cryptomys zechi	unknown	–
C. h. hottentotus (common mole-rat)	2–14/social	Davies & Jarvis 1986; Bennett 1988
C. h. natalensis	2–3/social	Hickman 1979
C. h. amatus	6–10/social	Jarvis, Bennett, Aguillar, & Faulkes, unpubl. data
C. darlingi	5–9/social	Bennett et al. 1994
C. damarensis (Damaraland mole-rat)	12–41/eusocial	Jarvis & Bennett 1993
Heterocephalus glaber (naked mole-rat)	25–300/eusocial	Jarvis 1985; Brett 1991a

tionary history when the bathyergids split off from their hystricomorph ancestors (Allard & Honeycutt 1992). According to Allard and Honeycutt's phylogeny, the Damaraland mole-rat has a more recent evolutionary history,
strongly suggesting that eusociality evolved independently in the naked mole-
rat and the Damaraland mole-rat, presumably in response to similar environmental pressures.

Because of the range of sociality they display and their comparatively close
phylogenetic relationships, the Bathyergidae represent a unique mammalian
family within which to investigate the relationships between ultimate and
proximate environmental factors regulating reproductive success and the evolution of social behavior and reproductive suppression. Comparison of social
group size and environment has led to the hypothesis that rainfall pattern and
subsequent aridity of the habitat is correlated with the degree of sociality

(Jarvis & Bennett 1991, 1993; Jarvis et al. 1994). As aridity increases, the energetic cost of burrowing becomes greater. Also, the distribution of roots and tubers, the staple diet of African mole-rats, becomes more dispersed and patchy, increasing the risk of unsuccessful foraging for solitary animals or small groups (Lovegrove & Wissel 1988; Brett 1991b; Jarvis & Bennett 1991). These constraints would bestow a selective advantage on groups of animals cooperatively foraging, leading ultimately to the evolution of behavioral and reproductive division of labor within colonies. Thus, solitary species are found only in mesic zones, social species are found in both mesic and arid zones, and eusocial species are restricted to the most harsh and arid habitats. In the latter, although rainfall is unpredictable, food in the form of underground tubers is continuously present, and both the Damaraland and the naked mole-rat breed continuously throughout the year. Most of the bathyergids inhabiting more mesic areas, where climatic changes are more predictable, appear to be seasonal breeders (Jarvis & Bennett 1991, 1993).

11.5 Naked Mole-Rats

11.5.1 Sexual and Agonistic Interactions in Captive Colonies

In colonies of naked mole-rats, sexual interactions and reproduction are restricted entirely to the queen and one, two, or sometimes three breeding males (Jarvis 1981, 1991; Lacey & Sherman 1991). A strong behavioral bond is apparent between the queen and the breeding male(s), as is illustrated by the large amounts of time they spend in close bodily contact within the nest chamber and the high frequencies of mutual anogenital nuzzling and sniffing, activities not observed among nonbreeders (Jarvis 1991). The breeding animals are usually the largest (e.g., body weights, mean ± SEM: queens (nonpregnant: $n = 7$), 56.6 ± 2.6 g; breeding males ($n = 13$) 40.4 ± 2.1 g; nonbreeding females ($n = 39$) 32.6 ± 1.1 g; nonbreeding males ($n = 41$) 31.2 ± 0.9 g) and are also the most dominant members of the colony. Among nonbreeders, dominance status is more difficult to ascertain, because of the absence or low frequency of agonistic interactions such as shoving, biting, tugging, and incisor fencing (as defined by Lacey et al. 1991). However, "passing in tunnels" is a commonly observed naked mole-rat behavior. It occurs when two animals meet face to face in a tunnel, engage in mutual sniffing of the facial area, and then one animal (higher ranking) passes over the top of the other. Quantification of this over/under behavior has shown it to be remarkably consistent between animals, and calculations of dominance rank based on this behavior produced a linear hierarchy according to

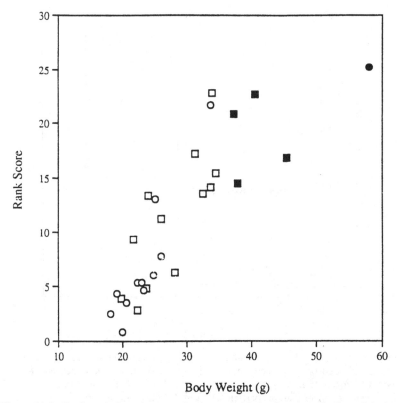

Body Weight (g)

Figure 11.1. Rank score based on over/under tunnel behavior versus body weight in a colony of 28 naked mole-rats, • = queen; o = nonbreeding females; ■ = breeding males; □ = nonbreeding males. $R = 0.91$, $P < 0.001$ (Spearman's rank correlation coefficient). From Faulkes, Clarke, Murray, and Waters in prep.

Landau's index (Chase 1974). Furthermore, rank score based on over/under interactions correlated significantly with both body weight (Figure 11.1) and urinary testosterone in nonbreeding males and females including the queen (Faulkes, Clarke, Murray, & Waters in prep.).

Attainment of queen status following the death or experimental removal of the previous breeder is often, but not always, achieved by fighting among female contenders who become reproductively active. There is good evidence that queen succession depends on dominance status within the colony (Margulis, Saltzman, & Abbott 1995; Faulkes et al. in prep.). An example of this is shown in Figure 11.2, where, prior to queen removal, the individual succeeding the queen was the next highest-ranking female in the dominance hierarchy and also had the highest urinary testosterone concentrations

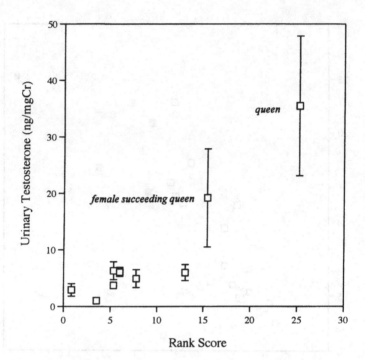

Figure 11.2. Mean ± SEM concentrations of urinary testosterone versus dominance rank score based on over/under tunnel-passing behavior, for the queen and 8 non-breeding females from a captive colony of 28 naked mole-rats. $R = 0.80$, $P < 0.01$ (Spearman's rank correlation coefficient). From Faulkes, Murray, and Waters in prep.

among nonbreeding females. During the periods of social unrest when queen succession occurs, deaths often result, not only among the competing females but also among higher ranking males, including breeders (Jarvis 1991; Lacey & Sherman 1991; Faulkes unpublished data). Because removal of a queen produces elevated urinary testosterone concentrations in all males in a naked mole-rat colony (see Figure 11.9), the increased testosterone may induce male aggression and fighting, resulting in the male deaths observed. Social order is generally restored in the colony when a new female has taken over as queen; agonistic interactions among colony members then become relatively rare and subtle events. For example, in a colony of 15 individuals, the frequency of agonistic interactions (shoving, biting, and incisor fencing) per 10-minute behavioral observation was 0.28 ± 0.07 when the colony was stable with an established queen (1,620 minutes of observation over 56 days) and 0.55 ± 0.09 when the colony was unstable during a period of 72 days after the established queen was removed (1,760 minutes of observation). During this time the breeding male and another high-ranking male were killed following fighting with the new queen (Faulkes et al., in prep.).

Perhaps the most obvious sign of aggression is that of shoving, a behavior initiated by the queen or occasionally by breeding males and directed significantly more often at larger individuals of both sexes (Reeve & Sherman 1991; Faulkes et al. in prep.). In colonies where the queen has been removed, shoving behavior develops and begins to be expressed in the succeeding queen around the time that she becomes reproductively active, illustrating the relationship between reproductive status and the expression of this behavior (Margulis et al. 1995; Faulkes et al. in prep.). To date, little is known about the ontogeny of dominance behavior and the attainment of breeder status in colonies of naked mole-rats.

The existence of clear dominance hierarchies within colonies of naked mole-rats is intriguing, given their lack of genotypic variability (Faulkes, Abbott, & Mellor 1990; Reeve et al. 1990). The major influence in determining dominance status and ultimately which animals will become breeders may involve subtle developmental effects. For example, the intrauterine position (IUP) of the fetus (i.e., how an individual male or female is positioned in utero with respect to the sex of adjacent fetuses) is known to affect various behavioral and morphological characteristics in the Mongolian gerbil, *Meriones unguiculatus,* and the house mouse, *Mus musculus.* In the former, IUP affects reproductive success, scent-marking behavior, and adult plasma testosterone concentrations in males (Clarke, Tucker, & Galef 1992), whereas in the latter, IUP affects timing of puberty, sensitivity to and production of pheromones (Vom Saal 1989), and aggression in females (Gandelman, vom Saal, & Reinisch 1977). In mice, a female fetus positioned between two males in utero receives greater androgen exposure than do females positioned in utero next to one male or next to only females (vom Saal & Bronson 1978). Such females exhibit a degree of genital and behavioral virilization, including increased aggression (Gandelman et al. 1977). It would be interesting to ascertain whether similar intrauterine effects operate in naked mole-rats (in which single litters can number up to 27 offspring; Jarvis 1991) and whether female naked mole-rat aggressive behavior is linked to prenatal androgen exposure.

On attainment of the dominant position, breeding female naked mole-rats not only become heavier than nonbreeding females and express different behavioral and reproductive characteristics, but they undergo specific alterations to their physique. Jarvis, O'Riain, and McDaid (1991) have shown that the elongated shape of a queen is due to 40 percent growth in the length of her vertebrae. They speculate that such specific phenotypic change may be adaptive in coping with large litter sizes and in accommodating the passage of the pregnant female through the narrow tunnel system. Whether such specific bone growth is dependent on elevated circulating ovarian hormones found in breeding females, but not in nonbreeding females, or is dependent on growth hormone or insulin-mediated events, remains to be determined.

11.5.2 Physiology of Reproductive Suppression in Nonbreeding Males and Females

Naked mole-rats are also unusual among mammals (and, perhaps, singular cooperative breeders) in that a clear physiological block to reproduction is apparent in nonbreeders of both sexes. Although suppression of reproductive physiology has been reported in nonbreeding female marmosets and other callitrichid primates (Abbott & Hearn 1978; French this volume), dwarf mongooses (Creel et al. 1992; Creel & Waser this volume), and Damaraland mole-rats (Bennett et al. 1993), in most species reproductive suppression among males appears to be predominantly behavioral (Keverne et al. 1984; Sapolsky 1993). Endocrine deficiencies are not usually observed among nonbreeding males (but see Abbott 1993 and Perret 1992 on subordinate male common marmosets and lesser mouse lemurs, respectively), and suppression is thought to be due to exclusion from mating as a result of interactions with more dominant individuals—for example, dwarf mongooses (Creel et al. 1992; this volume) and Damaraland mole-rats (Bennett et al. 1993).

In nonbreeding naked mole-rats of both sexes, reproductive suppression is apparently mediated by changes in the secretion of hypothalamic GnRH, resulting in inadequate hormonal stimulation of the gonads by pituitary gonadotropins and ultimately in a state of infertility. When compared with breeders, both male and female nonbreeding naked mole-rats have reduced concentrations of plasma LH, and their pituitaries are less responsive to exogenous GnRH, suggesting a lack of priming of the pituitary gland as a consequence of altered secretion of endogenous GnRH. The lack of sensitivity to exogenous GnRH in both nonbreeding males and females is reversible after four repeated hourly doses, as shown in Figure 11.3 for females (Faulkes, Abbott, Jarvis & Sherriff 1990; Faulkes & Abbott 1991; Faulkes, Abbott, & Jarvis, 1991). In females, this disruption of GnRH secretion ultimately results in anovulation (Faulkes, Abbott, Jarvis, & Sherriff 1990), and the reproductive tract and ovaries of suppressed nonbreeders remain in a pre-

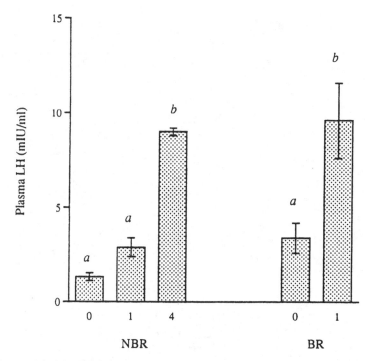

Figure 11.3. Mean ± SEM plasma LH concentrations in nonbreeding (NBR) and breeding (BR) naked mole-rats ($n = 4 - 6$ in each column) before (0) and 20 minutes after an subcutaneous injection of 0.1 mg GnRH (1), and 20 minutes after the last of four 0.1 mg injections given at hourly intervals (4). Significant differences between columns are indicated by different letters ($P < 0.01$; Duncan's multiple range test following ANOVA for repeated measures). Adapted from Faulkes, Abbott, Jarvis, Sheriff et al. 1990.

pubescent state (Figure 11.4). Histological examination of ovaries removed from nonbreeding females has shown an absence of corpora lutea or corpora albicantia and only the very occasional presence of a preovulatory follicle, in contrast to breeding females (Faulkes 1990). As Figure 11.4 suggests, the mass of the ovary and the whole reproductive tract was vastly greater in breeders than in nonbreeders (ovary: breeder ($n = 8$) 0.79 ± 0.24 mg/g body weight [BW] vs. nonbreeder ($n = 14$) 0.25 ± 0.03 mg/g BW, p < 0.05; reproductive tract: breeder ($n = 7$) 27.1 ± 8.9 mg/g BW vs. nonbreeder ($n = 59$) 1.5 ± 0.1 mg/g BW, p < 0.01) (Faulkes 1990).

Among male naked mole-rats as well, altered secretion of hypothalamic GnRH and pituitary LH gives rise to clear physiological differences between breeders and nonbreeders, with the latter having lower concentrations of uri-

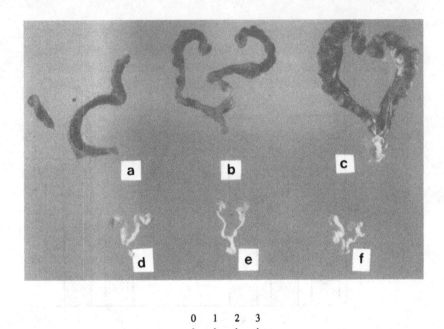

0   1   2   3
└──┴──┴──┘
cm

Figure 11.4. Reproductive tracts and ovaries from female naked mole-rats. a, b, and c are from breeding females having nonpregnant body weights of 52 g, 56 g, and 45 g, respectively. d, e, and f are from nonbreeding females having body weights of 76 g, 33 g, and 46 g, respectively.

nary testosterone compared with the former (mean ± SEM: 23.8 ± 2.3 vs. 5.2 ± 1.4 ng/mg Cr, respectively; P < 0.001; Faulkes & Abbott 1991). Histological examination of naked mole-rat testes showed that nonbreeding males (captive and wild-caught) had fewer Leydig cells (testosterone secreting) than did breeding males (Faulkes 1990). Probable nonbreeding males captured from colonies in the wild also showed a poor in vitro Leydig cell response to an ovine LH stimulation teste (Onyango et al. 1991). Although the reproductive and dominance status of wild-caught male naked mole-rats was unknown, they were all assumed to be nonbreeders because the mass of their testes and reproductive tracts resembled those of captive nonbreeders (Faulkes 1990). In captive colonies, breeding and nonbreeding males were identified, respectively, on the basis of which males did or did not consort with the queen (Faulkes & Abbott 1991). While these observations suggest that suppression of reproductive physiology occurs in non-breeding male

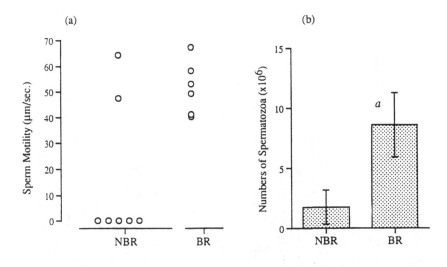

Figure 11.5. Sperm motility (a) and numbers of spermatozoa in one half of the reproductive tract (b) in nonbreeding (NBR) and reproductively active (BR) male naked mole-rats. (a) individual data points showing NBR, $n = 7$; BR, $n = 6$. (b) Mean I, NBR, $n = 7$; BR, $n = 7$. SEM values shown; $a = P < 0.05$ vs. nonbreeders (students' t test). Adapted from Faulkes et al. 1994.

naked mole-rats, levels of reproductive hormones are sufficient to support some spermatogenesis, and these males apparently produce mature spermatozoa (Faulkes & Abbott 1991; Jarvis 1991). However, closer investigation of spermatozoa in nonbreeders has shown that these males produce significantly lower numbers of sperm compared with breeders and that in most nonbreeding males these sperm are not motile (Figure 11.5; Faulkes et al. 1994).

The neuronal mechanisms that bring about socially induced suppression of GnRH secretion in nonbreeding naked mole-rats remain unknown. In socially suppressed female common marmoset monkeys, GnRH attenuation is mediated by a mechanism involving both opioid peptides and an increased sensitivity to the negative feedback effects of estradiol (Abbott, George, Barrett, et al. 1990). A similar system is thought to operate in the seasonally anoestrus ewe (Robinson, Radford, & Karsch 1985; Yang et al. 1988), suggesting that selection during the course of evolution may have favored a common neuroendocrine mechanism in the modulation of reproduction by environmental cues, at least in spontaneous ovulators (Abbott 1988). Our studies on the naked mole-rat also suggest a specific neuroendocrine regulatory mechanism controlling repro-

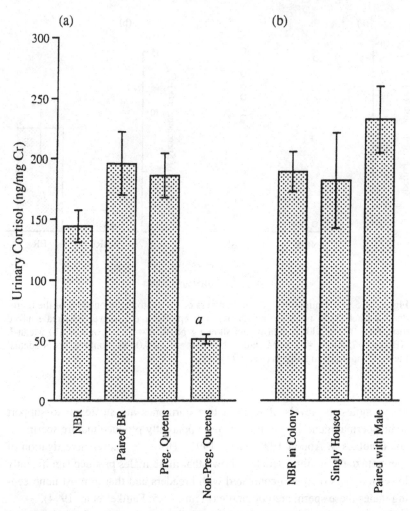

Figure 11.6. Concentrations of urinary cortisol (mean ± SEM). (a) In nonbreeding females (NBR: 28 samples from 15 females from 3 colonies), breeding females in male/female pairs (paired BR: 21 samples from 4 females), pregnant queens in colonies (preg. queens: 19 samples from 4 females), and nonpregnant queens in colonies (nonpreg. queens: 6 samples from 3 females). $a = P < 0.05$ vs. pregnant queens, paired BR and NBR (Duncan's multiple range test following ANOVA). (b) In two females while they were nonbreeders in a colony, when singly housed for 6 weeks and when paired with a male ($n = 6$, 10 and 5 samples, respectively).

duction and so far have ruled out a causal link between elevated urinary levels of cortisol and reproductive suppression. Although nonpregnant breeding queens had significantly lower concentrations of urinary cortisol than did non-breeding females and pregnant queens (Figure 11.6a), when nonbreeding females were separated from their colonies, their concentrations of urinary cortisol remained high, but they still ovulated and became reproductively active (Figure 11.6b; Faulkes 1990). In males, too, breeders had significantly lower urinary cortisol concentrations than did nonbreeders (73.8 ± 7.4 vs. 164.9 ± 19.7 ng/mg Cr, respectively), but when nonbreeders were separated from their colonies, they became reproductively active even though urinary cortisol concentrations increased from 165.7 ± 20.6 to 234.1 ± 46.3 ng/mg Cr (Faulkes 1990). These observations are similar to findings in nonhuman primates, where differences in glucocorticoids between individuals of different social rank do not seem to be directly involved in the physiological suppression of reproduction (Mendoza et al. 1979; Saltzman et al. 1994).

In female naked mole-rats, a possible role for prolactin in the mechanism of suppression has proved difficult to ascertain because of the lack of a suitable prolactin assay (Faulkes 1990). However, circumstantial evidence suggests that this hormone may not be directly implicated: The breeding queen has a postpartum estrus, which may often occur when she is still lactating (8–11 days postpartum) and therefore at a time when prolactin levels would be expected to be elevated (Jarvis 1991). This is similar to the situation suggested for another aseasonal, cooperatively breeding mammal, the common marmoset (McNeilly et al. 1981), in which prolactin has not been functionally linked to anovulation in female subordinates (Abbott et al. 1981). A simple, direct relationship between chronic physiological stress and reproductive inhibition does not seem to be operative in the naked mole-rat, but more definitive data are still required.

In both male and female nonbreeding mole-rats, reproductive suppression is readily reversible if the social cues maintaining reproductive suppression are removed. For example, if nonbreeding females are removed from their colonies and either paired with a male or housed singly and then paired with a male, urinary progesterone concentrations rise for the first time to levels indicative of a luteal phase of an ovarian cycle after approximately eight days (Figure 11.7; Faulkes, Abbott, & Jarvis 1990). Likewise, if males are similarly separated from their colonies, concentrations of plasma LH and urinary testosterone increase significantly, with urinary testosterone reaching levels comparable to those of breeding males after approximately five days (Faulkes & Abbott 1991). Separation experiments such as these, together with studies following the removal or death of a queen, prove conclusively that the breeding queen plays the central role in bringing about suppression of reproduction in both male and female nonbreeding naked mole-rats. The suppression among females by the queen is readily apparent, because when a queen dies she is replaced by a former nonbreeder (Jarvis 1991; Lacey & Sherman 1991).

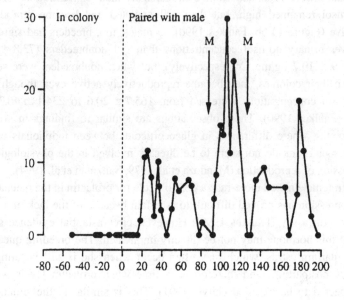

(a) Female 21

(b) Female 29

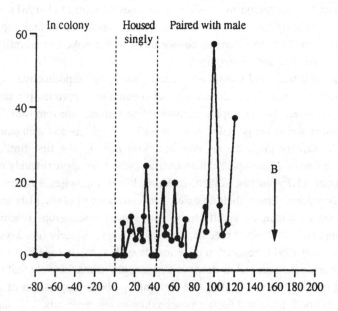

Days from Separation

11.5.3 Regulation of Reproductive Physiology in Breeding Males

What has not been clear until recently is the extent of the control of the whole colony, including both breeding and nonbreeding males, by the queen. Even the breeding males have little or no autonomy over their reproductive function within the colony. This phenomenon can be seen both in a colony situation and in male–female pairs. In the latter case, when urinary testosterone profiles of reproductively active males paired with females were plotted relative to the mate's ovarian cycle, testosterone concentrations in the male peaked during the follicular phase of the cycle, just before estrus (Figure 11.8). The central role of the breeding queen in imposing reproductive suppression on both breeding and nonbreeding males within a colony was seen when measurements of male urinary testosterone were made before and after the removal of the breeding queen. An example is shown in Figure 11.9. Upon removal of the queen, urinary testosterone concentrations rose significantly in both breeders and nonbreeders, indicating that reproductive suppression was released and the hypothalamic–pituitary–gonadal axis had become active in the nonbreeding males and was no longer being modulated by the queen in breeding males. An explanation for the higher urinary testosterone levels achieved by nonbreeding males in comparison to breeding males following queen removal is still being determined (see Figure 11.9).

The physiological manipulation of reproductive function by the queen may take its toll on the male consorts. Jarvis (1991) showed that the alpha breeding male (the consort that mated and anogenitally nuzzled with the queen most frequently in captive colonies) consistently lost an average of 16 to 34 percent of its body mass (over 1 to 7 years) from its heaviest body weight attained before gaining alpha status. Such consistent deterioration in body size and condition of alpha males does not occur in nonbreeding males. It may reflect the consequences of physiological stress in alpha males from queen-induced stimulation (around estrus) and inhibition (when the queen is not approaching estrus) of their reproductive endocrinology and behavior. On the other hand, the periods of high testosterone concentrations induced in

---

(Facing page)
Figure 11.7. Urinary progesterone profiles for two nonbreeding females removed from their colonies and (a) paired directly with a male, (b) housed singly for 6 weeks before pairing with a male, showing the rapid commencement of ovarian cyclicity after separation, with or without the presence of a male. M = observations of mating; B = birth of litter. Adapted from Faulkes, Abbott, and Jarvis 1990.

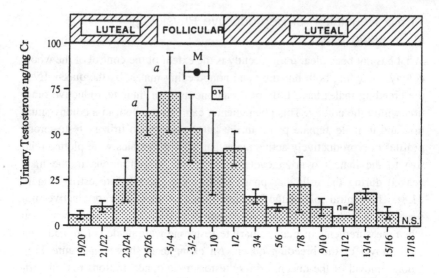

Day of female mate's cycle

Figure 11.8. Concentrations of urinary testosterone in 68 samples collected from five male naked mole-rats, housed in male–female pairs, over 12 female ovarian cycles. Data are shown as the mean ± SEM, plotted relative to the ovarian cycle of the female mate in 2-day intervals, assuming a mean total cycle length of 34 days (Faulkes et al. 1990). M: mean ± SEM day of mating, $n = 6$ observations. NS: no samples collected from this period. ov: presumed day of ovulation. * = $P < 0.05$ vs. days 15–16, 19–20, and 21–22. Based on Duncan's multiple-range test following one-way ANOVA for repeated measures.

alpha males by the queen, probably more frequently and for longer than in the other male consorts, may result in chronic immune suppression. Certainly, seasonally induced high testosterone levels in free-living male hopping mice, *Notomys alexis* (Breed 1976), have been implicated in the annual demise of these males following mating. Given the latter scenario, it would be advantageous for alpha male naked mole-rats to experience high-circulating testosterone levels for the minimum time.

11.5.4 Cues Regulating Male and Female Reproduction

Socially induced suppression of reproduction in male and female naked mole-rats appeared to be mediated by a mechanism involving direct contact with the breeding queen, rather than by primer pheromones (Faulkes & Abbott 1993). The lack of obvious primer pheromone effects in the inhibition of naked mole-rat reproduction was surprising because of the unusual communal toilet habits of naked mole-rats (Lacey et al. 1991). Each colony has one or

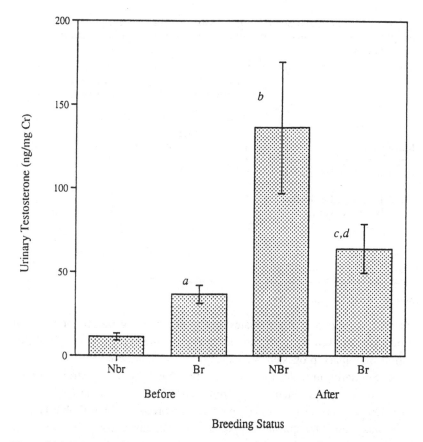

Figure 11.9. Mean concentrations of urinary testosterone in breeding and nonbreeding naked mole-rats from a colony of 28. A total of 222 samples (from 6 nonbreeding and 4 potential breeding males) were collected up to 126 days before and up to 52 days after removal of the breeding queen. $a = P < 0.05$ vs. NBr males before queen removal; $b = P < 0.001$ vs. NBr before queen removal; $c = P < 0.05$ vs. Br before queen removal; $d = P < 0.01$ vs. NBr after queen removal. From Duncan's Multiple Range Test following ANOVA.

more toilet chambers where animals usually urinate and defecate. Following urination or defecation, naked mole-rats roll around in the soiled litter, thereby facilitating transfer of any potential urinary or secreted pheromones. Given this large potential for cohesive colony transmission of chemical signals, it was assumed that primer pheromones played an integral part in reproductive inhibition (Jarvis 1981; Abbott 1988). However, reproductive activation could not be prevented or delayed in nonbreeding males and females separated from, but maintained in olfactory contact with their colonies by daily transfer of bedding and litter from the nest, food, and toilet chambers

Table 11.2. *Time from separation to reproductive activation in male and female naked mole-rats in control and bedding transfer separation experiments*

Manipulation	Days to start of luteal phase (females)		Days to increase of urinary testosterone (males)	
	Control	Bedding transfer	Control	Bedding transfer
Animals removed & paired	5	9	9	9
	18		1	
Animals removed & housed	7	15	7	1
singly before pairing	4	8	5	3
	5	3	3	1
		12		8
Mean ± SEM	7.8 ± 2.6	9.4 ± 2.0	5.0 ± 1.4	4.4 ± 1.7
	$n = 5$	$n = 5$	$n = 5$	$n = 5$

Source: Adapted from Faulkes and Abbott, 1993.

(Faulkes & Abbott 1993). Females undergoing bedding transfer and control procedures showed no significant difference in time to the first sustained elevation of urinary progesterone, indicative of a luteal phase of the ovarian cycle. Similarly, in males there was no difference between control and bedding transfer groups in the time from separation to elevations of urinary testosterone to concentrations comparable with breeding males (Table 11.2).

Because odor cues appear unimportant in regulating reproductive inhibition in naked mole-rats, a suppression mechanism mediated by behavioral interactions may predominate, a hypothesis supported by animal transfer experiments. A nonbreeding female was removed from its colony and, in addition to daily transfer of soiled bedding and litter from the parent colony (to maintain olfactory contact), received groups of nonbreeding animals of both sexes that were rotated between the parent colony and the permanently separated female. Thus, the separated female was exposed not only to any potential suppressing chemosignals but also to behavioral cues from all the other colony members *except* the breeding queen and breeding male(s). In such experiments, the separated females became reproductively active and showed signs of ovarian activity after three days, again demonstrating that behavioral contact with the queen is the key factor in the imposition of reproductive suppression (Smith 1994; Smith, Faulkes, & Abbott in prep.).

All the available evidence, therefore, points to a behavioral mechanism initiated by the queen as the primary cue in suppression of reproduction in nonbreeding male and female naked mole-rats. But what is the exact nature of such cues, and how do they bring about suppression? Our current hypothesis is that overt and subtle agonistic interactions between the queen and the non-breeders cause a neuroendocrine response in the latter, the physiological result of which is inhibition of gonadal function and infertility. Overt aggression in naked mole-rat colonies is rare except during queen succession. It seems likely that such events are critical periods when the social hierarchy is rearranged, and the new queen enforces her dominance over the other colony members and reestablishes reproductive inhibition. After succession, only subtle cues from the new queen may be required to maintain suppression in most colonies. One possible behavior implicated in the process of suppression is shoving. As mentioned previously, the queen initiates shoving with the highest frequency, and she directs this behavior at larger individuals of both sexes (Reeve & Sherman 1991; Faulkes, Clarke, Murray, & Waters in prep.). Also, in captive colonies containing animals of mixed kinship, the queen shoves less closely related individuals more frequently (Reeve & Sherman 1991).

Currently, there are two hypotheses regarding the function of shoving behavior. The threat-reduction hypothesis proposes that shoving reduces the threat of challenges for the queen's reproductive status. The activity–incitation hypothesis proposes that shoving serves to incite activity in "lazy workers" (Reeve & Sherman 1991; Reeve 1992). These two hypotheses are by no means mutually exclusive, and behavioral data support both. The lack of sex bias in the recipients of shoving from the queen and the fact that the queen also shoves the breeding male(s) led Reeve and Sherman (1991) to favor the activity–incitation hypothesis, because males seemingly posed no threat to the reproductive status of the queen. However, our physiological studies suggest that cues from the queen suppress reproduction in both male and female non-breeding naked mole-rats and modulate reproductive function in breeding males, thus offering an explanation as to why all these individuals are recipients of shoves. However, a causal link between changes in shoving behavior by the queen toward breeding males and elevation of urinary testosterone levels in these males around the time of estrus in the queen remains to be demonstrated. Although males may not be a direct threat to the queen per se, if physiological suppression were relaxed within the colony and testosterone levels in male colony members increased sufficiently to cause increased levels of aggression, then the social hierarchy could collapse and the queen could well find herself threatened. This may explain why, during some queen successions, breeding males engage in fighting and are often killed by the new queen (Reeve & Sherman 1991; Faulkes, Clarke, Murray, & Waters in prep.).

Thus, it seems likely that shoving behavior functions in both reproductive suppression threat reduction and activity incitation. From an evolutionary point of view, maintaining social order, inhibiting reproduction in subordinates of both sexes, and encouraging work-related activity would all be expected ultimately to increase the reproductive success of the queen.

11.5.5 Relevance of Reproductive Adaptations to Cooperative Breeding

Naked mole-rats lie at one extreme of the continuum of mammalian cooperative breeding strategies. However, they share many similarities with other singular cooperative breeders with respect to the ultimate and proximate factors involved in the evolution and maintenance of their social system. The main difference that sets naked mole-rats apart from other African mole-rats and all other mammals is the intense degree of inbreeding, giving rise to very high levels of intracolony genetic relatedness. Nucleotide sequence variation in the mitochondrial DNA D-loop region, which is normally highly variable (Harrison 1989), was absent within and between adjacent colonies in Mtito Andei, Kenya (Faulkes, Abbott, O'Brien, et al. in prep.). Multilocus minisatellite DNA fingerprints have produced band-sharing coefficients as high as 1.0 within some wild colonies from the same region of Kenya (Faulkes, Abbott, O'Brien, et al. in prep.), with average relatedness within colonies being very high ($r = 0.81$, compared with 0.5 in siblings of an outbred species; Reeve et al. 1990). The high level of inbreeding results from and is enforced by the high cost of dispersal for nonbreeders (Jarvis et al. 1994). In other singular cooperative breeders, such as the Damaraland mole-rat and the dwarf mongoose, intragroup relatedness is less than that found in the naked mole-rat. However, the opportunity for dispersal is greater among the former two species than the latter. In Damaraland mole-rats, inbreeding is avoided and dispersal occurs after periods of heavy rainfall. In 13 out of 14 newly formed colonies studied in Namibia, the breeding male and female were from different colonies. Nevertheless, even with such a dispersal phase, 90 percent of nonbreeders do not get an opportunity to reproduce (Jarvis & Bennett 1993). In the dwarf mongoose, average intragroup relatedness (r) equals 0.33 (Creel & Waser 1991), which is relatively high for a mammal. Subordinate dwarf mongooses of both sexes (usually higher-ranking individuals) occasionally rear their own offspring and thus obtain direct fitness, but the amount of indirect (inclusive) fitness obtained by helping to rear the offspring of the breeding female is far greater (Keane et al. 1994).

Reproductive Dictatorship by a Single Female*325*

Given the high degree of intracolony relatedness in colonies of naked mole-rats, why are social dominance and aggression so important, given the possibility of death from fighting? In terms of gene transmission and the relative benefits to individuals from inclusive versus individual fitness, the actual individuals breeding should not matter. The answer to the question may lie in the importance of social dominance in initiating and maintaining reproductive suppression and social order within colonies and in the specialized responses of colony members to dominance-related cues. Without such adaptations, altruism and cooperation could not occur, and the niche that naked mole-rats fill could not be successfully exploited.

The extreme breeding system of naked mole-rats also gives rise to some interesting questions regarding the interaction between genes and the environment and in particular the heritability of traits that contribute to attainment of dominance. Such traits have been shown to be heritable in male deer mice. *Peromyscus maniculatus* (Dewsbury 1990). In naked mole-rats, how does phenotypic variation in traits that contribute to attainment of dominance and to attainment of breeding status develop against a background of reduced genetic variability? Such phenotypic variation may occur as a result of stochastic developmental and environmental effects, like different fetal intrauterine position (vom Saal 1989; Clarke et al. 1992), when female fetuses positioned in utero between two males might develop more aggressive behavior than female fetuses adjacent to only one male or only females.

Although naked mole-rat colonies for the most part appear to be stable, harmonious societies, in certain circumstances social order may break down. For example, the breeding queen may be deposed by a nonbreeding female that has become increasingly dominant and reproductively active. In one such case, in a captive colony at the Institute of Zoology in London, an acyclic, nonbreeding female increased dramatically in body weight (from 35 g to 50 g over a 70-day period), fought and killed the queen, and became the new breeding female. There are also reported cases in captive colonies of a second queen emerging and coexisting with the original queen, and one case of a captive colony having three breeding queens (Jarvis 1991). It is not known if such phenomena are a laboratory artifact, or if they also occur in the wild. There is good evidence to suggest that the queen's control over the colony may lapse around the time of late pregnancy and parturition, because nonbreeding females may develop perforate vaginas at this time (Jarvis 1991). It has also been noted that the rate of shoving by the queen declines at this time, again suggesting a causal link between this behavior and reproductive suppression (Reeve & Sherman 1991).

A plurality of breeding females in captive naked mole-rat colonies may reflect attempts by nonbreeding females to reproduce in their own right. In the wild, a violent uprising among females might readily lead to fissioning of a colony and the "budding off" of a new founder colony from the parent one. Jarvis (1991) indeed showed that in four out of six of her captive colonies where more than one breeding female was present at any one time, the appearance of a second breeding female was preceded by a period of good recruitment of pups to the colonies. In captivity, where spontaneous fissioning of colonies cannot occur, reduced reproductive success can be a consequence of a plurality in breeding females in a colony. When two or even three breeding females were tolerated within a captive colony, pup survival was poor (0% to 3%; 0–16 pups; three colonies: Jarvis 1991) in comparison to when only one breeding female was present (18% to 40%; 10 to 179 pups; the same three colonies: Jarvis, 1991). Often such plurality in breeding females was terminated violently (see preceding paragraph; Jarvis 1991; Lacey & Sherman this volume). These findings would be consistent with the proposed occurrence of fission–dispersal phases in naked mole-rat colonies (Jarvis 1991).

Such deviations from a single breeding female in a naked mole-rat colony may also reflect the competitive attempts of a few reproductive hopefuls or high-ranking nonbreeding animals to gain reproductive status themselves *within* their parent colony with a breeding female already resident. The emergence of subordinate females as active breeders within their current social group has been found in other cooperative species: common marmosets (captivity: Abbott 1984; wild: Digby & Ferrari 1994), lion tamarin (wild: Dietz & Baker 1993; French this volume), and dwarf mongoose (wild: Keane et al., 1994). Such instances of competitive attempts to breed by subordinate females in the presence of a dominant breeding female may reflect natural selection in progress (Grafen 1988). In the naked mole-rat, common marmoset, and golden lion tamarin, such attempts are not particularly successful, either in terms of surviving offspring produced by the contending subordinate or in the survival of the contender itself. Such findings suggest that there is some selection pressure operating against subordinates that do not maintain behavioral and physiological suppression in the presence of a dominant conspecific. On the other hand, the genetic typing data from wild dwarf mongooses suggested that older and high-ranking subordinates successfully reproduced and survived in their natal group in the presence of a dominant breeding female (Keane et al. 1994). Possibly, in this species, if *every* available group member is crucial to survival of all group members (e.g., in predator detection and avoidance) as well as in assisting in rearing offspring of the breeding female, and if migration is successful, then there may be little effective negative selection against older animals becoming refractory to reproductive suppression. Older animals could readily leave, and their absence from the group could be detrimental for survival and reproduction of the remaining breeding and nonbreeding animals. This scenario for the dwarf mongoose, though, does not readily fit the conditions normally encountered by free-living naked mole-rats. Opportunities for dispersal from naked mole-rat colonies appear minimal, and group size is large (averaging 40 to 70 individuals) (Brett 1991a; Sherman et al. 1992; Jarvis et al. 1994).

11.6 Conclusions

As evident from the findings in this chapter and throughout this volume, subordinate, nonbreeding animals in singular, cooperatively breeding mammalian species are not phenotypically limited to a nonbreeding existence. Nonbreeders are highly responsive to changes in the social environment and to socially related sensory cues (see also Wasser and Barash 1983.) Even with dominant breeding animals resident, subordinate animals can occasionally mate and become pregnant. It remains to be determined what proximate changes occur to trigger the release of subordinates from physiological and behavioral suppression and what social or ecological circumstances determine the success or failure of such competitive attempts at breeding by subordinate animals in species that exhibit highly specialized cooperative breeding. Contributing factors might involve the immigration of unrelated animals of the opposite sex into their group (e.g., a novel male or female effect), increased group density, altered behavioral relationship with the dominant breeding animals, or reduced sensitivity to inhibitory social cues with increasing adult age (e.g., age-dependent changes in neurotransmitter release). Certainly, alterations in likelihood of (1) survival and successful breeding following dispersal, (2) predation, and (3) quality of home range may ultimately have an additional bearing on the degree of attempted reproduction by subordinates (MacDonald & Carr, 1989; Creel & Creel 1991; Waser et al. 1994).

Singular, cooperatively breeding mammals, and the naked mole-rat in particular, certainly provide excellent opportunities for determining the physiological, behavioral, and sensory adaptations mediating the proximate regulation of reproductive success. Unlike competitively breeding mammals, singular cooperative breeders are severely constrained: Successful reproduction can usually be timed only when social (or ecological) conditions permit. Reproduction by subordinates at inappropriate times (i.e., in the presence of a dominant breeding animal) may be disastrous in terms of survival and reproductive success (Jarvis 1991; Dietz & Baker 1993; Digby & Ferrari 1994; but see Keane et al. 1994). Only in such singular cooperative breeders has the social environment become so elevated in importance that it provides the predominant proximate cues regulating the timing of reproduction.

Acknowledgments

The authors thank M. Llovet and M. Gordon for excellent assistance in animal care; Dr. T. E. Smith and Miss H. O'Brien for technical assistance and aid in data collection; Drs. J. U. M. Jarvis and N. C. Bennett for supplying animals, tissues, and blood and urine samples; Drs. A. S. I. Loudon, M. S. Gosling, J. P. Hearn, W. Saltzman, N. J. Schultz-Darken, and J. A. French for criticism of earlier drafts of the manuscript; T. A. Dennett for photography; L. A. Smith, T. E. Thaker, and A. N. Norris for preparation of the manuscript. The naked mole-rat research reported here was supported by an SERC PhD studentship (CCF), grants from the Wellcome Trust (DHA, CGF), the Bonhote Bequest of the Linnean Society of London and the Royal Society of London (CGF), an MRC/AFRC Program Grant, and travel grants from the Journal of Reproduction and Fertility Ltd., SERC, National Geographic Society, The Royal Society, Wellcome Trust, Association for the Study of Animal Behavior, and the Society for Endocrinology. The writing of this paper was supported (in part) by NIH grant RR00167 to the Wisconsin Regional Primate Research Center. This chapter is publication number 34 – 020 of the WRPRC.

References

Abbott, D. H. (1984). Behavioral and physiological suppression of fertility in subordinate marmoset monkeys. *Amer. J. Primatol.* 6:169–186.
Abbott, D. H. (1988). Natural suppression of fertility. In *Symposia of the Zoological Society of London Number 60*, ed. G. R. Smith and J. P. Hearn, pp. 7–28. New York: Oxford University Press.
Abbott, D. H. (1993). Social conflict and reproductive suppression in marmosets and tamarin monkeys. In *Primate social conflict,* ed. W. A. Mason and S. P. Mendoza, pp. 331–372. Albany: State University of New York Press.
Abbott, D. H., Barrett, J., & George, L. M. (1993). Comparative aspects of the social suppression of reproduction in female marmosets and tamarins. In *Marmosets and tamarins: Systematics, behaviour, and ecology,* ed. A. B. Rylands, pp. 152–163. Oxford: Oxford University Press.
Abbott, D. H., & George, L. M. (1991). Reproductive consequences of changing social status in female common marmosets. In *Primate responses to environmental change,* ed. H. O. Box, pp. 294–309. London: Chapman and Hall.
Abbott, D. H., Barrett, J., & Faulkes, C. G. (1990). Pheromonal contraception. In *Chemical signals in vertebrates 5,* ed. D. W. MacDonald, D. Müller-Schwarze, and S. E. Natynczuk, pp. 169–183. New York: Oxford University Press.

Abbott, D. H., George, L. M., Barrett, J., Hodges, K. T., O'Byrne, K. T., Sheffield, J. W., Sutherland, I. A., Chambers, G. R., Lunn, S. F., & Ruiz de Elvira, M.-C. (1990). Social control of ovulation in marmoset monkeys: A neuroendocrine basis for the study of infertility. In *Socioendocrinology of primate reproduction,* ed. T. E. Ziegler & F. B. Barcovitch, pp. 135–158. New York: Wiley-Liss.

Abbott, D. H., and Hearn, J. P. (1978). Physical, hormonal and behavioral aspects of sexual development in the marmoset monkey *Callithrix jacchus. J. Reprod. Fertil.* 53:155–166.

Abbott, D. H., McNeilly, A. S., Lunn, S. F., Hulme, M. J., & Burden, F. J. (1981). Inhibition of ovarian function in subordinate female marmoset monkeys *(Callithrix jacchus jacchus). J. Reprod. Fertil.* 63:335–345.

Allard, M. W., & Honeycutt, R. L. (1992). Nucleotide sequence variation in the mitochondrial 12S rRNA gene and the phylogeny of African mole-rats (Rodentia: Bathyergidae). *Mol. Biol. Evol.* 9:27–40.

Altmann, J., Hausfater, G., & Altmann, S. A. (1988). Determinants of reproductive success in savannah baboons, *Papio cynocephalus.* In *Reproductive success: Studies of individual variation in contrasting breeding systems,* ed. T. H. Clutton-Brock, pp. 403–418. Chicago: University of Chicago Press.

Barrett, J., Abbott, D. H., & George, L. M. (1993). Sensory cues and suppression of reproduction in subordinate female marmoset monkeys, *Callithrix jacchus. J. Reprod. Fertil.* 97:301–310.

Bennett, N. C. (1988). *The trend towards sociality in three species of South African mole-rats* (Bathyergidae): *Causes and consequences.* PhD dissertation, University of Cape Town.

Bennett, N. C., Jarvis, J. U. M. & Cotterill, F. P. D. (1994). The colony structure and reproductive biology of the afrotropical Mashona mole-rat, *Cryptomys darlingi. J. Zool., Lond.* 234:477–487.

Bennett, N. C., Jarvis, J. U. M., Faulkes, C. G., & Millar, R. P. (1993). LH responses of freshly caught female and male Damaraland mole-rats, *Cryptomys damarensis,* to single doses of exogenous GnRH. *J. Reprod. Fertil.* 99:81–86.

Breed, W. G. (1976). Effect of environment on ovarian activity of wild hopping mice *(Notomys alexis). J. Reprod. Fertil.* 47:395–397.

Brett, R. A. (1991a). The population structure of naked mole-rat colonies. In *The biology of the naked mole-rat,* ed. P. W. Sherman, J. U. M. Jarvis, & R. D. Alexander, pp. 97–136. Princeton: Princeton University Press.

Brett, R. A. (1991b). The ecology of naked mole-rat colonies: Burrowing, food, and limiting factors. In *The biology of the naked mole-rat,* ed. P. W. Sherman, J. U. M. Jarvis, and R. D. Alexander, pp. 137–184. Princeton: Princeton University Press.

Bronson, F. H. (1985). Mammalian reproduction: An ecological perspective. *Biol. Reprod.* 32:1–26.

Bronson, F. H. (1989). *Mammalian reproductive biology.* Chicago and London: University of Chicago Press.

Bronson, F. H., & Heideman, P. D. (1994). Seasonal regulation of reproduction in mammals. In *The physiology of reproduction,* 2nd ed. E. Knobil & J. D. Neill, pp. 541–583. New York: Raven Press.

Buffenstein, R., & Yahav, S. (1991). Is the naked mole-rat *Heterocephalus glaber* an endothermic yet poikilothermic mammal? *J. Therm. Biol.* 16:277–282.

Chase, I. D. (1974). Models of hierarchy formation in animal societies. *Behav. Sci.* 19:374–382.

Clarke, M. M., Tucker, L, & Galef, B. G. Jr. (1992). Stud males and dud males:

Intrauterine position effects on the reproductive success of male gerbils. *Anim. Behav.* 43:215–221.

Clutton-Brock, T. H., Albon, S. D., & Guinness, F. E. (1986). Great expectations: Dominance, breeding success and offspring sex ratios in red deer. *Anim. Behav.* 34:460–471.

Clutton-Brock, T. H., Guinness, F. E. & Albon, S. D. (1982). *Red deer: Behavioral ecology of two sexes.* Chicago: University of Chicago Press.

Cohen-Tannoudji, J., Locatelli, A., and Signoret, J. P. (1986). Non-pheromonal stimulation by the male of LH release in the anoestrous ewe. *Physiol. Behav.* 36:921–924.

Creel, S. R., & Creel, N. M. (1991). Energetics, reproductive suppression and obligate communal breeding in carnivores. *Beh. Ecol. Sociobiol.* 28:263–270.

Creel, S. R., Creel, N., Wildt, D. E., & Montfort, S. L. (1992). Behavioral and endocrine mechanisms of reproductive suppression in Serengeti dwarf mongoose. *Anim. Behav.* 43:231–246.

Creel, S. R., & Waser, P. M. (1991). Failures of reproductive suppression in dwarf mongooses (*Helogale parvula*): Accident or adaptation? *Behav. Ecol.* 2:7–15.

Davies, K. C., & Jarvis, J. U. M. (1986). The burrow systems and burrowing dynamics of the mole-rats *Bathyergus suillus* and *Cryptomys hottentotus* in the fynbos of the south western Cape, South Africa. *J. Zool. Lond.* 209:125–147.

Dewsbury, D. A. (1990). Fathers and sons: Genetic factors and social dominance in deer mice, *Peromyscus maniculatus. Anim. Behav.* 39:284–289.

Dietz, J. M., & Baker, A. J. (1993). Polygyny and female reproductive success in golden lion tamarins, *Leontopithecus rosalia. Anim. Behav.* 46:1067–1078.

Digby, L. J., & Ferrari, S. F. (1994). Multiple breeding females in free-ranging groups of *Callithrix jacchus. Int. J. Primatol.* 15:389–397.

Du Toit, J. T., Jarvis, J. U. M., & Louw, R. N. (1985). Nutrition and burrow energetics of the Cape mole-rat *Georychus capensis. Oecologia* 66:81–87.

Emlen, S. T. (1991). The evolution of cooperative breeding in birds and mammals. In *Behavioural ecology: An evolutionary approach,* ed. J. R. Krebs & N. B. Davies, pp. 301–337. Oxford, U.K.: Blackwell.

Epple, G., & Katz, Y. (1980). Social influences on first reproductive success and related behaviors in the saddle-back tamarin (*Saguinus fuscicollis,* Callitrichidae). *Int. J. Primatol.* 1:171–183.

Faulkes, C. G. (1990). *Social suppression of reproduction in the naked mole-rat* Heterocephalus glaber. PhD thesis, University of London.

Faulkes, C. G., & Abbott, D. H. (1991). Social control of reproduction in both breeding and non-breeding male naked mole-rats, *Heterocephalus glaber. J. Reprod. Fertil.* 93:427–435.

Faulkes, C. G., & Abbott, D. H., (1993). Evidence that primer pheromones do not cause social suppression of reproduction in male and female naked mole-rats. *Heterocephalus glaber. J. Reprod. Fertil.* 99:225–230.

Faulkes, C. G., Abbott, D. H., & Jarvis, J. U. M. (1990). Social suppression of ovarian cyclicity in captive and wild colonies of naked mole-rats. *Heterocephalus glaber. J. Reprod. Fertil.* 88:559–568.

Faulkes, C. G., Abbott, D. H., & Jarvis, J. U. M. (1991). Social suppression of reproduction in male naked mole-rats, *Heterocephalus glaber. J. Reprod. Fertil.* 901:593–604.

Faulkes, C. G., Abbott, J. H., Jarvis, J. U. M., & Sheriff, F. E. (1990). LH responses of female naked mole-rats, *Heterocephalus glaber,* to single and multiple doses of exogenous GnRH. *J. Reprod. Fertil.* 89:317–323.

Faulkes, C. G., Abbott, D. H., Liddell, C. E., George, L. M., & Jarvis, J. U. M. (1991).

Hormonal and behavioral aspects of reproductive suppression in female naked mole-rats. In *The biology of the naked mole-rat,* ed. P. W. Sherman, J. U. M. Jarvis, & R. D. Alexander, pp. 426–445. Princeton: Princeton University Press.

Faulkes, C. G., Abbott, D. H., & Mellor, A. L. (1990). Investigation of genetic diversity in wild colonies of naked mole-rats (*Heterocephalus glaber*) by DNA fingerprinting. *J. Zool. Lond.* 221:87–97.

Faulkes, C. G., Clarke, F. M., Murray, J., Waters, T. (in prep.). Hormonal and behavioral correlates of dominance and queen succession in captive colonies of the eusocial naked mole-rat, *Heterocephalus glaber.* For submission to *Animal Behavior.*

Faulkes, C. G., O'Brien, H. P., Bruford, M. W., Wayne, R. K., Lau, L., Roy, M., & Abbott, D. H. (in prep.). Minisatellite DNA and mitochondrial d-loop sequence variation within and between populations of naked mole-rats. *Heterocephalus glaber.* For submission to *Molecular Ecology.*

Faulkes, C. G., Trowell, S. N., Jarvis, J. U. M., & Bennett, N. C. (1994). Investigation of numbers of spermatozoa and sperm motility in reproductively active and socially suppressed males of two eusocial African mole-rats, the naked mole-rat (*Heterocephalus glaber*), and the Damaraland mole-rat (*Cryptomys damarensis*). *J. Reprod. Fertil.* 100:411–416.

Gandelman, R., vom Saal, F. S., & Reinisch, J. M. (1977). Contiguity to male foetuses affects morphology and behaviour of female mice. *Nature (Lond.)* 266:722–724.

Grafen, A. (1988). On the uses of data on lifetime reproductive success. In *Reproductive success: Studies of individual variation in contrasting breeding systems,* ed. T. H. Clutton-Brock, pp. 454–471. Chicago: University of Chicago Press.

Haigh, G., Cushing, B. S., & Bronson, F. H. (1988). A novel post-copulatory block of reproduction in white-footed mice. *Biol. Reprod.* 38:623–626.

Hamilton, W. D. (1964). The genetical evolution of social behavior, I., II. *J. Theoret. Biol.* 7:1–52.

Harrison, R. G. (1989). Animal mitochondrial DNA as a genetic marker in population and evolutionary biology. *Trends Ecol. Evol.* 4:6–11.

Hickman, G. C. (1979). Burrow system structure of the bathyergid. *Cryptomys hottentotus* in Natal, South Africa. *J. Saugetierkd.* 46:293–298.

Jarvis, J. U. M. (1981). Eu-sociality in a mammal – cooperative breeding in naked mole-rat *Heterocephalus glaber* colonies. *Science N.Y.* 212:571–573.

Jarvis, J. U. M. (1985). Ecological studies on *Heterocephalus glaber,* the naked mole-rat, in Kenya. *Natl. Geogr. Soc. Res. Rep.* 20:429–437.

Jarvis, J. U. M. (1991). Reproduction of naked mole-rats. In *The biology of the naked mole-rat,* ed. P. W. Sherman, J. U. M. Jarvis, & R. D. Alexander, pp. 384–425. Princeton: Princeton University Press.

Jarvis, J. U. M., & Bennett, N. C. (1991). Ecology and behavior of the family Bathyergidae. In *The biology of the naked mole-rat,* eds. P. W. Sherman, J. U. M. Jarvis, & R. D. Alexander, pp. 66–96. Princeton: Princeton University Press.

Jarvis, J. U. M., & Bennett, N. C. (1993). Eusociality has evolved independently in two genera of bathyergid mole-rats – but occurs in no other subterranean mammal. *Behav. Sociobiol.* 33:253–260.

Jarvis, J. U. M., & Sale, J. B. (1971). Burrowing and burrow patterns of East African mole-rats *Tachyoryctes, Heliophobius,* and *Heterocephalus. J. Zool. Lond.* 163:451–479.

Jarvis, J. U. M., O'Riain, M. J., Bennett, N. C., & Sherman, P. W. (1994). Mammalian eusociality: A family affair. *Trends Ecol. Evol.* 9:47–51.

Jarvis, J. U. M., O'Riain, J., & McDaid, E. (1991). Growth and factors affecting body size in naked mole-rats. In *The biology of the naked mole-rat,* ed. P. W. Sherman, J. U. M. Jarvis, and R. D. Alexander, pp. 358–383. Princeton: Princeton University Press.

Jennison, M. D., & MacDonald, D. W. (1994). Cooperative breeding in mammals. *Trends Ecol. Evol.* 9:89–93.

Karsch, F. J., Bittman, E. L., Foster, D. L., Goodman, R. L., Legan, S. J., & Robinson, J. E. (1984). Neuroendocrine basis of seasonal reproduction. *Rec. Prog. Hor. Res.* 40:185–232.

Keane, B., Waser, P. M., Creel, S. R., Creel, N. M., Elliot, L. F., & Minchella, D. J. (1994). Subordinate reproduction in dwarf mongooses. *Anim. Behav.* 47:65–75.

Keverne, E. B., Eberhardt, J. A., Yodyingyuad, V., & Abbott, D. H. (1984). Social influences on sex differences on the behavior and endocrine state of talapoin monkeys. In *Progress in brain research* Vol. 61, ed. G. J. Vries, J. P. C. DeBruin, H. B. M. Uylinjs, & M. A. Carver, pp. 331–345. Amsterdam: Elsevier.

Lacey, E. A., Alexander, R. D., Braude, S. H., Sherman, P. W., & Jarvis, J. U. M. (1991). An ethogram for the naked mole-rat: Non-vocal behaviours. In *The biology of the naked mole-rat,* ed. P. W. Sherman, J. U. M. Jarvis, & R. D. Alexander, pp. 209–242. Princeton: Princeton University Press.

Lacey, E. A., & Sherman, P. W. (1991). Social organization of naked mole-rat colonies: Evidence for a division of labor. In *The biology of the naked mole-rat,* ed. P. W. Sherman, J. U. M. Jarvis, & R. D. Alexander, pp. 275–336. Princeton, N.J.: Princeton University Press.

Lincoln, G. A. (1981). Seasonal aspects of testicular function. In *The testis,* ed. H. Burger and P. de Kretser, pp. 255–302. New York: Raven Press.

Lovegrove, B. G., & Wissel, C. (1988). Sociality in mole-rats: Metabolic scaling and the role of risk sensitivity. *Oecologia* 74:600–606.

McComb, K. (1987). Roaring by red deer stags advances the date of oestrus in hinds. *Nature (Lond.)* 330:648–649.

MacDonald, D. W., & Carr, G. M. (1989). Food security and the rewards of tolerance. In *Comparative socioecology,* ed. V. Standen & R. A. Foley, pp. 75–97. Oxford, U.K.: Blackwell.

McNeilly, A. S., Abbott, D. H., Lunn, S. F., Chambers, P. C., & Hearn, J. P. (1981). Plasma prolactin concentrations during the ovarian cycle and lactation and their relationship to return of fertility *post partum* in the common marmoset (*Callithrix jacchus*). *J. Reprod. Fertil.* 62:353–360.

Margulis, S. W., Saltzman, W., & Abbott, D. H. (1995). Behavioral and hormonal changes in female naked mole-rats (*Heterocephalus glaber*) following removal of the breeding female from a colony. *Horm. Behav.* 29:227–247.

Martin, G. B., Oldham, C. M., & Lindsay, P. R. (1980). Increased plasma LH levels in seasonally anovular merino ewes following the introduction of the rams. *Anim. Reprod. Sci.* 3:125–132.

Mendoza, S. P., Coe, C. L., Lowe, E. L., & Levine, S. (1979). The physiological response to group formation in adult male squirrel monkeys. *Psychoneuroendocrinol.* 3:221–229.

Michener, C. D. (1969). Comparative social behavior of bees. *Ann. Rev. Entomol.* 144:299–342.

Onyango, D. W., Otlanga-Owiti, G. E. Odirr-Okelo, D., & Makawiti, D. W. In vitro interstitial (Leydig) cell response to LH and concentrations of plasma testosterone and LH in the naked mole-rat (*Heterocephalus glaber, Rüppell*) *Afr. J. Ecol.* 29:76–85.

Perret, M. (1992). Environmental and social determinants of sexual function in the male lesser mouse lemur. *Folia Primatol.* 59:1–25.

Reeve, H. K. (1992). Queen activation of lazy workers in colonies of the eusocial naked mole-rat. *Nature (Lond.)* 358:147–149.

Reeve, H. K. & Sherman, P. W. (1991). Intracolony aggression and nepotism by the breeding female naked mole-rat. In *The biology of the naked mole-rat.*, ed. P. W. Sherman, J. U. M. Jarvis, & R. D. Alexander, pp. 384–425. Princeton: Princeton University Press.

Reeve, H. K., Westneat, D. F., Noon, W. A., Sherman, P. W., & Aquadro, C. F. (1990). DNA "fingerprinting" reveals high levels of inbreeding in colonies of eusocial naked mole-rat. *Proc. Natl. Acad. Sci. USA.* 87:2496–2500.

Rhine, R. J., Wasser, S., & Norton, G. W. (1988). Eight-year study of social and eco-logical correlates of mortality among immature baboons of Mikumi National Park, Tanzania. *Am. J. Primatol.* 16:199–212.

Robinson, J. E., Radford, M. H., & Karsch, F. J. (1985). Seasonal changes in pulsatile luteinizing hormone (LH) secretion in the ewe: Relationship of frequency of LH pulses to day length and response to estradiol negative feedback. *Biol. Reprod.* 33:324–334.

Saal, F. S., vom (1989). The production of and sensitivity to cues that delay and pro-long subsequent oestrous cycles of females in female mice are influenced by prior intrauterine position. *J. Reprod. Fertil.* 86:457–471.

Saal, F. S., vom & Bronson, F. (1978). In utero proximity of female house mouse fetuses to males: Effect on reproductive performance during later life. *Biol. Reprod.* 19:842–853.

Saltzman, W., Schultz-Darken, N., Scheffler, G., Wegner, F., & Abbott, D. H. (1994). Social and reproductive influences on plasma cortisol in female marmoset mon-keys. *Physiol. Behav.* 56:801–810.

Sapolsky, R. M. (1993). The physiology of dominance in stable versus unstable social hierarchies. In *Primate social conflict,* ed. W. A. Mason & S. P. Mendoza, pp. 171–204. Albany: State University of New York Press.

Sheffield, J. W., Abbott, D. H., & O'Shaugnessy, P. J. (1989). The effects of social rank on [3H] pregnenolone metabolism and androgen production by the common marmoset testis. *Serono Symposium Review* 20; 1:200.

Sherman, P. W., Jarvis, J. U. M., & Braude, S. H. (1992). Naked mole-rats. *Sci. Amer.* 267:72–78.

Signoret, J. P., & Lindsay, D. R. (1982). The male effect in domestic mammals: Effect on LH secretion and ovulation. Importance of olfactory cues. In *Olfaction and endocrine regulation,* ed. W. Briephol, pp. 63–72. London: IRL Press.

Smith, T. E. (1994). *Role of odour in the suppression of reproduction in female naked mole-rats* (Heterocephalus glaber) *and common marmosets* (Callithrix jacchus) *and the social organization of these two species.* PhD dissertation, University of London.

Smith, T. E., Faulkes, C. G., & Abbott, D. H. (in prep.). A behavioural mechanism involving direct contact with the queen plays a major role in the suppression of reproduction in subordinate female naked mole-rats (*Heterocephalus glaber*).

Sussman, R. W., & Garber, P. A. (1987). A new interpretation of the social organiza-tion and mating system of Callitrichidae. *Int. J. Primatol.* 8:73–92.

Waser, P. M., Creel, S. R., & Lucas, J. R. (1994). Death and disappearance: Estimat-ing mortality risks associated with philopatry and dispersal. *Behav. Ecol.* 5:135–141.

Wasser, S. K., & Barash, D. P. (1983). Reproductive suppression among female mam-mals: Implications for biomedicine and sexual selection theory. *Q. Rev. Biol.* 58:513–538.

Wasser, S. K., & Starling, A. K. (1988). Proximate and ultimate causes of reproductive suppression among female yellow baboons at Mikumi National Park, Tanzania. *Am. J. Primatol.* 16:97–121.

Wickings, E. J., & Dixson, A. F. (1992). Testicular function, secondary sexual development, and social status in male mandrills (*Mandrillus sphinx*). *Physiol. Behav.* 52:909–916.

Wilson, E. O. (1975). *Sociobiology*. Boston, Mass.: Harvard University Press.

Yang, K., Haynes, N. B., Lamming, G. E., Brooks, A. N. (1988). Ovarian steroid hormone involvement in endogenous opioid modulation of LH secretion in mature ewes during the breeding and nonbreeding seasons. *J. Reprod. Fertil.* 83:129–139.

12

Factors Influencing the Occurrence of Communal Care in Plural Breeding Mammals

SUSAN E. LEWIS and ANNE E. PUSEY

12.1 Introduction

Across group-living mammals, there is a continuum of reproductive skew from societies in which a single pair is responsible for all reproduction (singular breeders) to those in which all adults reproduce (plural breeders) (Keller & Reeve 1994). Factors influencing the degree of skew include extrinsic, ecological factors, such as saturation of available breeding sites, and intrinsic factors, such as the energetic costs of reproduction that limit the success of breeding attempts by subordinates (Vehrencamp 1983; Creel & Creel 1991; other chapters in this volume). Across this continuum, there is also variation in the degree to which members of the social group assist in the rearing of offspring of other members of the social group (i.e., communal care). Communal care that occurs at the singular-breeding end of the spectrum usually takes the form of nonbreeding helpers at the nest or den that defend and provision the offspring. Most chapters in this book focus on this kind of care. Communal care that occurs among plural-breeding species most frequently involves behaviors such as communal defense or provisioning of offspring, often by other breeding individuals within the group. In this chapter, we will examine the occurrence of communal care in plural-breeding mammals.

The difference between a system in which help is provided by only non-breeding individuals and one in which help is provided by breeding individuals is profound, because the former involves helpers sacrificing their direct reproductive efforts in the short term, whereas the latter does not necessarily imply a loss of current direct fitness. By not differentiating these two systems in reviews that examine the distribution and functions of communal care in mammals (e.g., Riedman 1982; Gittleman 1985; Emlen 1991; Jennions & Macdonald 1994), potential differences in the causes and consequences of the communal behaviors become obscured. In the chapters of this book that focus on communal care in singular breeders, the primary questions under investigation are (1) why not disperse? (2) why delay breeding? and (3) why help? The first two questions highlight issues of particular concern to singular breeding mammals. As reproductive skew decreases, questions regarding dispersal and delayed breeding become less important, but questions regarding the value of breeding in groups rather than alone become more important. The key questions then become why live and breed in groups? and why help? In this chapter, we will discuss (1) why animals live and breed in groups, (2) the types of communal care provided by breeding females in plural breeding systems and proposed evolutionary explanations for this care, and (3) nonoffspring nursing as an example of a specific communal behavior for which ecological correlates of its occurrence are better understood.

12.2 Definitions and Ultimate Explanations for Communal Care

A wide variety of behaviors have been classified as communal care in the literature on social birds and mammals. Gittleman (1985, p. 188) distinguished between "direct" forms of communal care, such as "feeding, carrying, grooming, playing, and actively protecting the young" and "indirect" forms, such as "territorial defense of critical resources, scent marking, herding movements, or long-distance vocalizations that facilitate the spacing of individuals or groups." This broad classification of communal care results in nearly all social mammals being considered as communal breeding species. Jennions and Macdonald (1994, p. 90) limit their definition of communal (helping) behavior by labeling Gittleman's indirect forms as "group size effects," "fitness consequences of variation in group size which are not due to changes in the amount of costly help provided." However, the dichotomy between direct care and indirect care, or group size effects, is often not clear. For example, if several females chase a predator away from a nursery group, this would seem to be a clear example of direct care because the costly actions of the females help young that are not their own. But if each female chases the predator, even if she is alone with only her own young, then the more effective defense of the young by the group is an example of a "group size effect." In this chapter we consider all behavior that occurs as a consequence of the presence of young in the group and that benefits the young of other individuals to be a potential form of communal care. This definition parallels Clutton-Brock's (1991, p. 339) definition of parental care as "any form of parental behavior that appears likely to increase the fitness of a parent's offspring" in that it carries no implications about costs in terms of energy or fitness.

The ultimate causes of communal care have been widely discussed in the literature on helping in birds and mammals and are dealt with at length in several chapters of this volume (e.g., Riedman 1982; Gittleman 1985; Brown 1987; Emlen 1991; Jennions & Macdonald 1994; Mumme this volume; Tardif this volume). In singular breeders, helping by nonbreeders may occur because of positive effects on the helper's future direct fitness, its present indirect fitness, or its future indirect fitness (Mumme this volume). Among plural breeders there is the additional possibility that communal care brings immediate and often mutualistic benefits to the direct fitness of the participants. Alternatively, communal care may occur in spite of neutral or negative effects on the alloparent's inclusive fitness if the costs of avoiding giving such care are even greater.

In this chapter, we concentrate on care given by breeding females because care given by breeding females is unique to the plural breeding situation. Males and juveniles may also give care in plural breeding species, but the explanations for such care may be more similar to those for singular breeding species, which are dealt with in other chapters of this volume. A list of the common and scientific names of species listed in this chapter can be found in Table 12.1.

12.3 Why Live and Breed in Groups?

Except for the need to live with a mother during lactation and to come together with a mate for reproduction, mammals are theoretically capable of living an entirely solitary life. Yet mammals live in a diverse variety of social groups. Several authors have reviewed the benefits and costs associated with living in groups (e.g., Alexander 1974; Koenig 1981). Benefits of grouping are generally classified as those that tend to decrease the risk of predation, such as the dilution effect or increased vigilance, and those that tend to increase the ability to secure and hold resources, such as enhanced territorial defense or foraging efficiency (Clode 1993). It is evident, however, that the costs and benefits of group living may vary, depending on the sex, age, and reproductive condition of each group member. The composition and stability of groups often varies over time to reflect these differential benefits. The dynamics of group composition when dependent offspring are present provide an indicator of whether the benefits of grouping apply primarily to the adults in the group or whether grouping is particularly advantageous to raising dependent offspring.

In many species, females with young associate more closely than they do at other times of the year and form cohesive nursery groups of females and

Table 12.1. *Common and scientific names of species identified in the chapter*

Common name	Scientific name
Monotremes	Marsupials
Eastern gray kangaroo	*Macropus giganteus*
Lagomorphs	
European rabbit	*Oryctolagus cuniculus*
Chiropterans	
Pallid bat	*Antrozous pallidus*
Fringed myotis	*Myotis thysanodes*
Evening bat	*Nycteceius humeralis*
Rodrigues fruit bat	*Pteropus rodricensis*
Mexican free-tailed bat	*Tadarida brasiliensis*
Primates	
Mexican spider-monkey	*Ateles geoffroyi*
Vervet monkey	*Cercopithicus aethiops*
Patas monkey	*Erythrocebus patas*
Ring-tailed lemur	*Lemur catta*
Rhesus macaque	*Macaca mulatta*
Barbary macaque	*Macaca sylvanus*
Chimpanzee	*Pan troglodytes*
Hanuman langur	*Presbytis entellus*
Nilgiri langur	*Presbytis johnii*
Capped langur	*Presbytis pileata*
Carnivores	
Coyote	*Canis latrans*
Timber wolf	*Canis lupus*
Spotted hyena	*Crocuta crocuta*
Domestic cat	*Felis catus*
Dwarf mongoose	*Helogale parvula*
Brown hyena	*Hyaena brunnea*
Banded mongoose	*Mungos mungo*
Coati	*Nasua narica*
Raccoon dog	*Nyctereutes procyonoides*
African lion	*Panthera leo*
Red fox	*Vulpes vulpes*

young in close proximity. These groups sometimes form because of the positive, mutualistic advantages to all young that result, and such grouping may be considered as a form of communal care. Species in which nursery groups apparently form because of advantages to the young include some antelope, a variety of rodents, most bats, some pinnipeds, and some carnivores. As an example, although colonial living in sea lions is associated with increased disease transmission and higher female–female aggression (Vedros et al. 1971; Francis 1987), females most often group during the breeding season. Nursery

Table 12.1. (*cont.*)

Common name	Scientific name
Pinnipeds	
Grey seals	*Halichoeris grypus*
Hawaiian monk seal	*Monachus schauinslandi*
Northern elephant seal	*Mirounga angustirostris*
Southern sea lion	*Otaria flavescens*
Rodents	
Spiny mouse	*Acomys cahirinus*
Black-tailed prairie dog	*Cynomys ludovicianus*
Mara	*Dolichotis patagonum*
Southern flying squirrel	*Glaucomys volans*
Fat dormouse	*Glis glis*
Capybara	*Hydrochoerus hydrochaeris*
Plains viscacha	*Lagostomus maximus*
California vole	*Microtus californicus*
Meadow vole	*Microtus pennsylvanicus*
House mouse	*Mus musculus*
Arctic ground squirrel	*Spermophilus parryi*
Artiodactyls	
Elk	*Cervus elaphus*
Wildebeest	*Connochaetes taurinus*
Giraffe	*Giraffa camelopardalis*
Sable antelope	*Hippotragus niger*
Oryx	*Oryx gazella*
Musk-ox	*Ovibus moschatus*
Warthog	*Phacochoerus aethiopicus*
African buffalo	*Syncerus caffer*
Collared peccary	*Tayassu tajacu*
Cetaceans	
Sperm whale	*Physeter macrocephalus*
Bottle-nosed dolphin	*Tursiops truncatus*
Proboscideans	
African elephant	*Loxodonta africana*

crèches formed by southern sea lions have a significant impact on decreasing pup mortality (Campagna et al. 1992). Pups of females living in breeding aggregations with nursery crèches are better protected than are those of lone females from infanticidal young males (Campagna, LeBoeuf, & Cappozzo 1988; Campagna et al. 1992). These crèches also facilitate reunions of females and their pups, decreasing pup death from starvation following separation.

In other species, however, females may be forced to raise their young in groups when essential resources such as protected breeding areas are spatially or temporally limited, even though their young do not benefit from communal rearing. Northern elephant seals breed in groups of up to several hundred females and a few breeding males. Two factors have been suggested for the occurrence of plural breeding in this species: physical constraints due to the limited number of beaches protected from terrestrial predators, and the advantages females receive from mating with the most dominant male in the group (LeBoeuf & Briggs 1977). Elephant seal pup mortality is directly proportional to the density of the female group (Reiter, Stinson & LeBoeuf 1978). Most infant mortality results from separation of the mother and infant, resulting in starvation, injury by being bitten by alien females when attempting to nurse, or trampling by males (LeBoeuf & Briggs 1977). European rabbits also frequently breed in groups despite negative effects on offspring born into the group (Cowan 1987). Nest sites of this species are very localized, resulting in frequent formation of plural breeding groups. This grouping increases disease transmission, predation, and intraspecific aggression, especially that directed toward nondescendant offspring. The reproductive success of individuals in plural breeding groups is lower than that of singular breeders.

   There are also examples of females that live in groups throughout the year but that separate themselves from the social group for a period when vulnerable, dependent offspring are present (e.g., eastern gray kangaroos: Jarman & Southwell 1986; African lions: Pusey & Packer 1994a; brown hyenas: Owens & Owens 1984; giraffes: Pratt & Anderson 1982; warthog, oryx, and sable antelope: Estes 1991; southern flying squirrels: Layne & Raymond 1994; plains viscacha: Branch 1993). This suggests that females benefit from living in groups but that these benefits are surpassed by costs associated with having dependent offspring within those groups. For example, black-tailed prairie dogs benefit from sociality by decreasing their risk of predation (Hoogland 1981). However, intragroup aggression increases during the period when females are pregnant and lactating, for reproductive females establish exclusive territories within the coterie (Hoogland 1981). A variety of behaviors of coterie members, such as stealing nest material, trampling neonates, interfering with suckling, and killing offspring, may provide a selective pressure for females to withdraw from the social group when dependent offspring are present (Michener & Murie 1983). Infanticide, primarily by other resident lactating females, may be especially important, for it is the major cause of juvenile mortality in prairie dogs (Hoogland 1985). Other reasons to separate young from the group include the avoidance of competition with older juveniles (e.g., lions: Pusey & Packer 1994a), and avoidance of predators that may be more likely to find and catch young helpless animals in a group (e.g., several species of ungulates: Jarman 1974).

12.4 Types of Alloparental Care

In this section, we present examples of the types of alloparental care that have been observed in plural breeding species. In each case, we discuss possible evolutionary explanations for the behavior. In most cases, insufficient data are available to fully understand the distribution of such care across species or the evolutionary explanations for it, but in the case of nonoffspring nursing, more progress has been made in these directions. The benefit from any given type of alloparental care may be a primary factor leading to raising young in a social group, or it may be a secondary benefit of grouping for other reasons. When possible, these alternatives are distinguished.

12.4.1 Formation of Nursery Groups for the Benefit of Dilution

Although protection from predation by the dilution effect is often an important benefit for group living by adults, it may be particularly important for younger, more vulnerable animals. The advantages of dilution appear to be a primary reason that some animals pool their young in nursery groups, thus gaining mutual benefits. For example, common wildebeests live in large, loosely structured aggregations throughout the year. Calving in wildebeests is highly synchronous, and during this time pregnant and lactating females form more cohesive groups than are found at other times of year. When calves are young and feeble, survival rates are highest in larger nursery groups where younger calves benefit by being able to hide among older calves (Estes 1976; Estes & Estes 1979). The structured nursery groups also have advantages over the larger, loosely structured aggregations because calves are less likely to become separated from their mothers (Estes & Estes 1979). Dilution has also been suggested to be one of the reasons for communal rearing in black-tailed prairie dogs (Hoogland, Tamarin, & Levy 1989) and is probably important in many other species.

12.4.2 Formation of Nursery Groups for the Benefit of Thermoregulation

Nursery crèches may also enhance social thermoregulation. Again, these advantages should be of mutual benefit to all participating breeding females. In bats, raising young in large groups provides thermoregulatory benefits that can be critical to young with poorly developed thermoregulatory capabilities. In pallid bats, roosting in groups can significantly decrease energy expenditure (Trune & Slobodchikoff 1976). McCracken (1984) suggested that the immense groups formed in Mexican free-tailed bat maternity colonies elevate and stabilize the body temperature of roosting bats. The selective pressure of thermoregulation apparently balances the pressure to roost in smaller groups in which costs such as nonoffspring nursing and parasite and disease transfer might be lower. Thermoregulation may also explain why many species of rodents nest communally during both breeding and nonbreeding seasons (e.g., meadow voles: McShea & Madison 1984; California voles: Ostfeld 1986; fat dormice: Pilastro 1992; house mice: Wilkinson & Baker 1988; maras: Taber & Macdonald 1992).

12.4.3 Communal Defense of Young

In many plural breeders, groups of females actively defend their own and other young against infanticidal conspecifics and predators by threatening, chasing, or attacking them. In some species, nursery groups or crèches appear to be formed primarily for reasons of defense. Lions live in permanent social groups known as prides, which are fission–fusion units. Females give birth alone and keep their young hidden until they are about 6 weeks old. Then they mix them with other cubs in the pride and associate much more constantly with the other mothers than they do when they do not have cubs (Packer, Scheel, & Pusey 1990; Pusey & Packer 1994b). Females gain no benefits from crèching in terms of feeding efficiency, but females fiercely defend their young against potentially infanticidal extragroup males and are significantly more successful in defending cubs when the females are in groups of two or more than when alone (Packer et al. 1990; Pusey & Packer 1994a). They thus gain a strong mutual benefit from pooling their young in crèches. Similarly, the pups of house mice that share communal nests are less likely to die by infanticide than those raised by single females (Manning et al. 1995). Communal defense against infanticide has been recorded in several other plural breeding species (e.g., ring-tailed lemurs: Periera & Izard 1989; coatis: Russell 1981; collared peccaries: Packard et al. 1990; arctic ground squirrels: Michener 1983). In many of these species, as in lions, communal defense is by females that defend the young of other individuals in the process of defending their own. There are some cases, however, in which females seem to take additional risks on behalf of nonoffspring. For example, old female hanuman langurs without infants were seen to try to defend the infants of other females against infanticidal males (Hrdy 1977). Hrdy interpreted this behavior in terms of kin selection. More data are needed to understand the evolutionary basis of similar instances in other species.

Communal defense of dependent young from predators has been observed in a variety of taxa that live in permanent, cohesive groups, but it is unlikely that this is a primary reason for group living. For example, although group living in capybaras appears primarily to be a function of resource availability, adults also cluster together with juveniles in the center of the group when threatened by a predator. Females have also been observed to carry the infants of other group members on their back while swimming from danger (Macdonald 1981; Herrera & Macdonald 1987). More aggressive defense has been observed in several species (e.g., ring-tailed lemurs: Periera & Izard 1989; African elephants: Dublin 1983; Lee 1987; Kingdon 1979; African buffalo: Estes 1991; Sinclair 1977; collared peccaries: Byers & Bekoff 1981; banded mongoose: Rood 1975; musk-oxen: Tener 1954). Some species, such as spotted hyenas, are thought to keep their young in communal dens because of the advantages of communal defense against both potentially infanticidal conspecifics and predators (East, Hofer, & Turk 1989).

12.4.4 "Babysitting"

Whereas communal defense involves simultaneous behavior by several group members that are all present with the young, in other cases one or a few females remain in the proximity of the young of other females while the mother is at a distance or absent. This has been termed "babysitting." The extent to which the females actively care for the other young is poorly known in most instances. In pallid bats and fringed myotis, one or more adults have been observed to stay with the nursery crèche and possibly to retrieve young that fall from the roost (Beck & Rudd 1960; O'Farrell & Studier 1973). In domestic cats living in plural breeding groups, one adult is sometimes left with the kittens. It has been suggested that she may provide protection against infanticidal conspecifics when females are switching to a new nest (Feldman 1993). Similar protective behavior has also been reported in coatis (Kaufmann 1962), banded mongooses (Neal 1970), elk (Altmann 1956, 1963), and maras (Taber & Macdonald 1992).

In sperm whales, which forage at depths greater than calves can dive, calves are serially accompanied by different members of their social group at the surface while the majority of the group is foraging. Groups containing calves stagger their dives more than those that do not, and this nonsynchronous diving has been interpreted as babysitting (Whitehead, in press). Unlike most other species, in which there are probably other advantages to group living besides babysitting, Whitehead (in press) suggests that babysitting to protect young from predators has been a primary evolutionary force for sociality in female sperm whales.

In primates, babysitting can take the form of one or more alloparental adults carrying the offspring of a female in the group. The extent to which this

344   S. E. Lewis and A. E. Pusey

represents alloparental care is variable (Hrdy 1976; Nicolson 1987). In some species (e.g., chimpanzees: Nishida 1983; vervet monkeys: Lancaster 1971; Fairbanks 1990; rhesus macaques: Berman 1993a, b) carrying behavior is primarily restricted to nulliparous females and has been suggested to be a means of learning maternal behavior (Hrdy 1976; Nicolson 1987; Fairbanks 1990; Tardif this volume). There is evidence that carrying by inexperienced females or other members of the social group may present a significant danger to the offspring (Hrdy 1976; Silk 1980), and females may be reluctant to allow others to carry their young. However, in captive vervet monkeys, females whose infants were carried more by nulliparous females gained the advantage of shorter interbirth intervals (Fairbanks 1990). In other primate species (e.g., capped langurs: Stanford 1992; niligiri langurs: Poirier 1968; Barbary macaques: Small 1990; patas monkeys: Zucker & Kaplan 1981; ring-tailed lemurs: Gould 1992) carrying behavior is most common among parous females, and the risk of injury to offspring is low. Small (1990) and Stanford (1992) suggest that carrying by parous females is a low-cost communal behavior that is advantageous for the mother because it allows her more time for foraging and may benefit the alloparent by enhancing her social relationships within the group. Zucker and Kaplan (1981) also suggest that young patas monkeys benefit from carrying because unrelated females will be willing to carry them when fleeing from danger.

12.4.5 Adoption

Adoption of dependent offspring by other members of the social group has been observed in a variety of taxa (reviewed by Spencer-Booth 1970; Riedman 1982), including many primates (Thierry & Anderson 1986; Agoramoorthy & Rudran 1992), northern elephant seals (Riedman & LeBoeuf 1982), brown hyenas (Owens & Owens 1984), lions (Schaller 1972), and red fox (von Schantz 1984). Adoption is generally a rare occurrence and is unlikely to be a primary reason for communal rearing. However, some authors have suggested that behaviors such as carrying in primates and allo-mothering in elephants promote adoption if the mother dies (Douglas-Hamilton & Douglas-Hamilton 1975; Hrdy 1976; Vogel 1984). In some cases, the individuals adopting are relatives of the infant, but this is probably not the case in elephant seals (Riedman & LeBoeuf 1982) or several cases of adoption by adult female primates (Thierry & Anderson 1986; Goodall 1990). Some cases of adoption by female primates have been attributed to "reproductive error" arising out of attraction to young infants (Thierry & Anderson 1986). More study of the evolutionary basis of this behavior is required.

12.4.6 Group Provisioning

Group provisioning of young is frequently reported in the social carnivores (Macdonald & Moehlman 1982). However, it most often takes the form of nonbreeding adults or subadults provisioning the offspring of a singular breeding female, and it is rare in plural breeders. One of the few cases observed involves related female brown hyenas provisioning cubs of other females within the social group, regardless of whether they also have dependent cubs (Owens & Owens 1984).

12.4.7 Helper-Assisted Birth

There are a few reports in the literature of individuals within a social group providing assistance to a female during and following parturition. For example, Kunz, Allgaier, Seyjagat, and Caligiuri (1994) recently observed a captive female Rodrigues fruit bat attend to a second female before, during, and after parturition. The attendant female: "(1) intermittently groomed [the mother's] anovaginal region; (2) grasped her with partially outstretched wings; (3) 'tutored' her in a feet-down birthing posture; (4) groomed the emerging pup; and (5) physically assisted the mother by maneuvering the pup into a suckling position" (Kunz et al. 1994, p. 691). Helper-assisted birth has been reported in a few additional species (reviewed in Kunz et al. 1994). In captive dolphins, females were observed to attend and model exaggerated straining movements in front of females experiencing difficult labor (Essapian 1963). In captive spiny mice, when labor is protracted, perhaps because of the large size of the young, females gave "obstetrical help" (Dieterlin 1962). And in captive raccoon dogs, a male groomed his mate and nosed and licked the anogenital region while she was in labor (Yamamoto 1987). More information on the frequency of such behavior in the wild and on the relationships between the females is necessary in order to understand the evolutionary, or nonadaptive alternative, explanations of such behavior.

12.4.8 Nonoffspring Nursing

Nonoffspring nursing, when females nurse young that are not their own, is perhaps the most extreme example of communal care among plural breeding mammals. Lactation is costly (Hanwell & Peaker 1977), so the diversion of milk to the young of others is an example of a behavior with clear costs to the female's immediate direct fitness. In the literature on mammalian social behavior from the 1960s to the early 1980s, it is not uncommon to read reports of indiscriminate nursing among plural breeding mammals (e.g.,

Davis, Herreid, & Short 1962; Rood 1974; Macdonald 1981). Although it may not occur as widely or as indiscriminately as early accounts suggested, nonoffspring nursing has been reported in a variety of mammals, including bats (e.g., Watkins & Shump 1981; McCracken 1984; Wilkinson 1992), primates (e.g., Periera, Klepper, & Simons 1987; O'Brien 1988), carnivores (e.g., Russell 1983; Macdonald et al. 1987; Pusey & Packer 1994a), rodents (e.g., King 1963; Rood 1972; König 1993; French 1994; Manning et al. 1995), pinnipeds (e.g., Fogden 1971; Riedman & LeBoeuf 1982) and ungulates (e.g., Bradley 1968; Byers & Bekoff 1981; Jensen 1988). In addition to its occurrence in plural breeders, nonoffspring nursing has been reported among typically singular breeders if more than one female in the social group has bred (e.g., coyotes: Camenzind 1978; timber wolves: Paquet, Bragdon, & McCusker 1982; red fox: Macdonald 1979). There is also evidence that nonoffspring nursing may occur as the result of induced lactation in some mammalian species (e.g., ring-tailed lemurs: Periera & Izard 1989; Mexican spider monkeys: Estrada & Paterson 1979; rhesus macaques: Holman & Goy 1980; timber wolves: Fentress & Ryon 1982; banded mongooses: Rood 1980; dwarf mongooses: Creel et al. 1991).

There are several possible evolutionary explanations for nonoffspring nursing. It is often assumed to have mutualistic advantages because of increased levels of nutrition for all young. However, evidence in support of this has been found only in captive rodents. Some of the most detailed studies of nonoffspring nursing have been conducted on house mice. Early laboratory studies indicated that offspring raised in communal litters were nursed indiscriminately by both females, had higher weaning weights than did similar offspring raised with a single female, and had the same pup-to-dam ratio (Sayler & Salmon 1969, 1971; Werboff, Steg, & Barnes 1970). Sayler and Salmon (1971) demonstrated that this was due to increased milk production by females. Furthermore, König (1993) found that females paired with their sisters weaned more and heavier offspring than did females paired with an unfamiliar female or females rearing their pups alone in laboratory experiments. These results were attributed to the combined effects of communal nursing and cooperative brood care. Interestingly, König (1994) found that when a female joined a second female that had pups, the new female frequently killed a few of the existing pups, perhaps to ensure her own pups access to a more plentiful milk supply.

In these studies, food was available on an *ad libitum* basis, and increased milk production presumably involved an increase in food intake. However, under field conditions, milk is more likely to be a limited resource such that by nursing additional young a female either reduces the milk available to her own offspring or must increase her foraging time to compensate for the losses. The redirection of milk to nonoffspring under these conditions may still benefit the female in various ways. First, if the recipients of her milk are close kin, the female may improve her indirect fitness. Communally nursing females are often closely related. For example, in a field study of house mice, Wilkinson and Baker (1988) found that communally nesting females were significantly more closely related to each other than expected by chance (see also lions, Section 12.6). Second, females may enhance their direct fitness by improving the survival of individuals that eventually help them or their offspring through better thermoregulation (Sayler & Salmon 1969), cooperation (Bertram 1976), or the dilution effect (Hoogland et al. 1989; Wilkinson 1992). Third, nonoffspring nursing may reduce the variance in milk intake if the behavior is reciprocated by other mothers (Caraco & Brown 1986). If individual mothers spend long periods away from their young or show temporal variation in their foraging success, reciprocal nonoffspring nursing could shorten the interval between meals and provide a more constant, if not more plentiful, milk supply. Finally, nonoffspring nursing may provide a mechanism for decreasing the energetic costs or risk of infection of carrying unconsumed milk, as has been suggested for evening bats (Wilkinson 1992).

It is also possible that nonoffspring nursing has no advantages for the mother and occurs because she is constrained to rear her young in close proximity to others for nonnutritional reasons. Several authors have described nonoffspring nursing as parasitism of unwilling females by nonoffspring (Mexican free-tailed bats: McCracken 1984; Northern elephant seals: Reiter et al. 1978), or as misdirected care (grey seals: Fogden 1971; Hawaiian monk seals: Boness 1990). This explanation is similar to Jamieson's (1989) explanation for provisioning by alloparents in communally breeding birds.

12.5 Comparative Study of Nonoffspring Nursing

To investigate the factors influencing the distribution of nonoffspring nursing across mammals, Packer, Lewis, and Pusey (1992) carried out a literature review and questionnaire survey on 100 species of group-living mammals in 14 orders. Such interspecific comparisons are often hampered by the possibility that related species show similar adaptations through shared phylogeny (Harvey & Mace 1982; Grafen 1989; Pagel & Harvey 1989). For example, nonoffspring nursing is ubiquitous in canids, suggesting that the trait was present in a common ancestor. Therefore, Packer, Lewis, and Pusey (1992) used phylogenetic regression and evolutionary covariance regression models (Grafen 1989; Pagel & Harvey 1989) to locate and compare independent evo-

348   S. E. Lewis and A. E. Pusey

Table 12.2. *Factors influencing the frequency of nonoffspring nursing across taxa[a]*

Independent variable	Increase	Decrease	P (sign test)
Captivity	16	5	0.026
Increase in median litter size	13	1	0.022
Increase in maximum litter size	14	2	0.004
Increase in degree of milk theft			
Monotocous species	11	1	0.006
Polytocous species	2	5	ns
			$P < 0.02$ (Fisher test)
Increase in number of lactating females per group			
Monotocous species	9	3	ns
Polytocous species	0	8	0.008
			$P < 0.01$ (Fisher test)

[a]Numbers indicate the number of evolutionary changes in the frequency of nonoffspring nursing that were associated with a concomitant change in the independent variable. An increase indicates that the frequency of nonoffspring nursing increased with an increase in the independent variable. A decrease indicates that nonoffspring nursing decreased with an increase in the independent variable. The comparisons also indicate that milk "theft" is more likely to occur in monotocous species than in polytocous species and that nonoffspring nursing is likely to increase with group size in monotocous species but decrease with group size in polytocous species (Fisher test comparisons; adapted from Packer et al. 1992).

lutionary events. These programs identify evolutionary changes in the dependent variable (nonoffspring nursing) at any taxonomic level and specify whether these changes are associated with a concomitant change in the independent variable of interest (e.g., litter size or group size).

Packer, Lewis, and Pusey (1992) investigated the effects of a variety of ecological, demographic, and behavioral factors on the frequency of nonoffspring nursing and made four significant findings, which are summarized in Table 12.2. First, nonoffspring nursing is more common in captivity. This may be because crowding and disturbance increase the chances of misdirected maternal care, but it may also be because captive animals have access to unlimited food, thus decreasing the costs of nonoffspring nursing. Subsequent analyses excluded data from captive studies. Second, nonoffspring nursing increases with litter size across taxa. Third, in monotocous species (average litter size = 1) nonoffspring nursing is more likely to be classified as milk "theft," as described earlier for Mexican free-tailed bats and Northern elephant seals, and it is more likely to occur in species that continue to produce milk after the loss of their own offspring. In contrast, in polytocous species (average litter size > 1), nonoffspring nursing is less likely to be classified as milk theft, occurring in situations in which females either cannot or do not discriminate between young in mixed litters (e.g., Manning et al. 1995) or in which females are able to discriminate their own from foreign young but allow foreign young to nurse (e.g., lions, see Section 12.6). Fourth, whereas nonoffspring nursing tends to occur more often in monotocous species when social group size is large, in polytocous species it is most common when group size is small and decreases significantly as group size increases.

Why should nonoffspring nursing be more common and better tolerated in polytocous species? One possible reason is that as litter size increases, the potential for confusion and misdirected care increases. However, nonoffspring nursing actually declines with group size in polytocous species, even though the potential for confusion is likely to be highest in large groups. Instead, Packer, Lewis, and Pusey (1992) suggest that the costs of nonoffspring nursing are lower in polytocous than monotocous species. Monotocous females are presumably adapted to produce enough milk for just one young. They would need either to increase their food intake considerably above the optimum to sustain the additional demands of a nonoffspring or to continue lactation after the loss of their own young and thus delay future reproduction (e.g., Altmann, Altmann, & Hausfater 1978; Louden, McNeily, & Milne 1983). But a polytocous female, adapted to providing milk for several offspring, should be able to nurse an additional young at little cost, especially if she has a smaller than average litter size or has lost young during a temporary food shortage.

The fact that nonoffspring nursing in polytocous species is most common in taxa with small group sizes is consistent with models of cooperation based on kin selection, mutualism, and reciprocity. In mammals, where females are usually philopatric, small groups are likely to have the highest average coefficients of relatedness (Bertram 1976; Murray 1985; Wilkinson 1987), and kin selection may be the most plausible explanation for truly communal nursing. However, mutualistic advantages from cooperation are also highest in small groups (Packer & Ruttan 1988; Lima 1989), and reciprocity is also most likely to evolve in small groups (Boyd & Richerson 1988). Packer, Lewis, and Pusey's (1992) survey did not provide sufficiently detailed data to allow assessment of the relative importance of each of these processes. These researchers were able, however, to examine the idea that reciprocal communal nursing might minimize the interval between meals (Caraco & Brown 1986). If this were important, we would expect nonoffspring nursing to be more common in taxa in which females spent long periods away from their young or when nursing by females in the group was asynchronous, but this was not the case (Packer et al. 1992).

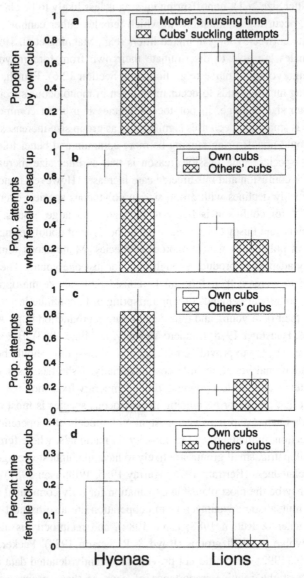

Figure 12.1. Comparisons of the behavior of female hyenas and lions and their cubs. Vertical lines indicate standard errors. (a) Proportion of mothers' total nursing (the sum of the time spent nursing each cub divided by the total time watched) that went to their offspring (n_1 = 9 hyenas, n_2 = 25 lions, Z = − 4.392, P < 0.0001), and proportion of total suckling attempts received by females that were from their offspring (Z = − 0.49, ns). (b) Proportion of suckling attempts received by females from offspring and other cubs in which the female's head was up during the attempt (species difference with offspring, n_1 = 9 hyenas, n_2 = 22 lions, U = 6, P = 0.0001; species differ-

12.6 Comparison of Nonoffspring Nursing in Lions and Hyenas

Packer, Lewis, and Pusey's (1992) comparative study of nonoffspring nursing indicated some general factors that affect the occurrence of nonoffspring nursing, but the study was limited by the quality and quantity of information available across species. A more detailed comparison of nonoffspring nursing in African lions and spotted hyenas (Pusey & Packer 1994a) supports the importance of litter size and kinship in the distribution of nonoffspring nursing across taxa and points to a further factor – the costs of vigilance against milk theft.

Both lions and hyenas live in permanent social groups and keep their young in nursery crèches during lactation, probably because of the advantages of communal defense against infanticidal conspecifics and possibly predators (see page 342 Section 12.4.3). They differ in that hyenas keep their cubs in a communal den with burrows that only young hyenas can enter, whereas lions do not. This appears to afford hyenas greater opportunity to rest away from their cubs. Lions remain with their cubs throughout the period of lactation, leaving them only briefly to hunt, whereas hyenas spend little time at the communal den, except when they come to nurse their cubs. Lions nurse nonoffspring about 30 percent of the time, whereas hyenas very rarely nurse cubs other than their own (Figure 12.1).

Pusey and Packer (1994a) suggest that two key factors contribute to the difference in the frequency of nonoffspring nursing in these species: the costs of discriminating against nonoffspring and the relative costs of milk loss. Lions and hyenas receive equal numbers of suckling attempts by nonoffspring, but hyenas are more vigilant and resist these attempts, whereas lions are more likely to accept nonoffspring (see Figure 12.1). Because hyenas may visit the

---

ences with other cubs, $n_1 = 8$ hyenas, $n_2 = 21$ lions, U = 8.4, $P = 0.01$; difference between offspring and other cubs: hyenas, T = 4, $n = 7$ females, ns; lions, T = 39, $n = 21$ females, $P = 0.008$, Wilcoxon matched pairs test). (c) Proportion of suckling attempts from offspring and other cubs resisted by females (species difference in resistance of offspring, $n_1 = 9$ hyenas, $n_2 = 24$ lions, U = 25, $P = 0.0009$, other cubs, $n_1 = 8$ hyenas, $n_2 = 23$ lions, U = 4.5, $P = 0.0001$; difference between offspring and other cubs: hyenas, T = 0, $n = 8$ females, $P = 0.01$; lions, T = 62, $n = 22$ females, $P = 0.04$). (d) Percentage of time females licked their offspring and the cubs of other females (species difference in licking offspring, $n_1 = 9$ hyenas, $n_2 = 21$ lions, U = 87, ns, others' cubs, U = 18, $P = 0.0006$; difference between licking offspring and other cubs: hyenas, T = 0, $n = 7$ females, $P = 0.02$; lions, T = 10, $n = 20$ females, $P = 0.0004$). Reprinted from Pusey and Packer, 1994a, by permission of Oxford University Press.

den primarily to nurse their young, the energetic costs of remaining vigilant and resisting nursing attempts may be more manageable than the costs incurred by lions that are approached frequently throughout the day by their own and foreign cubs. The differences in the relative costs of milk loss between the two species may stem from three causes. First, hyenas have a smaller litter size, producing a maximum of two cubs and sometimes lacking the ability to feed more than one, whereas lions have a mean litter size of two (range two to four). Second, hyenas nurse for a longer period and are dependent entirely on milk for a much longer period than are lions. Finally, because hyenas live in larger social groups, average kinship between group members is lower than that of lions, thus increasing the costs of losing milk to nonoffspring.

12.7 Factors Influencing Variation in Nonoffspring Nursing in Lions

Pusey and Packer (1994a) carried out a detailed study of 13 nursery crèches of lion cubs and four single mothers and their cubs to investigate the causes of nonoffspring nursing in lions. They found that females discriminated their own cubs from nonoffspring, nursing and licking each of their own cubs more than each nonoffspring and rejecting them less (Figure 12.1). Cubs also behaved differently toward their own mother and other females. They made more suckling attempts on their own mother and were more surreptitious in approaching other females to suckle, waiting to approach them until other cubs were suckling and being less likely to announce themselves to the female (Figure 12.2).

Several factors influenced the frequency of nonoffspring nursing among females. First, female lions nursed the same amount regardless of litter size but gave a higher proportion of their total nursing to nonoffspring when their own litter size was small (Figure 12.3). This result is consistent with Packer, Lewis, and Pusey's (1992) explanation for the finding that nonoffspring nursing increases with litter size across taxa: Females that are adapted to nursing a number of cubs can nurse others at little extra cost, especially if their own litter is smaller than average. Second, females in groups where average coefficients of kinship were highest nursed more indiscriminately than those with low kinship (Figure 12.4). In addition, within groups where coefficients of kinship varied, females nursed the cubs of their closest relatives most. Third, females gave proportionately more of their nursing to nonoffspring as the age of their own young increased (Pusey & Packer 1994a).

Pusey and Packer (1994a) found little evidence for general nutritional benefits of nonoffspring nursing in lions. First, cubs raised with single females

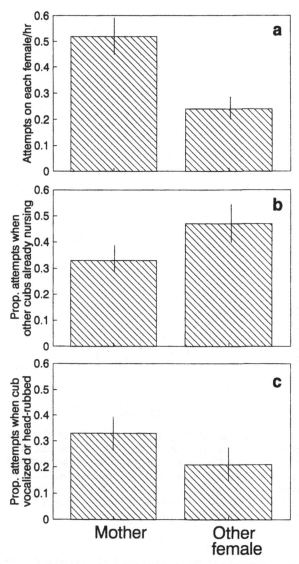

Figure 12.2. (a) Number of independent nursing attempts per hour by lion cubs on their mother and on other females (Z = − 5.8, n = 53 cubs of known maternity, P < 0.00001). (b) Proportion of independent suckling attempts by lion cubs on their mother or on other females in which other cubs were already suckling from the female (Z = − 3.1, n = 45 cubs, P < 0.01). (c) Proportion of independent suckling attempts by cubs on their mother or on other females in which the cub vocalized or head-rubbed the female immediately before the attempt (Z = − 3.4, n = 45 cubs, P < 0.001). Reprinted from Pusey and Packer, 1994a, by permission of Oxford University Press.

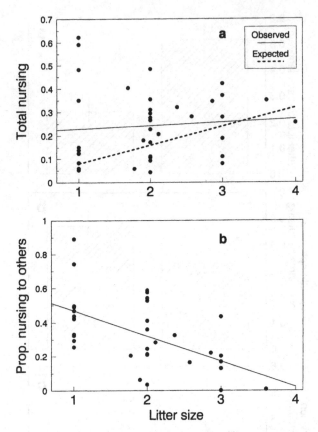

Figure 12.3. (a) Rate of total nursing by females with different litter sizes. The solid line shows the regression line of total nursing against own litter size ($R^2 = 0.03$, $n =$ 35 females, ns). The dashed line shows the expected rate of nursing if females nursed only their own offspring and nursed each at the average rate observed. (b) The proportion of a female's total nursing that goes to other cubs plotted against her own litter size ($R^2 = 0.34$, $P = 0.0002$, $n = 35$ females). Reprinted from Pusey and Packer, 1994a, by permission of Oxford University Press.

appeared to gain access to as much milk as did cubs raised in crèches. Indeed, there is evidence from other sources that females raising their cubs in crèches actually obtain lower levels of food intake and thus are likely to have less milk than single females (Packer 1986; Packer et al. 1990). Second, there was no evidence that nonoffspring nursing shortened the interval between meals or reduced the variance in meal size. Females left and returned to their cubs synchronously and usually experienced similar levels of food intake, so there was little opportunity for cubs to nurse from other females when their mother was absent or had little milk.

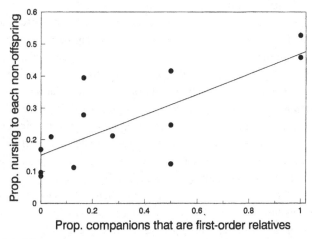

Prop. companions that are first-order relatives

Figure 12.4. The mean per capita proportion of nursing by all females in each crèche that went to other cubs plotted against the probability that all females in the crèche are first-order relatives ($R^2 = 0.58$, $P = 0.003$, n = 13 crèches). Per capita proportion of nursing is the mean time that a female nursed each other cub divided by this number plus the mean time she nursed each of her own cubs. This would be 0 if females nursed their own cubs exclusively, 1 if they exclusively nursed other cubs, and 0.5 if nursing was apportioned equally across all cubs. Note that this correlation remains significant when a variety of potentially confounding variables are taken into account (e.g., litter size, group size, age). Reprinted from Pusey and Packer 1994a, by permission of Oxford University Press.

Pusey and Packer (1994a) concluded that lions raise their young in crèches because of the advantages of defense against infanticidal males rather than because of nutritional benefits from nonoffspring nursing. Because females rest with their cubs for most of the day, they are subject to frequent nursing attempts from nonoffspring, and some nonoffspring nursing occurs as an inevitable consequence of group rearing. However, nonoffspring nursing occurs least when the costs are the highest. It is least common when litter size is large and females are least likely to have milk to spare, when kinship is low, and when the females' cubs are younger and more dependent on milk.

12.8 Conclusions

There is a great diversity in the incidence of communal care among plural breeding species. This diversity reflects in part the diversity of reasons that females raise young in groups. In some species, females are constrained to be in groups for reasons unrelated to the welfare of their young, such as a shortage of suitable breeding sites or to gain protection against sexual harassment by males. In such cases, young raised in large groups may suffer increased mortality, and what alloparental care takes place may best be described as parasitism, as in the case of nonoffspring nursing in Northern elephant seals. In other species, females form nursery groups because of positive benefits for the young. Such benefits include dilution, thermoregulation, and group defense against infanticidal conspecifics or predators. In these cases, the females often gain mutual benefits from grouping to care for their young and thus increase their direct fitness. In other cases, females may perform apparently costly activities such as rescuing the young of other females, or babysitting. Here, explanations in terms of kinship and indirect fitness or reciprocity may be involved, although the possibility remains that these also occur because of misdirected care (Jamieson 1989; Tardif this volume).

The case of nonoffspring nursing is instructive because the factors influencing its occurrence have been better studied. This is clearly an example of costly alloparental care. The questions are to what extent it occurs as a by-product of females grouping for other reasons and to what extent it occurs because of positive benefits to females through the enhanced nutrition of young. The fact that nonoffspring nursing is usually described as milk theft in monotocous species and the fact that it tends to increase in frequency as group size increases suggest that it represents parasitism or misdirected parental care in these species and occurs as a by-product of grouping for other reasons. In polytocous species, nonoffspring nursing often occurs with little opposition from the female and occurs most in small groups, suggesting that females have control over the amount that occurs. The possibility exists that in at least some of these cases, nonoffspring nursing brings some of the benefits discussed earlier – and may even be a cause of grouping. However, in the detailed studies of lions (Pusey & Packer 1994a) and free-ranging house mice (Manning et al. 1995), the evidence suggests that females group and rear their young communally because of the benefits of communal defense against infanticide rather than because of nutritional advantages of communal nursing. The costs of the inevitable nonoffspring nursing that occurs are minimized by restricting it to situations in which the costs are least, either by choice of close kin as nest mates in the case of mice, or by regulating it depending on kinship, litter size, and age of cubs in the case of lions.

12.9 Directions for Future Research

As has been advocated in other chapters in this book, future studies of communal care in plural breeders require a careful evaluation of the ecological reasons for grouping in each species and a consideration of the evolutionary explanations for the alloparental care shown by each group member. To what extent do the relative costs and benefits of alloparental care vary because of factors such as the age of young or the size of the social group? Does grouping occur because of the advantages of alloparental care, or vice versa? If the former is true, which alloparental behaviors can best be considered the primary factors leading to grouping? It is also important to note that although grouping may initially evolve for other reasons, alloparental care may then evolve and add to the benefits of grouping (Emlen 1994). There is a need for comparative studies like the one on nonoffspring nursing to enable us to understand the distribution of other forms of alloparental care in plural breeders, such as the occurrence of communal defense of the young. Also, there is a need for detailed studies of additional species that show extensive nonoffspring nursing to determine whether nonoffspring nursing ever brings real nutritional benefits under natural conditions.

Acknowledgments

Jeffrey French and Craig Packer provided helpful critiques of earlier versions of this chapter. Nicole Ribbens and Matt Pawlowski assisted with the preparations of Table 12.1. Anne Pusey was supported by NSF grant BSR 8507087 during preparation of this manuscript.

References

Agoramoorthy, G., & Rudran, R. (1992). Adoption in free-ranging red howler monkeys, *Alouatta seniculus* of Venezuela. *Primates* 33:551–555.
Alexander, R. D. (1974). The evolution of social behavior. *Ann. Rev. Ecol. Syst.* 5:325–388.
Altmann, J., Altmann, S. A., & Hausfater, G. (1978). Primate infant's effects on mother's future reproduction. *Science* 201:1028–1030.
Altmann, M. (1956). Patterns of herd behavior in free-ranging elk of Wyoming, *Cervus canadensis nelsoni. Zoologica (New York).* 41:65–71.
Altmann, M. (1963). Naturalistic studies of maternal care in moose and elk. In *Maternal behavior in mammals,* ed. H. L. Rheingold, pp. 210–225. New York: Wiley.
Beck, A. J., & Rudd, R. L. (1960). Nursery colonies in the pallid bat. *J. Mammal.* 41:266–267.
Berman, C. M. (1983a). Differentiation of relationships among rhesus monkey infants. In *Primate social relationships,* ed. R. A. Hinde, pp. 89–93. Oxford, U.K.: Blackwell.
Berman, C. M. (1983b). Early differences in relationships between infants and other group members based on mother's status: their possible relationship to peer – peer rank acquisition. In *Primate social relationships,* ed. R. A. Hinde, pp. 154–156. Oxford, U.K.: Blackwell.

358   S. E. Lewis and A. E. Pusey

Bertram, B. C. R. (1976). Kin selection in lions and evolution. In *Growing points in ethology,* ed. P. P. G. Bateson & R. A. Hinde, pp. 281–301. Cambridge: Cambridge University Press.

Boness, D. J. (1990). Fostering behavior in Hawaiian monk seals: Is there a reproductive cost? *Behav. Ecol. Sociobiol.* 27:113–122.

Boyd, R., & Richerson, P. (1988). The evolution of reciprocity in sizeable groups. *J. Theoret. Biol.* 132:337–356.

Bradley, R. M. (1968). *Some aspects of the ecology of the warthog* (Phacochoerus aethiopicus *Pallas) in Nairobi National Park.* MS thesis, University of East Africa.

Branch, L. C. (1993). Social organization and mating system of the plains viscacha (*Lagostomus maximus*). *J. Zool. (Lond.)* 229:473–491.

Brown, J. L. (1987). *Helping and communal breeding in birds: Ecology and evolution.* Princeton: Princeton University Press.

Byers, J. A., & Bekoff, M. (1981). Social, spacing, and cooperative behavior of the collared peccary, *Tayassu tajacu. J. Mammal.* 62:767–785.

Camenzind, F. J. (1978). Behavioral ecology of coyotes on the National Elk Refuge, Jackson, Wyoming. In *Coyotes: Biology, behavior, management,* ed. M. Bekoff, pp. 267–294. New York: Academic Press.

Campagna, C., Bisioli, C., Quintana, F., Perez, F., & Vila, A. (1992). Group breeding in sea lions: Pups survive better in colonies. *Anim. Behav.* 43:541–548.

Campagna, C., LeBoeuf, B. J., & Cappozzo, H. L. (1988). Pup abduction and infanticide in southern sea lions. *Behaviour* 107:44–60.

Caraco, T., & Brown, J. L. (1986). A game between communal breeders: When is food sharing stable? *J. Theoret. Biol.* 118:379–393.

Clode, D. (1993). Colonially breeding seabirds: Predators or prey? *Trends Ecol. Evol.* 8:336–338.

Clutton-Brock, T. H. (1991). *The evolution of parental care.* Princeton: Princeton University Press.

Cowan, D. P. (1987). Group living in the European rabbit (*Oryctolagus cuniculus*): Mutual benefit or resource localization? *J. Anim. Ecol.* 56:779–795.

Creel, S. R., & Creel, N. M. (1991). Energetics, reproductive suppression and obligate communal breeding in carnivores. *Behav. Ecol. Sociobiol.* 28:263–270.

Creel, S. R., Monfort, S. L., Wildt, D. E., & Waser, P. M. (1991). Spontaneous lactation is an adaptive result of pseudopregnancy. *Nature* 351:660–662.

Davis, R. B., Herreid, C. F., & Short, H. L. (1962). Mexican free-tailed bats in Texas. *Ecol. Monographs* 32:11–46.

Dieterlen, F. (1962). Geburt und Geburtshilfe bei der Stachelmaus, *Acomys cahirinus. Z. Tierpsychol.* 19:191–222.

Douglas-Hamilton, I., & Douglas-Hamilton, O. (1975). *Among the elephants.* Glasgow: Collins.

Dublin, H. T. (1983). Cooperation and competition among female elephants. In *Social behavior of female vertebrates,* ed. S. K. Wasser, pp. 2291–2313. New York: Academic Press.

East, M., Hofer, H., & Turk, A. (1989). Functions of birth dens in spotted hyaenas (*Crocuta crocuta*). *J. Zool. (Lond.)* 219:690–697.

Emlen, S. T. (1991). Cooperative breeding in birds and mammals. In *Behavioural ecology,* ed. J. R. Krebs & N. B. Davies. Oxford, U.K.: Blackwell.

Emlen, S. T. (1994). Benefits, constraints and the evolution of the family. *Trends Ecol. Evol.* 9:282–285.

Essapian, F. S. (1963). Observations on abnormalities of parturition in captive bottlenosed dolphins, *Tursiops truncatus,* and concurrent behavior of other porpoises. *J. Mammal.* 44:405–414.

Estes, R. D. (1976). The significance of breeding synchrony in the wildebeest. *East African Wildlife Journal* 14:135–142.

Estes, R. D. (1991). *The behavior guide to African mammals.* Berkeley: University of California Press.

Estes, R. D., & Estes, R. K. (1979). The birth and survival of wildebeest calves. *Z. Tierpsychol.* 50:45–95.

Estrada, A., & Paterson, J. D. (1979). A case of adoption in a captive group of Mexican spider monkeys (*Ateles geoffroyi*). *Primates* 21:128–129.

Fairbanks, L. A. (1990). Reciprocal benefits of allomothering for female vervet monkeys. *Anim. Behav.* 40:553–562.

Feldman, H. N. (1993). Maternal care and differences in the use of nests in the domestic cat. *Anim. Behav.* 45:13–23.

Fentress, J. C., & Ryon, C. J. (1982). A long-term study of distributed feeding within a captive wolf pack. In *Wolves of the world: Perspectives of behavior, ecology, and conservation,* ed. F. H. Harrington & P. C. Paquet, pp. 209–222. Park Ridge, N.J.: Noyes Publications.

Fogden, S. C. L. (1971). Mother – young behaviour at grey seal breeding beaches. *J. Zool. (Lond.)* 164:61–92.

Francis, J. M. (1987). *Interfemale aggression and spacing in the northern fur seal* Callorhinus ursinus *and the California sea lion* Zalophus californianus. PhD dissertation, University of California, Santa Cruz.

French, J. A. (1994). Alloparents in the Mongolian gerbil: Impact on long-term reproductive performance of breeders and opportunities for independent reproduction. *Behav. Ecol.* 5:273–279.

Gittleman, J. L. (1985). Functions of communal care in mammals. In *Evolution: Essays in honour of John Maynard Smith,* ed. P. Greenwood, P. H. Harvey, & M. Slatkin, pp. 187–205. Cambridge: Cambridge University Press.

Goodall, J. (1990). *Through a window.* London: Weidenfeld & Nicolson.

Gould, L. (1992). Alloparental care in free-ranging *Lemur catta* at Berenty Reserve, Madagascar. *Folia Primatol.* 58:72–83.

Grafen, A. (1989). The phylogenetic regression. *Phil. Trans. Royal Soc. Lond.* 326:119–157.

Hanwell, A., & Peaker, M. (1977). Physiological effects of lactation on the mother. *Symp. Zool. Soc. Lond.* 41:297–312.

Harvey, P. H., & Mace, G. M. (1982). Comparisons between taxa and adaptive trends: Problems of methodology. In *Current problems in sociobiology,* ed. King's College Sociobiology Group, pp. 343–361. Cambridge: Cambridge University Press.

Herrera, E. A., & D. W. Macdonald. (1987). Group stability and the structure of a capybara population. *Symp. Zool. Soc. Lond.* 58:115–130.

Holman, G. D., & Goy, R. W. (1980). Behavioral and mammary responses of adult female rhesus to strange infants. *Horm. Behav.* 14:348–357.

Hoogland, J. L. (1981). The evolution of coloniality in white-tailed and black-tailed prairie dogs (Sciuridae: *Cynomys leucurus* and *C. ludovicianus*). *Ecology* 62:252–272.

Hoogland, J. L. (1985). Infanticide in prairie dogs: Lactating females kill offspring of their own kin. *Science* 230:1037–1040.

Hoogland, J. L., Tamarin, R. H., & Levy, C. K. (1989). Communal nursing in prairie dogs. *Behav. Ecol. Sociobiol.* 24:91–95.

Hrdy, S. B. (1976). Care and exploitation of non-human primate infants by conspecifics other than the mother. In *Advances in the study of behavior,* ed. J. Rosenblatt, R. A. Hinde, E. Shaw, & C. Beer, pp. 101–157. New York: Academic Press.

Hrdy, S. B. (1977). *The langurs of Abu*. Cambridge, Mass.: Harvard University Press.

Jamieson, I. G. (1989). Behavioral heterochrony and the evolution of birds' helping at the nest: An unselected consequence of communal breeding? *Am. Nat.* 133:394–406.

Jarman, P. J. (1974). The social organisation of antelope in relation to their ecology. *Behaviour* 48:215–267.

Jarman, P. J., & Southwell, C. J. (1986). Grouping, associations, and reproductive strategies in eastern grey kangaroos. In *Ecological aspects of social evolution*, ed. D. I. Rubenstein & R. W. Wrangham, pp. 399–428. Princeton: Princeton University Press.

Jennions, M. D., & D. W. Macdonald. (1994). Cooperative breeding in mammals. *Trends Ecol. Evol.* 9:89–93.

Jensen, P. (1988). Maternal behaviour and mother – young interactions during lactation in free-ranging domestic pigs. *Appl. Anim. Behav. Sci.* 20:297–308.

Kaufmann, J. H. (1962). Ecology and social behavior of the coati, *Nasua narica* on Barro Colorado Island, Panama. *University of California Publications in Zoology* 60:95–222.

Keller, L., & Reeve, H. K. (1994). Partitioning of reproduction in animal societies. *Trends Ecol. Evol.* 9:98–102.

King, J. A. (1963). Maternal behavior in *Peromyscus*. In *Maternal behavior in mammals*. ed. H. L. Rheingold, pp. 58–93. New York: Wiley.

Kingdon, J. (1979). East african mammals: An atlas of evolution in Africa. New York: Academic Press.

Koenig, W. D. (1981). Reproductive success, group size, and the evolution of cooperative breeding in the acorn woodpecker. *Am. Nat.* 117:421–423.

König, B. (1993). Maternal investment of communally nursing female house mice (*Mus musculus domesticus*). *Behav. Proc.* 30:61–74.

König, B. (1994). Components of lifetime reproductive success in communally and solitarily nursing house mice: A laboratory study. *Behav. Ecol. Sociobiol.* 34:275–283.

Kunz, T. H., Allgaier, A. L., Seyjagat, J., & Caligiuri, R. (1994). Allomaternal care: Helper-assisted birth in the Rodrigues fruit bat, *Pteropus rodricensis* (Chiroptera: Pteropodidae). *J. Zool. (Lond.)* 232:691–700.

Lancaster, J. (1971). Play-mothering: The relations between juvenile females and young infants among free-ranging vervet monkeys (*Cercopithecus aethiops*). *Folia Primatol.* 15:161–182.

Layne, J. N., & Raymond, M. A. V. (1994). Communal nesting of southern flying squirrels in Florida. *J. Mammal.* 75:110–120.

LeBoeuf, B. J., & Briggs, K. T. (1977). The cost of living in a seal harem. *Mammalia* 41:167–195.

Lee, P. C. (1987). Allomothering among African elephants. *Anim. Behav.* 35:275–291.

Lima, S. L. (1989). Iterated prisoner's dilemma: An approach to evolutionarily stable cooperation. *Am. Nat.* 134:828–834.

Louden, A. S. I., McNeilly, A. S., & Milne, J. A. (1983). Nutrition and lactational control of fertility in red deer. *Nature* 302:145–147.

Macdonald, D. W. (1979). Helpers in fox society. *Nature* 282:69–71.

Macdonald, D. W. (1981). Dwindling resources and the social behaviour of capybaras (*Hydrochoerus hydrochaerus*). (Mammalia). *J. Zool. Lond.*, 194:371–391.

Macdonald, D. W., & Moehlman, P. D. (1982). Cooperation, altruism, and restraint in the reproduction of carnivores. In *Perspectives in ethology*, Vol. 5, Ontogeny, ed. P. P. G. Bateson & P. H. Klopfer, pp. 433–467. New York: Plenum Press.

Macdonald, D. W., Apps, P. J., Carr, G. M., & Kerby, G. (1987). Social dynamics, nursing coalitions and infanticide among farm cats, *Felis catus. Adv. Ethol.* 28:1–66.

Manning, C. J., Dewsbury, D. A., Wakeland, E. K., & Potts, W. K. (1995). Communal nesting and communal nursing in house mice (*Mus musculus domesticus*). *Anim. Behav.* 50:741–751.

McCracken, G. F. (1984). Communal nursing in Mexican free-tailed bat maternity colonies. *Science* 223:1090–1091.

McShea, W. J., & Madison, D. M. (1984). Communal nesting between reproductively active females in a spring population of *Microtus pennsylvanicus. Can. J. Zool.* 62:344–346.

Michener, G. R. (1983). Kin identification, matriarchies, and the evolution of sociality in ground-dwelling sciurid. *Special Publications, American Society of Mammalogists* 7:528–572.

Michener, G. R., & Murie, J. O. (1983). Black-tailed prairie dog coteries: Are they cooperatively breeding units? *Am. Nat.* 121:266–274.

Murray, M. G. (1985). Estimation of kinship parameters: The island model with separate sexes. *Behav. Ecol. Sociobiol.* 16:151–159.

Neal, E. (1970). The banded mongoose. *Mungos mungos* Gmelin. *East African Wildlife Journal* 8:63–71.

Nicolson, N. A. (1987). Infants, mothers, and other females. In *Primate societies,* ed. B. B. Smuts, D. L. Cheney, R. M. Seyfarth, R. W. Wrangham, & T. T. Strusaker, pp. 330–342. Chicago: University of Chicago Press.

Nishida, T. (1983). Alloparental behavior in wild chimpanzees of the Mahale Mountains, Tanzania. *Folia Primatol.* 41:1–33.

O'Brien, T. G. (1988). Parasitic nursing behavior in the wedge-capped capuchin monkey (*Cebus olivaceous*). *Am. J. Primatol.* 16:341–344.

O'Farrell, M. J. & Studier, E. H. (1973). Reproduction, growth, and development in *Myotis thysanodes* and *M. lucifugus. Ecology* 54:18–30.

Ostfeld, R. S. (1986). Territoriality and mating system of California voles. *Journal of Animal Ecology* 55:691–706.

Owens, D. D., & Owens, M. J. (1984). Helping behaviour in brown hyenas. *Nature* 308:843–845.

Packard, J. M., Babbitt, K. J., Hannon, P. G., & Grant, W. E. (1990). Infanticide in captive collared peccaries (*Tayassu tajacu*). *Zoo Biology* 9:49–53.

Packer, C. (1986). The ecology of felid sociality. In *Ecological aspects of social evolution,* ed. D. Rubenstein & R. Wrangham, pp. 429–451. Princeton: Princeton University Press.

Packer, C., & Ruttan, L. (1988). The evolution of cooperative hunting. *Am. Nat.* 132:159–198.

Packer, C., Lewis, S., & Pusey, A. (1992). A comparative analysis of non-offspring nursing. *Anim. Behav.* 43:265–281.

Packer, C., Scheel, D., & Pusey, A. (1990). Why lions form groups: Food is not enough. *Am. Nat.* 136:1–19.

Pagel, M. D., & Harvey, P. H. (1989). Comparative methods for examining adaptation depend on evolutionary models. *Folia Primatol.* 53:203–220.

Paquet, P. C., Bragdon, S., & McCusker, S. (1982). Cooperative rearing of simultaneous litters in captive wolves. In *Wolves of the world: Perspectives of behavior, ecology, and conservation,* ed. F. H. Harrington & P. C. Paquet, pp. 223–237. Park Ridge, N.J.: Noyes Publications.

Periera, M. E., & Izard, M. K. (1989). Lactation and care for unrelated infants in forest-living ringtailed lemurs. *Am. J. Primatol.* 18:101–108.

Periera, M. E., Klepper, A. & Simons, E. L. (1987). Tactics of care for young infants by forest-living ruffed lemurs (*Varecia variegata variegata*): Ground nests, parking, and biparental guarding. *Am. J. Primatol.* 13:129–144.

Pilastro, H. (1992). Communal nesting between breeding females in a free-living population of fat dormouse (*Glis glis* L.). *Boll. Zool.* 59:63–68.

Poirier, F. E. (1968). The Nilgiri langur (*Presbytis johnii*) mother–infant dyad. *Primates* 9:45–68.

Pratt, D., & Anderson, V. H. (1982). Giraffe cow–calf relationships and social development of the calf in the Serengeti. *Z. Tierpsychol.* 51:233–251.

Pusey, A. E., & Packer, C. (1994a). Non-offspring nursing in social carnivores: Minimizing the costs. *Behav. Ecol.* 5:362–374.

Pusey, A. E., & Packer, C. (1994b). Infanticide in lions. In *Infanticide and parental care*, ed. S. Parmigiani, F. vom Saal & B. Svare, pp. 277–299. London: Harwood Academic Press.

Reiter, J., Stinson, N. L., & LeBoeuf, B. J. (1978). Northern elephant seal development: The transition from weaning to nutritional independence. *Behav. Ecol. Sociobiol.* 3:337–367.

Riedman, M. L. (1982). The evolution of alloparental care and adoption in mammals and birds. *Q. Rev. Biol.* 57:405–435.

Riedman, M. L., & LeBoeuf, B. J. (1982). Mother–pup separation and adoption in northern elephant seals. *Behav. Ecol. Sociobiol.* 11:203–215.

Rood, J. P. (1972). Ecological and behavioural comparisons of three genera of Argentine cavies. *Anim. Behav. Monogr.* 5:1–83.

Rood, J. P. (1974). Banded mongoose males guard young. *Nature* 248:176.

Rood, J. P. (1975). Population dynamics and food habits of the banded mongoose. *East African Wildlife Journal* 13:89–112.

Rood, J. P. (1980). Mating relationships and breeding suppression in dwarf mongoose. *Anim. Behav.* 28:143–150.

Russell, J. K. (1981). Exclusion of adult male coatis from social groups: Protection from predation. *J. Mammal.* 62:206–208.

Russell, J. K. (1983). Altruism in coati bands: Nepotism or reciprocity? In *Social behavior of female vertebrates*, ed. S. K. Wasser, pp. 263–290. New York: Academic Press.

Sayler, A., & Salmon, M. (1969). Communal nursing in mice: Influence of multiple mothers on the growth of young. *Science* 164:1309–1310.

Sayler, A., & Salmon, M. (1971). An ethological analysis of communal nursing by the house mouse (*Mus musculus*). *Behaviour* 40:62–85.

Schaller, G. B. (1972). *The Serengeti lion.* Chicago: University of Chicago Press.

Silk, J. (1980). Kidnapping and female competition among captive bonnet macaques. *Primates* 21:100–110.

Sinclair, A. R. E. (1977). The African buffalo: A study of resource limitation of populations. Chicago: University of Chicago Press.

Small, M. F. (1970). Alloparental behaviour in Barbary macaques, *Macaca sylvanus*. *Anim. Behav.* 39:297–306.

Spencer-Booth, Y. (1970). The relationship between mammalian young and conspecifics other than mothers and peers: A review. In *Advances in the study of behavior*, Vol. 3, ed. D. S. Lehrman, R. A. Hinde, & E. Shaw, pp. 120–180. New York: Academic Press.

Stanford, C. B. (1992). Costs and benefits of allomothering in wild capped langurs (*Presbytis pileata*). *Behav. Ecol. Sociobiol.* 30:29–34.

Taber, A. B., & Macdonald, D. W. (1992). Communal breeding in the mara, *Dolichotis patagonum. J. Zool. (Lond.)* 227:439–452.

Tener, J. S. (1954). A preliminary study of musk-oxen of Fosheim Penninsula, Ellesmere Island, NWT. *Canadian Wildlife Service, Wildlife Management Bulletin,* First Series, No. 9.

Thierry, B., & Anderson, J. R. (1986). Adoption in anthropoid primates. *Int. J. Primatol.* 7:191–216.

Trune, D. R., & Slobodchikoff, C. N. (1976). Social effects of roosting on the metabolism of the pallid bat (*Antrozous pallidus*). *J. Mammal.* 57:656–663.

Vedros, N. A., Smith, A. W., Shonewald, J., Migaki, G., & Hubbard, C. (1971). Leptospirosis epizootic among California sea lions. *Science* 172:1250–1251.

Vehrencamp, S. L. (1983). A model for the evolution of despotic versus egalitarian societies. *Anim. Behav.* 31:667–682.

Vogel, C. (1984). Cooperative and competitive aspects of allomothering within free-ranging langurs (*Presbytis entellus*). *Int. J. Primatol.* 5:390.

von Schantz, T. (1984). Female cooperation, male competition and dispersal in the red fox, *Vulpes vulpes. Oikos* 37:63–68.

Watkins, L. C., & Shump, K. A. Jr. (1981). Behavior of the evening bat *Nycteceius humeralis* at a nursery roost. *Am. Mid. Nat.* 105:258–268.

Werboff, J., Steg, J. M., & Barnes, L. (1970). Communal nursing in mice: Strain specific effects of multiple mothers on growth and behavior. *Psychon. Sci.* 19:269–271.

Whitehead, H. (in press). Babysitting, dive synchrony, and indications of alloparental care in sperm whales. *Behav. Ecol. Sociobiol.*

Wilkinson, G. S. (1987). Altruism and cooperation in bats. In *Recent advances in the study of bats,* ed. M. B. Fenton, P. Racey, & M. V. Rayner, pp. 299–323. Cambridge: Cambridge University Press.

Wilkinson, G. S. (1992). Communal nursing in the evening bat, *Nycteceius humeralis. Behav. Ecol. Sociobiol.* 31:225–235.

Wilkinson, G. S., & Baker, A. E. M. (1988). Communal nesting among genetically similar house mice. *Ethology* 77:103–114.

Yamamoto, I. (1987). Male parental care in the raccoon dog *Nyctereutes procyonoides* during the early rearing stages. In *Animal societies,* ed. Y. Eto, J. L. Brown & J. Kikkawa, pp. 189–196. Tokyo: Japan Societies Scientific Press.

Zucker, E. L., & Kaplan, J. R. (1981). Allomaternal behavior in a group of free-ranging patas monkeys. *Am. J. Primatol.* 1:57–64.

13

A Bird's-Eye View of Mammalian Cooperative Breeding

RONALD L. MUMME

13.1 Introduction

Cooperative breeding presents biologists with challenging questions. Under what ecological conditions does cooperative breeding evolve? How can we explain, at both proximate and ultimate levels, intra- and interspecific variation in the delayed dispersal, reproductive suppression, and alloparental care that is characteristic of cooperative breeders? Why is cooperative breeding prevalent in some taxa but rare or absent in others? Although difficult, these questions are not completely intractable; as should be evident from the other chapters in this volume, general answers are beginning to emerge, at least in broad outline. Definitive answers are still many years away but clearly within our reach.

This volume is devoted to cooperative breeding in mammals. Nonetheless, it may be useful to compare the chapters here with the relevant literature on cooperative breeding in birds. I have therefore written this chapter with three objectives in mind. My first goal is to provide a brief and somewhat selective review of avian cooperative breeding; more extensive reviews are readily available elsewhere (e.g., Brown 1987; Koenig & Mumme 1990; Stacey & Koenig 1990; Emlen 1991; Koenig et al. 1992). Second, by using other papers in this volume as a point of departure, I wish to compare what we now know about cooperative breeding in mammals with our current understanding of the phenomenon in birds. Third, I hope to identify profitable areas for future research, especially those areas where the study of mammalian cooperative breeding systems is likely to lead to important advances.

13.2 Fundamental Biological Differences between Birds and Mammals

Although birds and mammals differ in a number of respects, two differences seem especially relevant to any consideration of the evolution of cooperative breeding: (1) viviparity in mammals versus oviparity and incubation in birds, and (2) female lactation in mammals versus its absence in birds. For example, to the extent that nesting or denning behavior promotes the evolution of complex forms of sociality (Andersson 1984; Alexander, Noonan, & Crespi 1991; Seger 1991; Powell & Fried 1992; Solomon & Getz this volume), we might expect cooperative breeding to occur more frequently in the oviparous birds, where nesting behavior is virtually universal, than in the viviparous mammals, where nesting and denning behavior is common but not universal. In mammals, lactation places obvious constraints on the types of care that alloparents can provide to dependent young (e.g., Solomon & Getz this volume; Tardif this volume) and on the physiological mechanisms that underlie that care (e.g., Asa this volume; Carter & Roberts this volume). Such constraints, however, are largely absent in birds, and complex physiological adaptations may not be necessary for the expression of alloparental behavior in birds (Jamieson & Craig, 1987; Jamieson 1989).

These differences between birds and mammals are clearly important. Nonetheless, they are not so substantive that a priori we would expect avian and mammalian cooperative breeding systems to be fundamentally different. In fact, the three central issues that have been the focus of this volume on mammalian cooperative breeding are the very same issues that have guided the past two decades of research on avian systems. These are the issues of group living, reproductive skew, and alloparental care. First, what are the ecological and social factors that promote the formation of stable social units? Second, once these groups have formed, what determines who breeds? And third, why do some individuals provide alloparental care?

13.3 Why Do Groups Form?

Group living in cooperative breeders is usually a result of offspring delaying dispersal from their natal unit. The origin of delayed or reduced dispersal is therefore considered to be critical in setting the stage for helping behavior (Brown 1987) and is central to further understanding of the evolution of cooperative breeding (Lucas, Waser, & Creel this volume).

The first to propose a general hypothesis for the evolution of delayed dispersal was Selander (1964). In his monograph on the systematics of the *Campylorhynchus* wrens, a neotropical genus that includes several cooperative breeders, Selander wrote:

Assuming that adult mortality rates are relatively low and invariable from year to year in tropical species, it seems likely that there would be a greater tendency in tropical areas for the habitat of a species to be more fully and consistently occupied by breeding pairs of adults. Under such conditions, a young, inexperienced bird dispersing from the parental territory might have little chance of establishing an adequate territory and breeding. Possibly there would be greater advantage to the individual, in terms of total reproductive success, in delaying reproduction and remaining on the parental home area on the chance that the parental territory or an adjacent area with which the bird was familiar would become available. Chances for survival presumably would be greater and, by serving as a helper, the young bird would gain experience in parental activities which could enhance reproductive success later in life. (p. 206)

The foregoing passage is noteworthy because it explicitly recognizes the potential importance of limitations in the availability of high-quality territories (termed "habitat saturation" by later authors; Brown 1989, 1993), the relative inexperience of potential dispersers, and the benefits associated with philopatry and short-distance dispersal. Selander suggests that all these factors may be critical to the evolutionary origins of delayed dispersal and cooperative breeding. Subsequent authors have elaborated on these general themes, emphasizing either the high rates of occupancy of suitable territories and other constraints on independent breeding (e.g., Brown 1969, 1974; Koenig & Pitelka 1981; Emlen 1982) or the potential benefits associated with reduced dispersal and philopatry (Stacey & Ligon 1987, 1991; Waser 1988; Zack 1990). These differences in emphasis led Stacey & Ligon (1991, p. 842) to propose that "habitat saturation" and "benefits of philopatry" models for the origins of delayed dispersal "are based on fundamentally different conceptions of the ecological context in which either cooperative or noncooperative breeding occurs."

Recently, however, several authors have argued that habitat saturation and benefits of philopatry are not alternative models and that the distinctions between them are more apparent than real (Emlen 1991, 1994; Koenig et al. 1992; Brown 1993). In fact, all the various models for the evolution of delayed dispersal can be subsumed under the more general framework of the "delayed dispersal threshold model" proposed by Koenig et al. (1992). Thus, we should not ask whether habitat saturation or other ecological constraints on independent breeding are more or less important than are benefits of philopatry. Instead, we should be investigating the specific ecological, social, and historical factors that promote delayed dispersal in some individuals, populations, or species but not in others.

It is important to remember, however, that factors that cause within-population variation in the expression of delayed dispersal may be unimportant in contributing to interpopulation or interspecific variation. For example, in an elegant experimental study of the Seychelles warbler (*Acrocephalus sechellensis*), Komdeur (1992) showed that within-population variation in delayed dispersal was attributable to variation in the quality of the natal territory, whereas between-population variation was caused by variation in the degree of habitat saturation. Valuable though these results are, they do not necessarily tell us anything about interspecific variation; the question of why the Seychelles warbler is a cooperative breeder whereas other members of the genus *Acrocephalus* are not remains unanswered (Mumme 1992a). I will therefore consider the issue of delayed dispersal by explicitly recognizing three separate levels at which variation in this behavior can be expressed.

13.3.1 Within-Population Variation in Delayed Dispersal

What are the causes of within-population variation in the expression of delayed dispersal? For cooperatively breeding birds, this is a question with many different answers; experimental evidence from several different species has shown that dispersal behavior is influenced by a variety of social and ecological factors, including the availability of high-quality territories (Hannon et al. 1985; Zack & Rabenold 1989; Komdeur 1992; Walters, Copeyon, & Carter 1992), the availability of potential mates (Pruett-Jones & Lewis 1990), the age, experience, and competitive ability of potential dispersers (Hannon et al. 1985; Hunter 1987; Zack & Rabenold 1989), and the quality of the natal territory (Komdeur 1992, 1993).

Two negative experimental results, however, also are worth noting, because they demonstrate that neither habitat saturation nor variation in quality of potential breeding territories is universally important. First, in a study of cooperatively breeding purple gallinules (*Porphyrula martinica*), Hunter (1987) experimentally increased the availability of suitable breeding habitat. This newly created habitat was rapidly occupied by individuals that had previously been nonterritorial floaters, resulting in a marked decrease in the number of floaters in the study population. However, the habitat manipulation had no effect on the expression of delayed dispersal by juveniles, which continued to remain on their natal territories as nonbreeders. Thus, delayed dispersal in juvenile gallinules was not influenced by experimental manipulation of the degree of habitat saturation.

Second, in the superb fairy-wren (*Malurus cyaneus*), Pruett-Jones and Lewis (1990) used removal experiments to show that shortages of both breeding females and suitable breeding habitat prevent early dispersal by male nonbreeders. However, the authors found no evidence that variation in the quality of the natal territory or the breeding territory had any significant effect on dispersal behavior; essentially all nonbreeding males dispersed and became breeders when given an opportunity to do so.

Unlike the situation for birds, relatively few studies of cooperatively breeding mammals have explored likely sources of within-population variation in delayed dispersal. Nonetheless, themes similar to those described for birds are beginning to emerge (Solomon & Getz this volume). Perhaps the most complete analysis presently available for a cooperatively breeding mammal is the model of delayed dispersal in the dwarf mongoose (*Helogale parvula*) presented by Lucas, Waser, and Creel (this volume). In addition to providing a guide for future experimental work, their model clearly shows how subtle changes in factors affecting direct and indirect components of inclusive fitness can have profound effects on philopatry and optimal dispersal behavior.

13.3.2 Among-Population Variation in Delayed Dispersal

Little definitive work has examined the causes of among-population variation in dispersal behavior in either birds or mammals. An interesting avian example involves the cooperatively breeding Florida scrub jay (*Aphelocoma coerulescens*) and its almost exclusively noncooperatively breeding relatives from western North America. In Florida, young scrub jays delay dispersal from their natal territory for one to several years, and the breeding habitat is highly saturated (Woolfenden & Fitzpatrick 1984, 1990). However, a large nonbreeding surplus of young jays also exists in the noncooperative western populations; instead of remaining on their natal territories as nonbreeders, young western scrub jays live as floaters that inhabit either undefended marginal habitat or the periphery of established breeding territories (Atwood 1980; Carmen 1988). Thus, between-population variation in dispersal behavior of the scrub jay is caused by factors other than variation in the degree of habitat saturation (Koenig et al. 1992).

13.3.3 Interspecific Variation in Delayed Dispersal

What factors are primarily responsible for interspecific variation in the expression of delayed dispersal and cooperative breeding? Although several studies have searched for ecological correlates of cooperative breeding in birds (e.g., Dow 1980; Zack & Ligon 1985; Brown 1987; Ford et al. 1988), these efforts have met with relatively little success (Koenig et al. 1992). Delayed dispersal and cooperative breeding are observed in ecological, biogeographical, and taxonomic contexts that are extremely diverse.

However, an intriguing finding of Stacey and Ligon (1991) is noteworthy. These authors found that between-territory variation in reproductive success is considerably greater in two species of cooperatively breeding birds (green woodhoopoe *Phoeniculus purpureus* and acorn woodpecker *Melanerpes formicivorus*) than it is in an unrelated noncooperative species, the mountain chickadee (*Parus gambeli*). Stacey and Ligon suggested that high variation in territory quality may be characteristic of many cooperative breeders, whereas low variance in territory quality is more typical of noncooperatively breeding species. The delayed-dispersal threshold model also suggests that high variation in territory quality can promote philopatry (Koenig et al. 1992). It thus would be worthwhile to extend the analysis of Stacey and Ligon with additional comparative studies, preferably controlling for the potentially confounding effects of phylogeny by focusing on a single genus or family that comprises both cooperative and noncooperative breeders.

Phylogenetic influences cannot be ignored. Recent research has suggested that the distribution of cooperative breeding among both birds and mammals reflects historical factors and phylogenetic inertia as well as present-day ecology and current selective pressures (Russell 1989; Packer, Lewis, & Pusey 1992; Edwards & Naeem 1993; Ligon 1993; Lewis & Pusey this volume; Mohlman & Hofer this volume). Future analysis of interspecific variation in delayed dispersal and cooperative breeding must therefore rely on careful application of the comparative method (e.g., Packer et al. 1992; Edwards & Naeem 1993).

Among mammals, the subterranean rodents are clearly the most promising group for a comparative analysis of interspecific variation in delayed dispersal. The species that comprise the African family Bathyergidae span the entire range of animal sociality, from solitary, nonsocial forms (e.g., *Bathyergus, Heliophobius*) to fairly typical cooperative breeders (e.g., most species of *Cryptomys*) to the eusocial Damaraland and naked mole-rats (*Cryptomys damarensis* and *Heterocephalus glaber:* Jarvis et al. 1994). Furthermore, much of the interspecific variation in social organization of bathyergids can be explained by variation in habitat characteristics and the aridity–food-distribution hypothesis of Jarvis and her colleagues (1994) (see also Lacey & Sherman this volume). There are also several interesting and suggestive parallels between the bathyergids and South American subterranean rodents of the genera *Spalacopus* and *Ctenomys* (Lacey & Sherman this volume). Ongoing comparative work on the subterranean rodents thus promises to provide us with considerable insight into the causes of interspecific variation in delayed dispersal and cooperative breeding.

13.4 What Determines Who Breeds?

Once social groups have formed, either through delayed dispersal or other processes, what determines which individuals in these groups breed? Cooperative breeding birds and mammals show considerable variability in patterns of within-group reproduction. Many species are singular breeders that live in social units in which all or virtually all reproduction is performed by a single breeding pair, and alloparents are nearly always nonbreeders. Other species, however, are more egalitarian plural breeders in which reproductive roles are normally shared by many different group members, and alloparents are frequently breeders themselves.

As discussed by Creel and Waser (this volume), dichotomizing cooperative breeders as either singular or plural breeders is an oversimplification; singular and plural breeding are not so much distinct categories as they are the extremes of what can be called either the continuum of reproductive suppression (Creel & Waser this volume) or the eusociality continuum (Sherman et al. 1995; Lacey & Sherman this volume). Although cooperative breeders can be portrayed as occupying particular positions along the continuum (Figure 13.1), this, too, is an oversimplification, because it ignores important intraspecific variation that may exist in the degree of breeding suppression (see Figure 10.8 in Lacey & Sherman this volume; Moehlman & Hofer this volume). It is also important to remember that the term "reproductive suppression," as used throughout this chapter and in Figure 13.1, should not be viewed too narrowly; in other words, "reproductive suppression" (i.e., the failure of some individuals within a social group to reproduce) may indicate the absence of social or ecological stimuli normally required for breeding rather than the presence of specific physiological mechanisms that have evolved to suppress reproduction. Despite these caveats, Figure 13.1 illustrates the essential underlying question: How can we explain intra- and interspecific variation in the degree to which reproduction within social units is skewed?

13.4.1 Variation in Breeding Suppression: Proximate Causes

Few published studies have examined the proximate physiological mechanisms of reproductive suppression in cooperatively breeding birds (Reyer, Dittami, & Hall 1986; Mays, Vleck, & Dawson 1991; Schoech, Mumme, & Moore 1991; Wingfield, Hegner, & Lewis 1991). Although it would be premature to make any sweeping generalizations from this limited research, two early themes are worth noting.

First, in both the pied kingfisher, *Ceryle rudis* (Reyer et al. 1986), and the Harris' hawk, *Parabuteo unicinctus* (Mays et al. 1991), titers of sex steroids are low in nonbreeding males that are related to the breeding females in their group but are high in unrelated males. These observations suggest that reproductive suppression in these species may be endocrinological when the poten-

Singular Breeding
*(Complete suppression,
high skew in reproductive success)*

Florida scrub jay Red-cockaded woodpecker	Naked mole-rat Cotton-top tamarin
Stripe-backed wren Splendid fairy-wren	Dwarf mongoose Golden lion tamarin
Acorn woodpecker	Spotted hyena
Galapagos mockingbird Mexican jay	African lion

Plural Breeding
*(No suppression,
low skew in reproductive success)*

Figure 13.1. Representative species of cooperatively breeding birds and mammals on the continuum of reproductive suppression (Creel & Waser this volume) or on the eusociality continuum (Sherman et al. 1995; Lacey & Sherman this volume). The extremes of the continuum are complete suppression (high skew and singular breeding) and no suppression (low skew and plural breeding). Placement of each species is based on the findings presented earlier in this volume and in Stacey & Koenig (1990), Rabenold et al. (1990), and Haig, Walters, & Plissner (1994).

tial for incestuous matings exists but either absent or behavioral when the potential for incest is absent.

Although this conclusion is preliminary and based on data from only two species, it is consistent with genetic or behavioral evidence of incest-avoidance mechanisms in both the stripe-backed wren, *Campylorhynchus nuchalis* (Rabenold et al. 1990; Piper & Slater 1993) and the acorn wood-pecker, *Melanerpes formicivorus* (Koenig, Mumme, & Pitelka 1984). Clearly, however, considerably more data are needed on the relationship between incest avoidance and reproductive suppression in cooperatively breeding birds, particularly in species in which incestuous pairings appear to be rela-tively common, such as pukeko (*Porphyrio porphyrio:* Craig & Jamieson 1988, 1990), Galapagos mockingbirds (*Nesomimus parvulus:* Curry & Grant

1990), and splendid fairy-wrens (*Malurus splendens:* Rowley, Russell, & Brooker 1986; but see Brooker et al. 1990).

Inbreeding avoidance also appears to be important in many cooperatively breeding mammals. In both prairie voles (*Microtus ochrogaster*) and callitrichid primates, reproductive suppression of females is usually physiologically based when the potential for incestuous matings exists but either absent or behavioral when potential mates within the social unit are unrelated (Carter & Roberts this volume; French this volume). The view that inbreeding avoidance is important in shaping the proximate mechanisms of reproductive suppression is especially compelling for the callitrichids, where studies of both captive and wild populations have found evidence of inbreeding depression (Ralls & Ballou 1982; Dietz & Baker 1993; French this volume), a phenomenon that is yet to be demonstrated for any cooperatively breeding bird. Inbreeding avoidance, however, is by no means universal in cooperatively breeding mammals; incestuous matings occur frequently in field colonies of the naked mole-rat but are rare and usually avoided in the Damaraland mole-rat, at least in laboratory colonies (Jarvis et al. 1994; Faulkes & Abbott this volume; Lacey & Sherman this volume).

A second preliminary theme emerging from the work on the proximate mechanisms of reproductive suppression in cooperatively breeding birds is that in both Harris' hawks (Mays et al. 1991) and Florida scrub jays (Schoech et al. 1991), endocrinologically suppressed nonbreeders do not have elevated levels of corticosterone. This suggests that physiological suppression of nonbreeders is not caused by glucocorticoid-mediated stress imposed by aggression from behaviorally dominant breeders (e.g., Abbott 1987). Among mammals, glucocorticoid-mediated stress has also been excluded as a cause of physiological suppression, at least in naked mole-rats (Abbott & Faulkes this volume) and callitrichid primates (French this volume). It will be interesting to see if this preliminary generalization is corroborated by further research.

One additional theme emerging from studies of cooperative breeding mammals, but not evident from the limited number of relevant bird studies, is that the proximate mechanisms of reproductive suppression are typically stronger in females than in males (Abbott & Faulkes this volume). For example, in both the prairie vole and the dwarf mongoose, reproductive suppression is mediated by both behavioral and endocrinological mechanisms in females, but only by behavioral mechanisms in males (Carter & Roberts this volume; Creel & Waser this volume). The more rigidly physiological mechanisms of reproductive suppression characteristic of females have been hypothesized to reflect sex-based differences in both the cost of reproduction and confidence of parentage (Creel & Waser this volume).

13.4.2 Variation in Breeding Suppression: Ultimate Causes

There has been considerable interest in intraspecific variation in the degree of breeding suppression and reproductive skew within cooperative groups (Stacey 1982; Vehrencamp 1983; Brown 1987; Creel & Waser 1991; Emlen 1991; Jennions & Macdonald 1994; Keller & Reeve 1994). Empirical work has shown that patterns of within-group reproduction are influenced by a number of different factors, including avoidance of incestuous matings (Koenig et al. 1984; Rabenold et al. 1990; Piper & Slater 1993) and subtle conflicts of interest between males and females (Davies 1992) or between breeders and potential helpers (Creel & Waser 1991; Emlen & Wrege 1992, 1994). Although considerably more work remains to be done in this area, we now have an excellent framework for understanding the ultimate factors responsible for variation in reproductive suppression within natural populations.

But what accounts for interspecific variation in the degree of reproductive suppression? Our ability to answer this question is rudimentary, especially in cooperatively breeding birds. The rather sorry state of affairs is well illustrated by variation in the social organization of North American jays in the genus *Aphelocoma*.

The three species of *Aphelocoma* comprise what is probably the most thoroughly studied genus of cooperatively breeding birds in the world. One species, the Mexican jay (*A. ultramarina*, also known as the gray-breasted jay), breeds cooperatively throughout its range in the southwestern United States and Mexico. In Arizona it is a plural breeder that lives in groups of 5 to 13 individuals. In the Chisos Mountains of Texas, however, the Mexican jay is a singular breeder, living in groups of only 3 to 6 (Strahl & Brown 1987; Brown & Brown 1990). The Mexican jay's closely related congener, the scrub jay, is a singular cooperative breeder in Florida, where it lives in groups of 2 to 8 birds and where plural breeding occurs only rarely (Woolfenden & Fitzpatrick 1984, 1990). Although the scrub jay also appears to be a singular cooperative breeder in southern Mexico (Burt & Peterson 1993), it is noncooperative throughout its extensive range in western North America (Atwood 1980; Carmen 1988). The final member of the genus, the unicolored jay (*A. unicolor*), has been little studied in its limited range in the cloud forests of southern Mexico, but it appears to be intermediate between Florida scrub jays and Arizona Mexican jays in both group size and breeding system (Peterson & Burt 1992; Webber & Brown 1994).

How can we explain the considerable variation in social structure and reproductive suppression in this genus? Why is the Mexican jay a plural breeder in Arizona, whereas Mexican jays in Texas and scrub jays in Florida breed singly? Why do plurally breeding pairs of Mexican jays in Arizona co-defend a group territory with other pairs rather than defending separate individual breeding territories as do the other members of the genus? Despite the wealth of detailed ecological and behavioral data available for this genus, there are no clear answers to these questions. In fact, we don't have any particularly good ideas about how to begin looking for the answers to these questions. As pointed out by Brown & Brown (1990, p. 284), "we do not have a well-defined theory for the origin of plural breeding in jays" or, for that matter, in any other group of birds. This is a problem for which fresh, testable ideas are clearly needed.

Among cooperative breeding mammals, however, the situation is considerably more encouraging. Recent comparative studies by Gittleman (1989) and Creel and Creel (1991) suggest that among cooperatively breeding carnivores, strong reproductive suppression and singular breeding is more prevalent in species in which the cost of reproduction is especially high, whereas plural breeding and reduced suppression are more prevalent in carnivores with relatively low costs of reproduction. These relationships also are evident within the canids (Moehlman & Hofer this volume) and the callitrichid primates (French this volume).

A potential difficulty in making sense of the relationship between the cost of reproduction and reproductive suppression is that cause and effect are ambiguous (Creel & Creel 1991). One interpretation is that when costs of reproduction are high, subordinates are likely to accrue few net fitness gains from independent breeding and are therefore more likely to tolerate suppression by dominants. An alternative explanation, however, is also possible; once breeding suppression has become established in a species and subordinates are channeling their reproductive effort into providing alloparental care rather than breeding, dominant individuals may come under selection to produce larger litters of young that require prolonged periods of parental and alloparental care, thereby increasing the cost of reproduction (Creel & Creel 1991). Regardless of the difficulties of distinguishing cause and effect, the relationship between reproductive suppression and cost of reproduction provides an excellent base for future analyses of the evolution of variation in breeding suppression in cooperatively breeding mammals.

13.5 Why Do Some Individuals Provide Alloparental Care?

The defining characteristic of cooperative breeding systems is alloparental care. Some individuals, whether they be nonbreeders or breeders that are simultaneously caring for their own young, direct parentlike behavior toward young that are not their genetic offspring. How did this apparently altruistic behavior evolve? What determines which individuals become alloparents? What determines how much aid these individuals provide?

13.5.1 Alloparental Care: Proximate Causes

In cooperatively breeding birds, only two studies have explored the physiological mechanisms that underlie the expression of alloparental care. In the Harris' hawk, Vleck, Mays, Dawson, and Goldsmith (1991) found that nonbreeding males show an increase in circulating levels of the hormone prolactin at the time they begin provisioning nestlings. Similar results have been obtained recently for the Florida scrub jay; nonbreeders that provision nestlings (helpers) have higher levels of circulating prolactin than do nonprovisioning nonbreeders (Schoech, Mumme, & Wingfield in press). Although it is uncertain whether the relationships between prolactin levels and alloparental care in these two species is a causal one, they nonetheless suggest that evolutionary changes at the endocrinological level may underlie the expression of alloparental behavior in these species.

We know somewhat more about the proximate physiological basis of alloparental care in mammals. Probably the most complete data come from the canids, where alloparental care in females is regulated at a proximate level by a complex series of endocrinological changes associated with pseudopregnancy (Asa this volume; see also Creel et al. 1991). Male canids, though they become neither pregnant nor pseudopregnant, also show parental and alloparental care, and their care-giving behavior is associated with elevated levels of circulating prolactin. As discussed earlier, however, experimental studies are needed to demonstrate whether this relationship between elevated prolactin levels and alloparental behavior is a causal one (Asa this volume).

13.5.2 Alloparental Care: Ultimate Causes

The ultimate factors responsible for alloparental care have attracted considerable interest, particularly among researchers working with avian systems, and several recent and thorough reviews are available (Gittleman 1985; Brown 1987; Koenig & Mumme 1990; Emlen 1991; Emlen et al. 1991; Jennions & Macdonald 1994). Both adaptive and nonadaptive hypotheses have been proposed to explain the existence of alloparental behavior in cooperative breeders. The adaptive hypotheses all postulate that by caring for dependent young, alloparents enhance their fitness and are favored by natural selection. Using the framework suggested by Brown (1987), three general classes of adaptive hypotheses can be recognized, depending on the component of the alloparent's inclusive fitness that is enhanced by alloparental care.

13.5.2.1 Adaptive hypotheses based on future direct fitness

Following the early work of Skutch (1961) and Selander (1964), many authors have suggested that by caring for young, alloparents may enhance their prospects for successful independent reproduction in the future (i.e., increasing the future component of direct fitness: Brown 1987). This could be accomplished in any one of the number of different ways summarized in Table 13.1.

All the hypotheses listed in Table 13.1 have at least some empirical support. For avian cooperative breeders, however, the supporting evidence is correlative rather than experimental and therefore subject to alternative interpretations. For example, Lawton and Guindon (1981) have shown that the effectiveness of alloparents in the brown jay (*Cyanocorax morio*) increases with their age and experience. This result suggests that individuals gain valuable experience in caring for dependent young and that the experience they gain as alloparents may increase their own reproductive success when they become parents in the future. However, it is unclear whether the improvement in alloparental performance results from direct experience in alloparenting or merely from age-related improvements in foraging skill that would occur even in the absence of dependent young (see also Solomon & Getz this volume; Tardiff this volume). Experiments that control for the effects of age, such as those conducted with Mongolian gerbils by Salo and French (1989), would be required to discriminate between these alternatives.

13.5.2.2 Adaptive hypotheses based on present indirect fitness

If alloparental behavior increases the reproductive success of closely related breeders, it could be adaptive by enhancing the present indirect fitness of alloparents (Brown 1987). Of the many adaptive hypotheses for alloparental behavior, this one has received the strongest empirical support. For both avian and mammalian cooperative breeders there is extensive correlative evidence suggesting that alloparents increase the reproductive success of recipients (reviewed by Gittleman 1985; Brown 1987; Koenig & Mumme 1990; Emlen 1991; Jennions & Macdonald 1994). In addition, seven studies (four in birds, three in mammals) have examined the relationship between presence of alloparents and reproductive success experimentally, and five of the seven

Table 13.1. *Five adaptive hypotheses for alloparental behavior*

Hypotheses suggesting that alloparental behavior enhances future direct fitness

1. Alloparents gain access to group resources (Gaston 1978; Taborsky 1985).

2. Alloparental care increases reproductive success of recipients and augments group size, thereby increasing the alloparent's probability of becoming a breeder (Woolfenden & Fitzpatrick 1984).

3. Alloparental behavior promotes acquisition of future mates (Reyer 1984; Clarke 1989).

4. Alloparental behavior promotes formation of social bonds and acquisition of future alloparents (Ligon & Ligon 1983; Clarke 1989).

5. Alloparental care provides experience in caring for dependent young and promotes and increases future reproductive success (Skutch 1961; Lawton & Guindon 1981; Salo & French 1989).

Sources: Modified from Koenig & Mumme (1990) and Emlen (1991).

demonstrated significant enhancement of reproduction, at least under some circumstances (Brown et al. 1982; Leonard, Horn, & Eden 1989; Solomon 1991, 1994; Mumme 1992b; Powell & Fried 1992; French 1994; Komdeur 1994). Strong evidence is also accumulating that alloparental care is often directed preferentially toward close kin (Curry 1988; Emlen & Wrege 1988; Clarke 1989; Packer et al. 1991; Mumme 1992b).

13.5.2.3 Adaptive hypotheses based on future indirect fitness

By lightening the load on closely related breeders, alloparents could increase the survival and future reproduction of those breeders, thereby increasing their own future indirect fitness (Brown 1987; Creel 1990). Once again, some correlative evidence in support of this hypothesis exists for cooperatively breeding birds (Mumme, Koenig, & Ratnieks 1989), but it is not terribly compelling. For example, survivorship of breeding females is higher in the presence of nonbreeders in the splendid fairy-wren and pied kingfisher (e.g., Reyer 1984; Russell & Rowley 1988). However, this higher survival may be attributable to the effects of living in groups rather than the effects of alloparental behavior per se (Koenig & Mumme 1990). Among cooperatively breeding mammals, there is also some evidence that alloparental care can reduce work load on recipient breeders (e.g., Solomon & Getz this volume), but as pointed out by Moehlman and Hofer (this volume), more data are clearly needed before any definitive conclusions can be drawn.

13.5.2.4 Nonadaptive hypotheses for alloparental behavior

Among avian cooperative breeders, the leading nonadaptive hypothesis for the evolution of alloparental care is the one championed by Ian Jamieson and John Craig (Jamieson & Craig 1987; Jamieson 1989). These authors have proposed that helping behavior is an unselected consequence of group living and the neuroendocrine machinery responsible for the expression of parental care. In this view, "helping" is not a trait per se but the expression of a more general trait (parental care) in a nonparental context. Jamieson and Craig therefore suggest that alloparental behavior occurs in cooperative breeders simply because the social structure of these species provides an opportunity for parentlike behavior to be expressed in an alloparental context.

The views of Jamieson and Craig have been controversial, and their papers have stimulated a vigorous debate (Sherman 1989; Clarke 1989, 1990; Jamieson & Craig 1990; Koenig & Mumme 1990; Emlen et al. 1991; Jamieson 1991; Ligon & Stacey 1991; Mumme & Koenig 1991; White et al. 1991). At the heart of the debate lies the question of whether alloparental behavior and the proximate physiological machinery responsible for its expression have been shaped in any way by natural selection acting on potential alloparents. Because so little is known about the proximate basis of allo- parental care in birds, it is extremely difficult to answer this question.

In at least some cooperatively breeding birds, the generalized response to begging young does appear to have been modified in ways that would not be predicted by the unselected hypothesis. For example, white-fronted bee-eaters (*Merops bullockoides*) live in extended families of up to 16 individuals, within which one to three pairs breed at any one time. Even though most non- breeding group members are exposed to the stimuli of begging young, only about half of these individuals act as alloparents. In addition, care is directed preferentially toward close relatives (Emlen & Wrege 1988, 1989). These results and similar findings from other plural breeders (Curry 1988; Clarke, 1989, 1990) suggest that differences in exposure to begging young is often insufficient to explain variation in alloparental care and that alloparents facul- tatively change their behavior in ways that would not be predicted by the non- adaptive hypothesis of Jamieson and Craig. But for singular breeders, the data are equivocal (e.g., Mumme 1992b), and the unselected hypothesis is more difficult to reject (Mumme & Koenig 1991).

In mammalian cooperative breeders, there is more compelling evidence that the proximate mechanisms responsible for alloparental care have been shaped by natural selection, at least in some species. Probably the most persuasive case comes from rodents (Carter & Roberts this volume). In laboratory rats and mice, species that are not normally cooperative breeders, sexually naive males and females nonetheless can be induced to behave alloparentally by exposing them to infants over a period of several days, a process called "concaveation." Concaveation is consistent with the unselected hypothesis of Jamieson and Craig (1987) and with their contention that alloparental care may be nothing more than the nonadaptive expression of a general behavior (parental care) in a nonparental context (Jamieson 1989). However, in the cooperatively breeding prairie vole, sexually naive individuals are *spontaneously* alloparental, and a concaveation period is *not* required for the expression of alloparental care (Carter & Roberts this volume). Thus, prairie voles appear to have specific physiological adaptations that facilitate alloparental care.

Additional mammalian evidence of physiological adaptations that promote alloparental care comes from at least some viverrids and the canids, in which the expression of alloparental behavior in females is controlled proximately by unusual and elaborate endocrine mechanisms associated with pseudopregnancy (Creel et al. 1991; Asa this volume). It is difficult to imagine how the complex endocrine pathways that underlie alloparental care in these groups could have evolved nonadaptively.

13.5.2.5 Costs of alloparental care

The many adaptive hypotheses that have been proposed for alloparental behavior postulate that individuals gain either direct or indirect fitness benefits by caring for offspring of other individuals. As summarized earlier, there is also compelling empirical evidence that alloparental care can result in significant increases in either direct fitness (Salo & French 1989) or indirect fitness (Brown et al. 1982; Emlen & Wrege 1989; Mumme 1992b). However, we also have to consider the other side of the issue: What are the fitness costs of alloparental care? Somewhat surprisingly, very few studies have explicitly examined the costs of alloparental behavior per se. As argued by Tardif (this volume), a complete understanding of the effects of alloparental care on fitness requires data on the potential costs as well as the potential benefits. Careful analyses of the costs of alloparental care also promise to help explain intra- and interspecific variation in its expression (Creel & Creel 1991; Crick 1992; Pusey & Packer 1994; Lewis & Pusey this volume).

For avian cooperative breeders, the most complete analysis of costs is available from the pied kingfisher, in which alloparental care is associated with both increased energy expenditure and reduced survival of alloparents (Reyer 1984; Reyer & Westerterp 1985). Alloparental care has also been implicated in reduced survival in the stripe-backed wren (Rabenold 1990) and in loss of body mass during incubation in the white-winged chough (*Corcorax melanorhamphus:* Heinsohn, Cockburn, & Mulder 1990). Other studies, however, have not detected any appreciable cost of alloparental care (Curry & Grant 1990; Mumme 1992b).

13.6 Cooperative Breeding in Mammals: Future Directions

Does the comparison of cooperative breeding in birds and mammals presented here suggest anything about where future work in the field should be directed? I believe that it does. In particular, I see three areas in which research on mammals can help move the study of cooperative breeding past some key roadblocks that have impeded progress in the study of avian systems.

1. Proximate mechanisms and evolution. The physiological mechanisms that underlie cooperative breeding in birds are poorly known. This has slowed progress in understanding the evolution of cooperative breeding because, as Creel and Waser (this volume) have elegantly demonstrated, evolutionary and mechanistic approaches complement and enlighten each other (Sherman 1988). A perfect example of the complementarity of the two approaches is the debate over adaptive and nonadaptive explanations for alloparental care in birds. This controversy has persisted largely because the physiological mechanisms that underlie alloparental behavior in birds are virtually unknown, and we simply do not know if these mechanisms have been evolutionarily modified by natural selection acting on alloparents (Mumme & Koenig 1991).

In mammals, we know considerably more about the proximate mechanisms that underlie cooperative breeding (e.g., Asa this volume; Carter & Roberts this volume; Creel & Waser this volume; Faulkes & Abbott this volume; French this volume). By using these existing data as a base for further research and with judicious use of the comparative method, students of mammalian cooperative breeding systems are in an excellent position to explore not just evolutionary changes that have occurred in the outward manifestations of cooperative breeding, but also changes in the physiological mechanisms that exercise proximate control over delayed dispersal, reproductive suppression, and alloparental care.

2. Sources of interspecific variation and the comparative method. Researchers studying cooperative breeding in birds and mammals have used the comparative method to explore sources of interspecific variation in the expression of cooperative breeding. However, the comparative approach has met with much greater success in mammals. For example, comparative studies of mammals have shed light on the specific ecological factors that promote sociality in subterranean rodents (Jarvis et al. 1994; Lacey & Sherman this volume), on how the costs of alloparental care may explain interspecific variation in nonoffspring nursing (Packer et al. 1992; Pusey & Packer 1994; Lewis & Pusey this volume), and on how interspecific variation in reproductive suppression may be related to the cost of reproduction (Creel & Creel 1991; French this volume; Moehlman & Hofer this volume). Provocative conclusions such as these, however, have not emerged from any comparative study of cooperative breeding in birds. Whatever the reason (is phylogenetic inertia perhaps more prevalent in birds, thereby obscuring the ecological factors that initially promote the evolution of cooperative breeding?), it seems clear that further comparative studies of mammals are highly desirable, especially within groups that show considerable interspecific variation in social structure, such as the subterranean rodents (Lacey & Sherman this volume), callitrichid primates (French this volume), and voles (Solomon & Getz this volume). Mammalian studies will probably provide the best data on the causes of interspecific variation in delayed dispersal, reproductive suppression, and alloparental care.

3. Field and laboratory experimentation. Following the pioneering work of Brown et al. (1982), the study of the ecological factors that promote cooperative breeding is now moving into an experimental phase. Experimental studies of mammalian cooperative breeders have already made important contributions in this area (e.g., Salo & French 1989; Solomon 1991, 1994) and are likely to make many more, especially in species that can be maintained in captivity under seminatural conditions in which extraneous variables can be carefully controlled. Experiments on cooperatively breeding birds in captivity have not been attempted and in most cases are likely to be impractical.

Experimental manipulation of factors thought to promote delayed dispersal are clearly needed, especially when experiments are guided by models of delayed dispersal that are either general (e.g., Koenig et al. 1992) or make precise quantitative predictions (e.g., Lucas et al. this volume). Excellent comparative data on the ecological factors that affect dispersal are already available for mammals that are not cooperative breeders (e.g., Jones et al. 1988; Wolff 1992). As argued by Keller & Reeve (1994), experimental tests of models of reproductive suppression ("skew" models) are also needed. Most experimental analyses of the ecological factors that promote alloparental care have focused on its effects on current indirect fitness. With only a few exceptions (e.g., Salo & French 1989; Solomon 1994), the effects of alloparental care on future direct and indirect fitness have not been explored experimentally.

13.7 Conclusion: Toward "The New Synthesis"

Biologists have always been intrigued by comparisons between societies of invertebrates, especially insect societies, and those of vertebrates. They have dreamed of identifying the common properties of such disparate units in a way that would provide insight into all aspects of social evolution. . . . The goal can be expressed in modern terms as follows: when the same parameters and quantitative theory are used to analyze both termite colonies and troops of rhesus macaques, we will have a unified science of sociobiology. (Wilson 1975, p. 4)

In 1975, E. O. Wilson presented a vision of a unified science of social evolution in a book entitled *Sociobiology: The New Synthesis*. Now, some two decades later, it is worth asking if Wilson's vision has been realized. Clearly, the empirical base for a unified theory of social behavior is now tremendously richer than it was in 1975. Significant progress has also been made in shaping the broad outlines of the general theory that Wilson envisioned. But "the new synthesis" has not yet been achieved. Taxonomic boundaries that Wilson labored to overcome continue to divide the field, and there is still relatively little exchange of ideas between biologists studying vertebrate societies and their colleagues studying social insects. As a result, sociobiological theory has developed along two independent but parallel paths, one for insects and one for vertebrates (Keller & Reeve 1994; Sherman et al. 1995; Lacey & Sherman this volume).

"The formulation of a theory of sociobiology," Wilson wrote in 1975 (p. 5), "constitutes, in my opinion, one of the great manageable problems of biology for the next twenty or thirty years." Alas, 20 years have not been sufficient to produce the synthesis that Wilson envisioned. Will 30 be enough? Perhaps, but only if biologists studying cooperative breeding in vertebrates and eusociality in insects begin to work together to bridge the gulf that currently separates them (Keller & Reeve 1994). A general theory of social evolution must transcend the phylogenetic limitations and taxonomic biases of the biologists who build it.

Acknowledgments

I thank Nancy Solomon and Jeff French for inviting me to participate in the symposium on cooperative breeding in mammals at the 1992 meeting of the

Animal Behavior Society and, even after hearing my talk, inviting me to contribute a chapter to this volume. Thoughtful and constructive comments on the manuscript were provided by Walt Koenig, Steve Schoech, Paul Sherman, and Nancy Solomon.

References

Abbott, D. H. (1987). Behaviourally mediated suppression of reproduction in female primates. *J. Zool., (Lond.).* 213:455–470.
Alexander, R. D., Noonan, K. M., Crespi, B. J. (1991). The evolution of eusociality. In *The biology of the naked mole rat*, ed. P. W. Sherman, J. U. M. Jarvis, & R. D. Alexander, pp. 3–44. Princeton: Princeton University Press.
Andersson, M. (1984). The evolution of eusociality. *Ann. Rev. Ecol. Syst.* 15:165–189.
Atwood, J. L. (1980). Social interactions in the Santa Cruz Island scrub jay. *Condor* 82:440–448.
Brooker, M. G., Rowley, I. Adams, M., & Baverstorck, P. R. (1990). Promiscuity: An inbreeding avoidance mechanism in a socially monogamous species? *Behav. Ecol. Sociobiol.* 26:191–200.
Brown, J. L. (1969). Territorial behavior and population regulation in birds. *Wilson Bull.* 81:293–329.
Brown, J. L. (1974). Alternate routes to sociality in jays – with a theory for the evolution of altruism and communal breeding. *Am. Zool.* 14:63–80.
Brown, J. L. (1987). *Helping and communal breeding in birds: Ecology and evolution.* Princeton: Princeton University Press.
Brown, J. L. (1989). Habitat saturation and ecological constraints: Origin and history of the ideas. *Condor* 91:1010–1013.
Brown, J. L. (1993). Group territoriality and habitat quality: What are the issues? *Trends Ecol. Evol.* 8:187.
Brown, J. L., & Brown, E. R. (1990). Mexican jays: Uncooperative breeding. In *Cooperative breeding in birds: Long-term studies of ecology and behavior*, ed. P. B. Stacey & W. D. Koenig, pp. 269–288. Cambridge: Cambridge University Press.
Brown, J. L., Brown, E. R., Brown, S. D., & Dow, D. D. (1982). Helpers: Effects of experimental removal on reproductive success. *Science* 215:421–422.
Burt, D. B., & Peterson, A. T. (1993). Biology of cooperative-breeding scrub jays (*Aphelocoma coerulescens*) of Oaxaca, Mexico. *Auk* 110:207–214.
Carmen, W. J. (1988). *Behavioral ecology of the California scrub jay* (Aphelocoma coerulescens californica): *A noncooperative breeder with close cooperative relatives.* PhD dissertation, University of California, Berkeley.
Clarke, M. F. (1989). The pattern of helping in the bell miner (*Manorina melanophrys*). *Ethology* 80:292–306.
Clarke, M. F. (1990). The pattern of helping in the bell miner revisited: A reply to Jamieson and Craig. *Ethology* 86:250–255.
Craig, J. L., & Jamieson, I. G. (1988). Incestuous matings in a communal bird: A family affair. *Am. Nat.* 131:58–70.
Craig, J. L., & Jamieson, I. G. (1990). Pukeko: Different approaches and some different answers. In *Cooperative breeding in birds: Long-term studies of ecology and behavior*, ed. P. B. Stacey & W. D. Koenig, pp. 387–412. Cambridge: Cambridge University Press.

Creel, S. (1990). The future components of inclusive fitness: Accounting for interactions between members of overlapping generations. *Anim. Behav.* 40:127–134.

Creel, S. R., & Creel, N. M. (1991). Energetics, reproductive suppression and obligate communal breeding in carnivores. *Behav. Ecol. Sociobiol.* 28:263–270.

Creel, S. R., Montfort, S. L., Wildt, D. E., & Waser, P. M. (1991). Spontaneous lactation is an adaptive result of pseudopregnancy. *Nature. Lond.* 351:660–662.

Creel, S. R., & Waser, P. M. (1991). Failures of reproductive suppression in dwarf mongooses (*Helogale parvula*): Accident or adaptation? *Behav. Ecol.* 2:7–15.

Crick, H. Q. P. (1992). Load-lightening in cooperatively breeding birds and the cost of reproduction. *Ibis* 134:56–61.

Curry, R. L. (1988). Influence of kinship on helping behavior in Galapagos mockingbirds. *Behav. Ecol. Sociobiol.* 22:141–152.

Curry, R. L., & Grant, P. R. (1990). Galapagos mockingbirds: Territorial cooperative breeding in a climatically variable environment. In *Cooperative breeding in birds: Long-term studies of Ecology and Behavior,* ed. P. B. Stacey & W. D. Koenig, pp. 291–331. Cambridge: Cambridge University Press.

Davies, N. B. (1992). *Dunnock behaviour and social evolution.* Oxford: Oxford University Press.

Dietz, J. M., & Baker, A. J. (1993). Polygyny and female reproductive success in golden lion tamarins, *Leontopithecus rosalia. Anim. Behav.* 46:1067–1078.

Dow, D. D. (1980). Communally-breeding Australian birds with an analysis of distributional and environmental factors. *Emu* 80:121–140.

Edwards, S. V., & Naeem, S. (1993). The phylogenetic component of cooperative breeding in perching birds. *Am. Nat.* 141:754–789.

Emlen, S. T. (1982). The evolution of helping. I. An ecological constraints model. *Am. Nat.* 119:29–39.

Emlen, S. T. (1991). Cooperative breeding in birds and mammals. In *Behavioral ecology: An evolutionary approach,* 3rd ed., ed. J. R. Krebs & N. B. Davies, pp. 301–337. Oxford, U.K.: Blackwell.

Emlen, S. T. (1994). Benefits, constraints and the evolution of the family. *Trends Ecol. Evol.* 9:282–285.

Emlen, S. T., Reeve, H. K., Sherman, P. W., Wrege, P. H., Ratnieks, F. L. W., & Shellman-Reeve, J. (1991). Adaptive versus nonadaptive explanations of behavior: The case of alloparental helping. *Am. Nat.* 138:259–270.

Emlen, S. T., & Wrege, P. H. (1988). The role of kinship in helping decisions among white-fronted bee-eaters. *Behav. Ecol. Sociobiol.* 23:305–315.

Emlen, S. T., & Wrege, P. H. (1989). A test of alternate hypotheses for helping behavior in white-fronted bee-eaters of Kenya. *Behav. Ecol. Sociobiol.* 25:303–319.

Emlen, S. T., & Wrege, P. H. (1992). Parent–offspring conflict and the recruitment of helpers among bee-eaters. *Nature, Lond.* 356:331–319.

Emlen, S. T., & Wrege, P. H. (1994). Gender, status and family fortunes in the white-fronted bee-eater. *Nature, Lond.* 367:129–132.

Ford, H. A., Bell, H., Nias, R., & Noske, R. (1988). The relationship between ecology and the incidence of cooperative breeding in Australian birds. *Behav. Ecol. Sociobiol.* 22:239–249.

French, J. A. (1994). Alloparents in the Mongolian gerbil: Impact on long-term reproductive performance of breeders and opportunities for independent reproduction. *Behav. Ecol.* 5:273–279.

Gaston, A. J. (1978). The evolution of group territorial behavior and cooperative breeding. *Am. Nat.* 112:1091–1100.

Gittleman, J. L. (1985). Functions of communal care in mammals. In *Evolution: Essays in honour of John Maynard Smith,* ed. P. J. Greenwood, P. H. Harvey, & M. Slatkin, pp. 187–205. Cambridge: Cambridge University Press.

Gittleman, J. L. (1989). Carnivore group living: Comparative trends. In *Carnivore behavior, ecology and evolution,* ed. J. L. Gittleman, pp. 183–207. Ithaca: Cornell University Press.

Haig, S. M., Walters, J. R., & Plissner, J. H. (1994). Genetic evidence for monogamy in the cooperatively breeding red-cockaded woodpecker. *Behav. Ecol. Sociobiol.* 34:295–303.

Hannon, S. J., Mumme, R. L., Koenig, W. D., & Pitelka, F. A. (1985). Replacement of breeders and within-group conflict in the cooperatively breeding acorn woodpecker. *Behav. Ecol. Sociobiol.* 17:303–312.

Heinsohn, R. G., Cockburn, A., & Mulder, R. A. (1990). Avian cooperative breeding: Old hypotheses and new directions. *Trends Ecol. Evol.* 5:403–407.

Hunter, L. A. (1987). Acquisition of territories by floaters in cooperatively breeding purple gallinules. *Anim. Behav.* 35:402–410.

Jamieson, I. G. (1989). Behavioral heterochrony and the evolution of birds' helping at the nest: An unselected consequence of communal breeding? *Am. Nat.* 133:394–406.

Jamieson, I. G. (1991). The unselected hypothesis for the evolution of helping behavior: Too much or too little emphasis on natural selection? *Am. Nat.* 138:271–282.

Jamieson, I. G., & Craig, J. L. (1987). Critique of helping behaviour in birds: a departure from functional explanations. In *Perspectives in ethology,* Vol. 7, ed. P. P. G. Bateson & P. H. Klopfer, pp 79–98. New York: Plenum Press.

Jamieson, I. G., & Craig, J. L. (1990). Reply: Evaluating hypotheses on the evolution of helping behaviour in the bell miner, *Manorina melanophrys. Ethology* 85:163–167.

Jarvis, J. U. M., O'Riain, M. J., Bennett, N. C., & Sherman, P. W. (1994). Mammalian eusociality: A family affair. *Trends Ecol. Evol.* 9:47–51.

Jennions, M. D., & Macdonald, D. W. (1994). Cooperative breeding in mammals. *Trends Ecol. Evol.* 9:89–93.

Jones, W. T., Waser, P. M., Elliott, L. F., Link, N. E., & Bush, B. B. (1988). Philopatry, dispersal, and habitat saturation in the banner-tailed kangaroo rat, *Dipodomys spectabilis. Ecology* 69:1466–1473.

Keller, L., & Reeve, H. K. (1994). Partitioning of reproduction in animal societies. *Trends Ecol. Evol.* 9:98–102.

Koenig, W. D., & Mumme, R. L. (1990). Levels of analysis, functional explanations, and the significance of helping behavior. In *Interpretation and explanation in the study of animal behavior,* Vol. 1, ed. M. Bekoff & D. Jamieson, pp. 268–303. Boulder, Colo. Westview Press.

Koenig, W. D., Mumme, R. L., & Pitelka, F. A. (1984). The breeding system of the acorn woodpecker in central coastal California. *Z. Tierpsychol.* 65:289–308.

Koenig, W. D., & Pitelka, F. A. (1981). Ecological factors and kin selection in the evolution of cooperative breeding in birds. In *Natural selection and social behavior: Recent research and new theory,* ed. R. D. Alexander & D. W. Tinkle, pp. 261–280. New York: Chiron Press.

Koenig, W. D., Pitelka, F. A., Carmen, W. J., Mumme, R. L., & Stanback, M. T. (1992). The evolution of delayed dispersal in cooperative breeders. *Q. Rev. Biol.* 67:111–150.

Komdeur, J. (1992). Importance of habitat saturation and territory quality for evolution of cooperative breeding in the Seychelles warbler. *Nature, Lond.* 358:493–495.

Komdeur, J. (1993). Fitness-related dispersal. *Nature, Lond.* 366:23–24.

Komdeur, J. (1994). Experimental evidence for helping and hindering by previous offspring in the cooperative-breeding Seychelles warbler *Acrocephalus sechellensis. Behav. Ecol. Sociobiol.* 34:175–186.

Lawton, M. F., & Guindon, C. F. (1981). Flock composition, breeding success, and learning in the brown jay. *Condor* 82:27–33.

Leonard, M. L., Horn, A. G., & Eden, S. F. (1989). Does juvenile helping enhance breeder reproductive success. A removal experiment. *Behav. Ecol. Sociobiol.* 25:357–361.

Ligon, J. D. (1993). The role of phylogenetic history in the evolution of contemporary avian mating and parental care systems. *Curr. Ornithol.* 10:1–46.

Ligon, J. D., & Ligon, S. H. (1983). Reciprocity in the green woodhoopoe (*Phoeniculus purpureus*). *Anim. Behav.* 31:480–489.

Ligon, J. D., & Stacey, P. B. (1991). The origin and maintenance of helping behavior in birds. *Am. Nat.* 138:254–258.

Mays, N. A. Vleck, C. M., & Dawson, J. W. (1991). Plasma luteinizing hormone, steroid hormones, behavioral role, and nest stage in the cooperatively breeding Harris' hawk (*Parabuteo unicinctus*). *Auk* 108:619–637.

Mumme, R. L. (1992a). Delayed dispersal and cooperative breeding in the Seychelles warbler. *Trends Ecol. Evol.* 7:330–331.

Mumme, R. L. (1992b). Do helpers increase reproductive success? An experimental analysis in the Florida scrub jay. *Behav. Ecol. Sociobiol.* 31:319–328.

Mumme, R. L., & Koenig, W. D. (1991). Explanations for avian helping behavior. *Trends Ecol. Evol.* 6:343–344.

Mumme, R. L., Koenig, W. D., & Ratnieks, F. L. W. (1989). Helping behaviour, reproductive value, and the future component of indirect fitness. *Anim. Behav.* 38:331–343.

Packer, C., Gilbert, D. A., Pusey, A. E., & O'Brien, S. J. (1991). A molecular genetic analysis of kinship and cooperation in African lions. *Nature, Lond.* 351:562–565.

Packer, C., Lewis, S., & Pusey, A. (1992). A comparative analysis of non-offspring nursing. *Anim. Behav.* 43:265–281.

Peterson, A. T., & Burt, D. B. (1992). Phylogenetic history of social evolution and habitat use in the *Aphelocoma* jays. *Anim. Behav.* 44:859–866.

Piper, W. H., & Slater, G. (1993). Polyandry and incest avoidance in the cooperative stripe-backed wren of Venezuela. *Behaviour* 124:227–247.

Powell, R. A., & Fried, J. J. (1992). Helping by juvenile pine voles (*Microtus pinetorum*), growth and survival of younger siblings, and the evolution of pine vole sociality. *Behav. Ecol.* 3:325–333.

Pruett-Jones, S. G., & Lewis, M. J. (1990). Sex ratio and habitat limitation promote delayed dispersal in superb fairy-wrens. *Nature, Lond.* 348:541–542.

Pusey, A. E., & Packer, C. (1994). Non-offspring nursing in social carnivores: Minimizing the costs. *Behav. Ecol.* 5:362–374.

Rabenold, K. N. (1990). *Campylorhynchus* wrens: The ecology of delayed dispersal and cooperation in the Venezuelan savanna. In *Cooperative breeding in birds: Long-term studies of ecology and behavior,* ed. P. B. Stacey & W. D. Koenig, pp. 159–196. Cambridge: Cambridge University Press.

Rabenold, P. P., Rabenold, K. N., Piper, W. H., Haydock, J., & Zack, S. W. (1990). Shared paternity revealed by genetic analysis in cooperatively breeding tropical wrens. *Nature, Lond.* 348:538–540.

Ralls, K., & Ballou, J. (1982). Effects of inbreeding on infant mortality in captive primates. *Int. J. Primatol.* 3:491–505.

Reyer, H.-U. (1984). Investment and relatedness: A cost/benefit analysis of breeding and helping in the pied kingfisher (*Ceryle rudis*). *Anim. Behav.* 32:1163–1178.

Reyer, H.-U., Dittami, J. P., & Hall, M. R. (1986). Avian helpers at the nest: Are they psychologically castrated? *Ethology* 71:216–228.

Reyer, H.-U., & Westerterp, K. (1985). Parental energy expenditure: A proximate cause of helper recruitment in the pied kingfisher (*Ceryle rudis*). *Behav. Ecol. Sociobiol.* 17:363–369.

Rowley, I. C. R., Russell, E. M., & Brooker, M. G. (1986). Inbreeding: Benefits may outweigh costs. *Anim. Behav.* 34:939–941.

Russell, E. (1989). Co-operative breeding–a Gondwanan perspective. *Emu* 89:61–62.

Russell, E. M., & Rowley, I. C. R. (1988). Helper contributions to reproductive success in the splendid fairy-wren. *Behav. Ecol. Sociobiol.* 22:131–140.

Salo, A. L., & French, J. A. (1989). Early experience, reproductive success, and development of parental behaviour in Mongolian gerbils. *Anim. Behav.* 38:693–702.

Schoech, S. J., Mumme, R. L., & Moore, M. C. (1991). Reproductive endocrinology and mechanisms of breeding inhibition in cooperatively breeding Florida scrub jays (*Aphelocoma c. coerulescens*). *Condor* 93:354–364.

Schoech, S. J., Mumme, R. L., & Wingfield, J. C. (in press). Prolactin and helping behaviour in the cooperatively breeding Florida scrub jay (*Aphelocoma c. coerulescens*). *Anim. Behav.*

Seger, J. (1991). Cooperation and conflict in social insects. In *Behavioral ecology: An evolutionary approach.* 3rd ed., ed. J. R. Krebs & N. B. Davies, pp. 338–373. Oxford, U.K.: Blackwell.

Selander, R. K. (1964). Speciation in wrens of the genus *Campylorhynchus. Univ. Calif. Publ. Zool.* 74:1–224.

Sherman, P. W. (1988). The levels of analysis. *Anim. Behav.* 36:616–619.

Sherman, P. W. (1989). The clitoris debate and the levels of analysis. *Anim. Behav.* 37:697–698.

Sherman, P. W., Lacey, E. A., Reeve, H. K., & Keller, L. (1995). The eusociality continuum. *Behav. Ecol.* 6:102–108.

Skutch, A. F. (1961). Helpers among birds. *Condor* 63:198–226.

Solomon, N. G. (1991). Current indirect fitness benefits associated with philopatry in juvenile prairie voles. *Behav. Ecol. Sociobiol.* 29:277–282.

Solomon, N. G. (1994). Effect of the pre-weaning environment on subsequent reproduction in prairie voles, *Microtus ochrogaster. Anim. Behav.* 48:331–341.

Stacey, P. B. (1982). Female promiscuity and male reproductive success in social birds and mammals. *Am. Nat.* 120:51–64.

Stacey, P. B., & Koenig, W. D., ed. (1990). *Cooperative breeding in birds: Long-term studies of ecology and behavior.* Cambridge: Cambridge University Press.

Stacey, P. B., & Ligon, J. D. (1987). Territory quality and dispersal options in the acorn woodpecker, and a challenge to the habitat saturation model of cooperative breeding. *Am. Nat.* 130:654–676.

Stacey, P. B., & Ligon, J. D. (1991). The benefits of philopatry hypothesis for the evolution of cooperative breeding: Variance in territory quality and group size effects. *Am. Nat.* 137:831–846.

Strahl, S. D., & Brown, J. L. (1987). Geographic variation in social structure and behavior of *Aphelocoma ultramarina. Condor* 89:422–424.

Taborsky, M. (1985). Breeder-helper conflict in a cichlid fish with broodcare helpers: An experimental analysis. *Behaviour* 95:45–75.

Vehrencamp, S. L. (1983). A model for the evolution of despotic versus egalitarian societies. *Anim. Behav.* 31:667–682.

Vleck, C. M., Mays, N. A., Dawson, J. W., & Goldsmith, A. R. (1991). Hormonal correlates of parental and helping behavior in the cooperatively breeding Harris' hawk (*Parabuteo unicinctus*). *Auk* 108:638–648.

Walters, J. R., Copeyon, C. K., & Carter, J. H. III (1992). A test of the ecological basis of cooperative breeding in red-cockaded woodpeckers. *Auk* 109:90–97.

Waser, P. M. (1988). Resources, philopatry, and social interactions among mammals. In *The ecology of social behavior*, ed. C. N. Slobodchikoff, pp. 109–130. New York: Academic Press.

Webber, T., & Brown, J. L. (1994). Natural history of the unicolored jay in Chiapas, Mexico. *Proc. Western Foundation Vert. Zool.* 5:135–160.

White, C. S., Lambert, D. M., Millar, C. D., & Stevens, P. M. (1991). Is helping behavior a consequence of natural selection? *Am. Nat.* 138:246–253.

Wilson, E. O. 1975. *Sociobiology: The new synthesis*. Cambridge, Mass.: Harvard University Press.

Wingfield, J. C., Hegner, R. E., & Lewis, D. M. (1991). Circulating levels of luteinizing hormone and steroid hormones in relation to social status in the cooperatively breeding white-browed sparrow weaver, *Plocepasser mahali*. *J. Zool. Lond.* 225:43–58.

Wolff, J. O. (1992). Parents suppress reproduction and stimulate dispersal in opposite-sex juvenile white-footed mice. *Nature, Lond.* 359:409–410.

Woolfenden, G. E., & Fitzpatrick, J. W. (1984). *The Florida scrub jay: Demography of a cooperative-breeding bird*. Princeton: Princeton University Press.

Woolfenden, G. E., & Fitzpatrick, J. W. (1990). Florida scrub jays: A synopsis after 18 years of study. In *Cooperative breeding in birds: Long-term studies of ecology and behavior*, ed. P. B. Stacey & W. D. Koenig, pp. 241–266. Cambridge: Cambridge University Press.

Zack, S. (1990). Coupling delayed breeding with short-distance dispersal in cooperatively breeding birds. *Ethology* 86:265–286.

Zack, S., & Ligon, J. D. (1985). Cooperative breeding in *Lanius* shrikes. I. Habitat and demography of two sympatric species. *Auk* 102:754–765.

Zack, S., & Rabenold, K. N. (1989). Assessment, age and proximity in dispersal contests among cooperative wrens: Field experiments. *Anim. Behav.* 38:235–247.

Index